Global Megaprojects

Global Megaprojects

Lessons, Case Studies, and Expert Advice on
International Megaproject Management

Virginia A. Greiman
Boston University
Boston, MA, USA

Registered Office
John Wiley & Sons, Inc., 111 River Street, Hoboken, NJ 07030, USA

Editorial Office
John Wiley & Sons, Inc., 111 River Street, Hoboken, NJ 07030, USA

For details of our global editorial offices, customer services, and more information about Wiley products visit us at www.wiley.com.

Wiley also publishes its books in a variety of electronic formats and by print-on-demand. Some content that appears in standard print versions of this book may not be available in other formats.

Library of Congress Cataloging-in-Publication Data

Names: Greiman, Virginia, author.
Title: Global megaprojects : lessons, case studies, and expert advice on
    international megaproject management / Virginia A. Greiman, Boston
    University, MA.
Description: First edition. | Hoboken, NJ : Wiley, 2023. | Includes index.
Identifiers: LCCN 2022027391 (print) | LCCN 2022027392 (ebook) | ISBN
    9781119875208 (hardback) | ISBN 9781119875222 (adobe pdf) | ISBN
    9781119875215 (epub)
Subjects: LCSH: Project management–Case studies. | Risk management–Case
    studies. | Public works–Case studies. | Investments, Foreign–Case
    studies. | Globalization–Social aspects. | International cooperation.
Classification: LCC HD69.P75 G7524 2023 (print) | LCC HD69.P75 (ebook) |
    DDC 352.3/65–dc23/eng/20220801
LC record available at https://lccn.loc.gov/2022027391
LC ebook record available at https://lccn.loc.gov/2022027392

Cover Design: Wiley
Cover Images: © Fraser Hall/Getty Images

Set in 9.5/12.5pt STIXTwoText by Straive, Pondicherry, India
SKY10038449_111622

This book is dedicated to the creativity, imagination, and perseverance of all the engineers and construction workers building Innovative Megaprojects globally, and to all those who come after them in creating the great engineering megaprojects of the world.

Contents

List of Figures   *xiii*
List of Tables   *xv*
List of Boxes   *xvii*
Author's Perspectives   *xix*
Acknowledgments   *xxi*
Introduction to This Book   *xxiii*
Key Concepts and Objectives   *xxiii*
Pedagogy   *xxiv*
Book Structure   *xxv*
Overview of Book Chapters   *xxvi*
References   *xxviii*

1   Introduction to Global Megaprojects   *1*
Introduction   *1*
Globalization and Megaprojects   *2*
Characterization of Global Megaprojects   *3*
Megaprojects: The Global Timeline   *6*
Prehistory   *7*
The Middle Ages   *8*
The Golden Age of Globalization (1870–1914)   *9*
The Great Depression (1930–1940s)   *12*
The Great Megaproject Era (1950–1970s)   *14*
The Era of Great Tunnel, Energy, and Pipeline Projects: (1970s–2000)   *17*
The Millennium Projects   *23*
Why Study Megaprojects?   *26*
Summary   *33*
References   *35*

2 Megaproject Finance: Innovation and Value Driven Megaproject Management *43*

Introduction: Financing the World's Infrastructure *43*

Project Finance Definition and Characteristics *45*

Project Finance v. Corporate Finance *48*

Investment in the Developing Countries *49*

Infrastructure Financing in the United States *51*

Infrastructure Financing in the European Union *52*

Sources of Funding for Project Financing *54*

The Legal Framework: The Major Project Participants and Agreements *59*

Export Credit Agencies (ECAs) *65*

Multilateral Development Banks *67*

Public–Private Partnership Structures *69*

Evaluation Criteria for Project Finance *70*

Megaproject Evaluation: Beyond the Iron Triangle *77*

Looking Back and Looking Forward: Ex-ante and Ex-post Evaluations *78*

Evaluating Projects Through the Life Cycle *81*

Looking Back and Looking Forward and Qualitative Scoring *85*

Summary *86*

References *88*

3 The Multilaterals and World Development *95*

Introduction *95*

Growing Demand for Global Infrastructure *96*

The Role of the Multilaterals in Development *101*

Mobilization of Capital *107*

The Changing Landscape of Development Banks *116*

International Development Case Studies *118*

A Collective Action Perspective on the Planning of Global Megaprojects *121*

The Social, Economic and Institutional Value of Megaprojects *123*

Frameworks for Sustainable Development *124*

Summary *127*

References *129*

4 Leading Complex Global Projects *135*

Introduction *135*

Complexity in Relation to Uncertainty, Ambiguity, Conflict, and Risk *138*

Case Studies of Complex Projects *146*

Strategic Management of Complexity *151*
Summary *159*
References *161*

5 Global Megaproject Governance *167*
Introduction to Global Governance of Megaprojects *167*
Developing a Megaproject Governance Framework *174*
Project Governance Between Developed and Developing Countries:
Considerations for Improving Governance *188*
Case Studies in Governance: What Causes Governance Failure? *191*
Summary *198*
References *200*

6 Integrated Project Organizations and Public Private Partnerships *207*
Introduction *207*
Part I: Project Organization Integration: A New Mindset for Systems
Engineering and Program Management *209*
Part II: The Structure of Organizations as Systems of Systems *211*
Part III. Public–Private Partnerships: The Sharing of Risk and Opportunity *225*
Case Studies in Public–Private Partnership Development *236*
Emerging Trends and Social Considerations for PPP Development *241*
Summary *243*
References *249*

7 Managing the Megaproject Implementation and Delivery *255*
Introduction *255*
Phase 1: Initiation of the Megaproject *257*
Phase 2: Global Project Delivery Methodologies and Procurement *261*
Phase 3: Implementation *263*
Phase 4: Project Controls *279*
Phase 5: Transitioning a Megaproject to Operations *289*
Lessons Learned on Cost and Schedule *290*
Summary *294*
References *297*

**8 Megaprojects and Mega Risk: Opportunity, Risk, and Resilience
Management** *301*
Introduction *301*
Emerging Risks on a Global Scale Impacting Megaprojects *302*
Defining Risk and Risk Management *305*

Structure of Megaproject Risks *306*

Resilience and Risk Management *307*

What Is Project Risk Intelligence? *310*

Developing a Risk Management Framework: A Shared Vision of Risk *313*

Global Risk Factors in International Projects *325*

Business Continuity Planning in the Management of Risks *327*

Root Cause Analysis *327*

Characterizing Risk *330*

Normalization of Deviance Related to Risk Discovery *333*

Catastrophic Loss Potential: Natural and Manmade Disasters *336*

Summary *341*

References *344*

9 Megaprojects: Troubles and Triumphs *349*

Introduction *349*

What Is Meant by Success? Successful Megaproject Failures *350*

Characteristics of Failed Projects *356*

Characteristics of the Most Successful Projects *370*

Summary *372*

References *373*

10 Laws and Contracts in Global Megaprojects *381*

Introduction *381*

Planning for Procurement *382*

Concession Procurement *383*

Private Sector Procurement *383*

Megaproject Procurement Contracts in the United States *385*

Laws Governing International Megaproject Contracting *387*

Challenging Negotiations in Megaproject Contracts *391*

Collaborative Contracting *414*

Emerging Issues in Global Development Laws and Contracts *414*

Summary *419*

References *421*

11 Megaproject Innovation and Resilience *425*

Introduction *425*

Where Does Innovation Come From? *427*

What Is an Innovation Megaproject? *429*

Megaproject Innovation Programs *431*

What Is an Innovation Strategy? *435*

Enablers and Challenges to Innovation *439*
Best Practices for Innovation *443*
Summary *450*
References *452*

12 The Future of Global Megaprojects *459*
Introduction *459*
Blue Ocean Thinking *460*
Blue Economy Thinking *461*
Challenges of Future Megaprojects *463*
Meeting the Grand Challenges of the Twenty-first Century *469*
Global Change and International Cooperation *474*
Global Megaprojects and the Leaders of Tomorrow *474*
Megaprojects and the Growth of the Digital Economy *478*
Megaproject Management: Looking Back to Move the Future Forward *485*
Leadership for Megaprojects in the Emergent Era *486*
A New Principle Based Approach to Project Management *489*
Megaprojects and System Perspectives *490*
Summary *492*
References *494*

Glossary *501*
Acronyms *523*
Index *527*

List of Figures

Figure 1.1   Characteristics of a megaproject.   *4*

Figure 1.2   Megaprojects: the global timeline.   *6*

Figure 1.3   Global power sources.   *22*

Figure 2.1   Corporate v. project finance comparison.   *48*

Figure 2.2   Project company structure and legal agreements.   *55*

Figure 2.3   Borrower credit conditions.   *62*

Figure 2.4   Value capture projects.   *75*

Figure 2.5   Project evaluation phases.   *81*

Figure 2.6   Project finance evaluation ratings.   *84*

Figure 3.1   The 2030 agenda for sustainable development goals.   *96*

Figure 3.2   Private investment by financier.   *97*

Figure 3.3   Financing of Private Investment in Infrastructure Projects.   *98*

Figure 3.4   World Bank Global Commitments: $B in Loans, Grants, Equity Investments, & Guarantees.   *102*

Figure 3.5   Lending by sector (fiscal 2021 share of total of $15.6B).   *122*

Figure 4.1   Characteristics of complexity.   *141*

Figure 4.2   Complexity characteristics and effects.   *145*

Figure 5.1   Governance framework development.   *175*

Figure 5.2   Project organization governance frameworks.   *177*

Figure 6.1   High-level board integration of the International Space Station program.   *218*

Figure 6.2   Central Artery Tunnel Project Integrated Project Organization.   *222*

Figure 6.3   Crossrail integrated delivery structure.   *224*

Figure 6.4   Public–private partnership integrated roles.   *226*

Figure 6.5   PPP program and project development flowchart.   *231*

Figure 7.1   Life cycle in megaproject development.   *256*

Figure 7.2   Central Artery/Tunnel Project (CA/T) Oversight Coordination Commission (OCC).   *280*

Figure 7.3   Megaproject dispute resolution.   *287*

Figure 8.1   Megaproject risk structure.   *307*

Figure 8.2   Types of risk intelligence.   *311*

Figure 8.3   Risk management framework.   *313*

Figure 8.4   Megaproject risk model.   *314*

Figure 8.5   United Nations qualitative risk assessment.   *318*

Figure 8.6   Threat responses and opportunity responses.   *322*

Figure 8.7   Steps in root cause risk analysis.   *328*

Figure 8.8   Four categories of risk.   *331*

Figure 9.1   Sustainability pillars: the triple bottom line.   *352*

Figure 9.2   Case Studies in Megaproject Failures.   *360*

Figure 10.1 Project contract clauses.   *392*

Figure 10.2 Steps in claims dispute resolution at the Big Dig.   *405*

Figure 11.1 The Seven Global Innovation Index Pillars.   *447*

Figure 12.1 A System Perspective of Megaproject Management.   *491*

List of Tables

Table 1.1   The millennium megaprojects.   *24*
Table 2.1   Global sources of project finance.   *56*
Table 2.2   Key questions for the lender.   *61*
Table 2.3   Key questions for the equity sponsors.   *63*
Table 2.4   Export import (EXIM) banks.   *66*
Table 2.5   Value capture.   *74*
Table 3.1   Transition economies in Europe and the former Soviet Union and Asia (pre 2019).   *100*
Table 3.2   Role and contributions of the Multilateral Development Banks.   *104*
Table 3.3   Country and project requirements for lending.   *106*
Table 3.4   Sources of public and private financing for Global Development Projects.   *109*
Table 3.5   Overview of the World's Multilateral Development Banks.   *110*
Table 3.6   Select World Bank Projects from 2012 to 2021.   *119*
Table 3.7   Africa's mobilization of climate change funds.   *123*
Table 4.1   Complexity characteristics, concerns, and mitigation.   *139*
Table 4.2   Comparison of elements of complexity, risk, and uncertainty.   *142*
Table 4.3   Case studies: complex project characteristics, challenges, and solutions.   *147*
Table 5.1   Corporate governance v. project governance.   *169*
Table 5.2   Literature review on the characteristics of governance.   *170*
Table 5.3   Governance structures.   *176*
Table 5.4   Emerging models of sustainable governance.   *180*
Table 6.1   Structure of a public–private partnership (PPP).   *229*
Table 6.2   Advantages and challenges of a PPP.   *232*
Table 6.3   Public private partnership legislation.   *244*
Table 7.1   Crossrail and the Big Dig WBS comparison.   *270*
Table 7.2   NASA Advisory Council Meeting: reasons for cost growth in NASA projects.   *271*

Table 7.3   Boston's central artery/tunnel project timeline.   *274*

Table 7.4   Crossrail Ltd. project timeline.   *275*

Table 7.5   Litigation v. Arbitration comparison.   *286*

Table 8.1   Perception of risk across 10 countries.   *304*

Table 8.2   Selected sources of risk intelligence.   *312*

Table 8.3   Turning risks into opportunities.   *322*

Table 8.4   Department of Energy DoE root cause analysis top 10 risk-related issues and root causes.   *329*

Table 8.5   Major catastrophic disasters and losses.   *338*

Table 9.1   Critical factors in megaproject success or failure.   *358*

Table 9.2   Determinants of escalation.   *367*

Table 10.1   Megaproject law and policy initiatives.   *388*

Table 11.1   Crossrail and the central artery tunnel project innovation programs.   *432*

Table 11.2   Types of innovation.   *436*

Table 11.3   Internal and external drivers of sustainable innovation.   *440*

Table 11.4   Twenty-six enablers and challenges to innovation in megaprojects.   *441*

Table 11.5   Innovation in Asia.   *449*

Table 12.1   Projects of the future.   *460*

Table 12.2   Project management principles.   *489*

List of Boxes

Box 2.1   Types of Renewable Energy   45
Box 2.2   The Muara Laboh Geothermal Power Project in Indonesia   46
Box 3.1   Unlocking Private Investment in Infrastructure   105
Box 6.1   Insights on Organizational Attributes for Integration   212
Box 6.2   The F/A18E/F Super Hornet Case Study   215
Box 6.3   PPP Process in Ethiopia   229
Box 6.4   Airport Express Metro Line: Delhi Metro Rail Corporation,
           India   233
Box 6.5   Power Purchase Agreement   235
Box 6.6   Hong Kong PPP Development   236
Box 6.7   Unsolicited Proposal, Capital Beltway Project, Fluor Enterprises   240
Box 7.1   Best Value Delivery   263
Box 7.2   Resident Engineers (RE) Operations Manual: Typical Provisions   267
Box 7.3   Typical Responsibilities of the Resident Engineers   268
Box 7.4   Recommended Practices for Preventing or Reducing Schedule
           Delay   272
Box 7.5   The National Audit Organizations Review of Crossrail Delays   277
Box 7.6   The Central Artery/Third Harbor Tunnel (CA/T) Project Oversight
           Coordination Commission   280
Box 7.7   Key Questions That Must Be Answered in Designing a Quality
           Program Include the Following   282
Box 7.8   Selected Safety and Health Programs   283
Box 9.1   The Iridium Case Study   354
Box 9.2   Megaproject Success Factors   355
Box 10.1   Codes and Regulations   387
Box 10.2   SPECIFICATION Section 4.04 Differing Site Conditions   399
Box 10.3   Step-In Clause   408

Author's Perspectives

The motivation for this book came from working on global megaprojects commencing with my early experience on projects in Eastern and Central Europe after the fall of communism. Assisting economies in transition opened my eyes to the need for the rule of law to attract investment in implementing the many projects needed in these former war-torn countries. This experience created a desire to learn more about how successful projects are initiated, implemented, and transitioned to those who would operate these projects for many decades. Following my experience in these transition economies I had the opportunity to serve on the Executive Team of Boston's Central Artery/Tunnel Project (America's largest infrastructure project) which led to research and consulting opportunities at the Crossrail project in London and other megaprojects in Africa, Canada, Europe, North America and Southeast Asia.

The need for knowledge about megaprojects is apparent from every corner of the globe. Cities and countries worldwide are spending hundreds of billions of dollars annually to preserve the nation's infrastructure and construct the next generation of roads, bridges, tunnels, energy resources, and water supply. The Organization for Economic Co-operation and Development (OECD) is forecasting that global investment in infrastructure alone will cost as much as $70 trillion through 2030 (Mirabile et al. 2017). Thus, the need for development expertise is extensive and incorporates a broad range of disciplines such as project and program management, engineering, the environment, construction, research and development, competitive strategy, risk management, privatization, corporate responsibility, social and economic policy development, project finance, investment policy, business-government relations, sustainability, negotiations, development law, and ethics, to name a few.

This book is itself an ongoing research project and my goal is to not only share the lessons that have been learned from experienced practitioners, investors, and policy makers in the field but also to stimulate interest and encourage others to share their experiences so we are all better prepared to meet the challenges of the global megaprojects that the future will demand.

Reference

Mirabile, M., Marchal, V., and Baron, R. (2017). *Technical Note on Estimates of Infrastructure Investment Needs: Background Document to the Report Investing in Climate, Investing in Growth*. Paris, France: Organization for Economic Development (OECD).

Acknowledgments

This book results from a significant collaborative effort involving many organizations, research associations, and individuals. As I explored the tremendous amount of research on global megaprojects, I was truly grateful for the thoughtful analysis from so many diverse projects ranging – from important social endeavors to amazing scientific achievements to extraordinary infrastructure projects connecting cities, countries and regions of the world through transcontinental and transoceanic links.

I have many to thank for their tremendous contributions to this book including my former and present megaproject colleagues dedicated to understanding the tremendous challenges of building complex projects and seeking new ways to improve upon current practice.

To my former colleagues on the Big Dig, now working on other global projects for continuing to share their insights on the new obstacles and challenges they face. I am deeply appreciative of my colleagues across the globe including Wendy Wan and Chung-Leung Luk for our research collaboration, and adding additional perspective on megaproject development in Southeast Asia. I am grateful to the many friends I have worked with in years past at the World Bank, and more recently I have greatly appreciated the support of Peter Vincze and Erick Abiassi for their assistance in sharing the challenges of delivering projects in some of the world's poorest countries. To my colleagues at Harvard Law School thank you for your interest and support in advancing opportunities for students in the field of international development.

To Harry Dimitriou and his team at the Omega Centre for Mega Infrastructure and Development, at University College London for providing insights from their incredible work on transport policy-making and sustainable development on cities and regions in the developing world. To Kristina Gillin for providing understanding on the attributes of a sustainable nuclear decommissioning and sustainable repurposing of projects globally. To Emmanuelle Bernardin for our collaboration on global cyber projects and the challenges they present for the future. To Nicole Hunter for her advancement of professional women in construction.

To Elliott Sclar for our collaboration on research on the building of the Interstate Highway System, and to Alan Altshuler, Fred Salvucci, and Robert Paaswell in sharing your experiences in leading some of the largest transportation megaprojects in the United States and your support for infrastructure endeavors internationally. To Eric Rebentisch, PMI and the team of collaborators I had the pleasure to work with on the development of the book, *Integrating Program Management and Systems Engineering* which has broad applicability to global megaprojects.

To my graduate research assistants, Sanskriti Patwal, Meirgul Kaleshova, and Meruyert Nussambayeva a special thanks for keeping me motivated and excited about this project and for the collection and organization of extensive amounts of research. To Barry Unger for his unique view of megaprojects in the creative economy, to Keith Diggans, and Jim Hannon for sharing their insights on Agile in mega practice, and to Steve Bergstrom for his unique insights on the business and managerial side of global megaprojects.

Also, to my graduate students, and law students at Boston University whose enthusiasm for learning instill in me my passion for teaching. I am also thankful to my fellow professors and colleagues, especially those who provided insight and new direction. To Vijay Kanabar, Stephen Leybourne, Rich Maltzman, Irena Vodenska, John Sulllivan, Gerard Keegan, and to Bill Parker at the Defense Acquisition University for their support for my teaching and research on megaprojects and their ideas for improving the projects of the future. To Dean Zlateva and Sr. Associate Dean Chitkushev for their continuous support of my research efforts.

A special thanks to my publisher, John Wiley & Sons and my Production Editor, Sundaramoorthy Balasubramani and especially Kalli Schultea and Skyler Van Valkenburgh and the editorial team for your willingness to take on this project.

My apologies to anyone I may have forgotten to mention here.

Finally, to the readers of this book, I hope you will find value in learning about the many successes of global megaprojects and how they continue to make a difference in the lives of people all over the world.

Responsibility for errors or omissions in this book remains mine.

Introduction to This Book

> The need to operate as a Whole Team and be aligned on the key objectives cannot be understated. Leadership direction, collaborative attitudes, and contract terms must all encourage participants at every level to always share the truth and do what is in the best interest of the project.
>
> *– Keith Sibley, Program Manager (former leadership roles on multiple mega-projects)*

The goal of this book is to provide valuable lessons and best practices for policy makers, academics, students, scholars, urban planners, developers, architectural, engineering and construction professionals, project owners, sponsors and investors, non-profits, private industry, regulators, and government, environmental, energy, and transportation officials interested in infrastructure and urban and rural development. Better ways must be found to finance infrastructure projects, deliver economic and social benefits, and reduce the cost overruns, schedule delays, and catastrophic potential, and the overall cost of uncertainty, conflict and risk that plague so many global megaprojects.

Key Concepts and Objectives

This book provides an analysis of the difficulties in managing global megaprojects during each phase and throughout the life span of the project. Despite the huge volume of media articles, industry research, scholarly papers, and government analysis that has been published very little has been written about the day-to-day reality of managing these global projects from the perspectives of investors, lenders, government owners, project professionals and the ultimate beneficiaries of these projects. Global megaprojects have rarely been studied in depth for the lessons they offer in the management of megaproject complexity, uncertainty, and risk, particularly in interrelation to financial, economic, technical, legal, political, and social factors. Practicing professionals, policy makers, and students will find useful how these massive endeavors are conceived, developed, and implemented and how they manage to deliver benefits to society despite the trials and tribulations along the way. This book covers broadly the challenges of developing effective strategies and policies, improving project selection methodologies, overcoming emergence, uncertainty and complexity, finding opportunity and innovative solutions to impossible problems, and leading these global projects to long term success. The book offers illustrations and case studies from every corner of the globe to understand success factors, how to avoid project failure and how to create an innovative environment in which these projects can thrive.

The goal is to stimulate new ideas for future megaproject endeavors and to develop advanced thinking on how these projects can be better developed across companies, cultures and countries. The key themes in this book include: (i) selecting projects that have long term sustainability; (ii) building transparent frameworks; (iii) mobilizing capital and investment; (iv) integrating the project organization for more effective outcomes; (v) creating more resilient alliances and collaborative public private partnerships; (vi) building upon the United Nations sustainability goals in project planning and implementation; (vii) developing megaproject innovation and understanding the enablers and challenges to innovation; (viii) introducing blue ocean and blue economy strategies; (ix) reducing complexity and uncertainty and turning risk into opportunity; and (x) learning the lessons from the past and adapting them to the future to ensure more resilient projects that have the potential to develop an ecosystem of their own or become part of an existing ecosystem.

Pedagogy

In the last 50 years there has been a growing interest in Megaprojects the world over. Megaprojects bring together many disciplines and borrow from the fields of social science, psychology, biology, engineering, technology, energy, transportation, physics, politics, law, management, finance and economics to name a few. Megaprojects are finding answers to challenging problems such as medical treatments, water resourcing and disaster relief. Bent Flyvbjerg (2014) identifies the four "sublimes" – political, technological, economic, and aesthetic to explain the increased size and frequency of megaprojects.

There has been a call from the business, management, and project management communities for decades for more research on the history of project management. We need to learn from the past to do better in the future (Söderlund et al. 2017). Though there have been case studies on some of the larger industrial and engineering projects of our time there is a need for more studies to fill the gap of our understanding of best practices for megaprojects. Especially practices that seem to be critical for the success of projects and that are grounded on a contextual understanding of the particular project at hand such as planning techniques, coordination mechanisms, team structures, and interorganizational collaborations (Söderlund and Lenfle 2011).

As you will learn throughout this book megaproject theory assists policy makers, industry, project stakeholders including contractors, designers, engineers, and government in understanding better the challenges we face in selecting the right projects, implementing these large and complex endeavors and then ensuring that we deliver the benefits promised. Megaprojects can provide frameworks for structural decision making, strategy alignment, and investment choices that can be beneficial to all projects (Esty 2004). There is now a growing body of research on megaprojects that provides valuable insights that can be applied not only on megaprojects but provides valuable lessons for projects of all sizes large and small. The lessons learned will assist all those involved in the future challenges of megaproject management.

Since this book is designed to meet academic needs as well as the needs of professional practice, it is structured to provide pedagogical tools to enhance the learning experience.

Each chapter provides background on the major concepts as drawn from the extensive literature, from the major multilateral organizations, from the project sponsors, and most important from hundreds of interviews with many megaproject stakeholders and participants. Without the insights and extensive experience of policy makers, project managers, engineers, and contractors this book would not be possible.

Book Structure

This book contains 12 chapters. As described here, the chapters correspond to different aspects of global megaprojects from their initiation through implementation to transition to operations. Highlighted are the most important practices and each chapter explains how these practices and strategies can be implemented to achieve the approval of the project stakeholders and project management and, ultimately, to ensure project success.

Each chapter contains a similar framework and flow and includes the following elements:

1) *Introduction*: An overview of the topic covered and the goals of the chapter along with the role of the topic within the broader context of project and program management.
2) *Concepts:* An explanation of the fundamental elements that build upon the various disciplines and strategies, methodologies, and tools and techniques used in megaprojects generally. Relevant theoretical and empirical studies are highlighted in each chapter.
3) *Summary:* A review of the key points raised in the chapter.
4) *Lessons learned:* Highlights some of the most important lessons learned helping the reader understand how the concepts are applied in actual projects.
5) *Best practices:* Provides examples of strategies and tools and techniques to improve upon current practice.

6) *Discussion questions:* Discussion Questions incorporate case studies, critical thinking, and problem-solving exercises that afford students an opportunity to apply the concepts reviewed in each chapter.
7) *References:* References include a list of the author's research and the relevant literature on the chapter topic.

Overview of Book Chapters

Chapter 1: Introduction to Global Megaprojects

The first chapter provides an introduction to global megaprojects and includes a timeline of these projects from prehistory through to the present-day Millennium Projects. This chapter explains why the study of megaprojects is important and how these projects have influenced our world history and built a better society. The primary goal of this chapter is to set forth a framework for understanding the goals and benefits of these projects, and analyzes what makes megaprojects unique and worthy of future analysis and research.

Chapter 2: Megaproject Finance: Innovation and Value Driven Global Megaprojects

Chapter 2 covers the financing of global megaprojects from conception through project completion. It contrasts project finance with corporate finance and the advantages and disadvantages of each type of finance. It provides an overview of the challenges of infrastructure financing in the developing world, the United States and the European Union. The chapter explores the sources of funding, the legal framework, the major agreements and the role of the multilateral development banks and public private partnership structures. Finally, it emphasizes the importance of megaproject evaluation from both an ex-ante and ex-post approach and throughout the project life cycle.

Chapter 3: The Multilaterals and World Development

Chapter 3 provides an overview of the role of multilaterals in development and the mobilization of capital. It reviews the changing landscape of development banks and the financing of private investment in infrastructure projects. The chapter includes development case studies from around the world and shows the social, economic and institutional value of megaprojects. Finally, frameworks for sustainable development are presented along with a collective action perspective on the planning of global megaprojects.

Chapter 4: Leading Complex Global Projects

Chapter 4 introduces the characteristics of complex global projects and the importance of complexity in relation to uncertainty, ambiguity, conflict and risk. Several case studies of complex global projects are analyzed and strategies for reducing and managing complexity is explained.

Chapter 5: Global Megaproject Governance

Chapter 5 introduces how global megaprojects are governed through a framework that can differ from developed to developing countries. Case studies in governance gone right and governance gone wrong are introduced along with recommendations for building a more sustainable and resilient governance structure.

Chapter 6: Integrating Project Organizations and Public Private Partnerships

Chapter 6 focuses on the importance of project integration at the organizational level as well as integration of the disciplines of systems engineering and program management. Several studies in integration are presented along with the recent thinking concerning the structure of megaprojects as systems of systems. Emerging trends and social considerations in public private partnership development are analyzed through case studies from multiple countries and regions.

Chapter 7: Managing the Megaproject Implementation and Delivery

Chapter 7 traces the implementation of a megaproject from initiation through to the transitioning of a megaproject to operations. Two of the world's largest megaprojects serve as case studies and the lessons learned from these projects on cost and schedule failures are critically analyzed along with a discussion of solutions.

Chapter 8: Megaprojects and Mega Risk: Opportunity, Risk and Resilience Management

Emerging risks on a global scale are impacting megaprojects. Chapter 8 explores all aspects of risk management from turning risk into opportunity, to structuring projects for resilience and using risk intelligence to assist in risk mitigation and build more sustainable project structures. International projects are assessed in terms of global risk factors along with root cause analysis, normalization of deviance and reducing catastrophic loss potential.

Chapter 9: Megaprojects: Trouble and Triumphs

Global megaprojects are known for their amazing feats but also for some dramatic failures. Chapter 9 explores both by looking at the characteristics of failed projects as well as the characteristics of the world's most successful megaprojects and solutions to ensuring success and avoiding failure.

Chapter 10: Laws and Contracts in Global Megaprojects

Enforceable contracts are the backbone of successful global megaprojects. Chapter 10 includes planning for procurement both through public and private sector procurement. Laws governing international megaproject contracting are reviewed and challenging negotiations are explored. Emerging issues in global development laws and contracts are analyzed and the chapter concludes with an extensive checklist for contract development in emerging markets.

Chapter 11: Megaproject Innovation and Resilience

Megaprojects desperately need innovation to keep up with the rapidly changing technological and social environment. This chapter explains where innovation comes from and how to develop an innovation strategy that is unique to megaprojects. Enablers and challenges to innovation are reviewed along with best practices for innovation through the global innovation index pillars.

Chapter 12: The Future of Global Megaprojects

The concluding chapter introduces innovative thinking through the blue ocean and blue economy strategies. The chapter focuses on some of the Grand Challenges of the twenty-first century including global climate change, energy independence and food security for hundreds of millions of people around the globe. Global change and international cooperation are highlighted as important challenges of the coming decades as well as the impact of the new technologies on human rights and privacy. The leadership for megaprojects in the emergent era requires new approaches and skills along with a new principle-based approach.

References

Esty, B. (2004). *Modern Project Finance.* Hoboken, NJ: John Wiley & Sons.

Flyvbjerg, B. (2014). What you should know about megaprojects and why: An overview. *Project Management Journal* 45 (2): 6–19.

Mirabile, M., Marchal, V., and Baron, R. (2017). *Technical Notes on Estimates of Infrastructure Investment Needs: Background Note to the Report. Investing in Climate, Investing in Growth.* Paris, France: Organization for Economic Cooperation and Development (OECD).

Söderlund, J. and Lenfle, S. (ed.) (2011). Special issue: Project history. *International Journal of Project Management* 29 (5): 491–493.

Söderlund, J., Sankaran, S., and Biesenthal, C. (2017). The past and present of megaprojects. *Project Management Journal* 48 (6): 5–16.

1

Introduction to Global Megaprojects

It is not the strongest of the species that survives, nor the most intelligent, but rather the one most adaptable to change.

— Charles Darwin

Introduction

Megaproject infrastructure has long been an essential factor in socioeconomic development around the world, yet we still know so little about the phenomenon of these massive endeavors. The objective of this chapter is for readers to learn about the characteristics of global megaprojects and how these projects have evolved over centuries beginning with prehistory up to the present day. We look at the contributions as well as the challenges of these projects and the ways in which these projects are conceived and developed including the benefits of megaprojects to the larger eco system such as the expansion of trade and development globally. While the primary goal of this chapter is to set forth a framework for understanding the importance of megaprojects in the larger economic and social environment, it also analyzes what makes megaprojects unique and worthy of future analysis and research.

Infrastructure of all kinds has become a subject of great importance to governments in every corner of the world from the development of the $4.75B Hadron Collider at CERN to the Belt and Road Initiative in China (BRI), to the creation of the Smart Cities Projects in India, to the development of London's Crossrail Project, and the implementation of America's Interstate Highway System and the High-Speed Rail System with initial focus on California. The Organization for Economic Cooperation and Development (OECD) estimates global investment needs of $6.3 trillion per year from 2016 to 2030, without considering further

Global Megaprojects: Lessons, Case Studies, and Expert Advice on International Megaproject Management, First Edition. Virginia A. Greiman.

climate action (Mirabile et al. 2017). To meet this growing demand, there has been a recent call, within the megaproject scholarship, for a better understanding of "what goes on in megaprojects – how they are managed and organized, from within, by the managers who are tasked with bringing them to fruition" (Söderlund et al. 2017).

The goal of this book is to provide insight, based on experience, for more realistically approaching the complex and transformative process of infrastructure investment that the urban world of the twenty-first century demands (Greiman and Sclar 2019). We begin by proposing a definition of global megaprojects based on the extensive research on large-scale projects.

Megaprojects have existed since the beginning of recorded time and have been characterized by great monuments, feats of engineering, and lasting images of something more powerful than had been imagined in the past. Since prehistoric times we have been building megaprojects, so there is much to learn about these remarkable achievements of earlier generations. In this book, we look at projects not just from the traditional cost and schedule perspectives written about in the leading project management journals, but more significantly this book focuses on the great benefits and achievements of these projects so we can learn not just what went right or wrong; but more importantly what are the factors in achieving success and how these projects have evolved and been shaped over time. It is critical that we learn from the past, but we also must apply these important lessons to our megaprojects of the future. This book was written to demonstrate that megaprojects can and have accomplished major economic, social, and technical advancements thought impossible but achieved by successfully confronting the challenges of the time.

This chapter focuses on defining the unique complexity of global megaprojects, assessing the impact of cross-institutional differences, and identifying ways for its sponsors, managers, and other stakeholders to address these challenges.

Globalization and Megaprojects

> Globalization presumes sustained economic growth. Otherwise, the process loses its economic benefits and political support.
> — *Paul Samuelson, An American economist and the first American to win the Nobel Prize in Economic Sciences in 1970 known as the "father of modern economics"*

Before embarking on a review of global megaprojects over the ages, it is important to understand the influence that megaprojects have had not just on local communities but on globalization worldwide. The modern world is increasingly being defined by the term "globalization." Important changes in the global economy have become the main determinants of this phenomenon. Friedman (2000) defines globalization as "the inexorable integration of markets, nation states and technologies to a degree never witnessed before – in a way that is enabling individuals, corporations, and nation states to reach around the world farther, faster, deeper, and cheaper than ever before."

Globalization means the spread of free-market capitalism to virtually every country in the world. Globalization also has its own set of economic rules – rules that revolve around opening, deregulating, and privatizing your economy (Friedman 2000). Globalization is also used to describe the growing interdependence of the world's economies, cultures, and populations, brought about by cross-border trade in goods and services, technology, and flows of investment, people, and information (PIIE 2021).

Importantly, globalization encourages the development of megaprojects. First, globalization of financial capital and law facilitates cross-national partnerships. Second, technological innovations secure big, complex constructions (Kardes et al. 2013). Megaprojects occupy an important place in global relations and drive a number of explicit and implicit economic agendas (Pitsis et al. 2018). International public–private partnerships have allowed large-scale government projects, such as roads, bridges, or hospitals, to be completed with private funding and thus expanded opportunities for globalization through interconnectivity among cities, states, regions, and borders. The goal of this book is to provide insight, based on the experience of those who have contributed to these projects, for more realistically approaching the complex and transformative process of infrastructure investment that the urban world of the twenty-first century demands (Greiman and Sclar 2019).

The next generation megaprojects will advance science and technology in a way that has never been done before. The International Space Station and the Hadron Collider Project at CERN are just two examples of unparalleled projects that will increase our knowledge and further our understanding about the universe for centuries to come.

Characterization of Global Megaprojects

As shown in Figure 1.1, megaprojects have been characterized by size, duration, uncertainty, ambiguity, complexity, cultural environments, dynamic governance, large-scale policy making, and significant political and external influences (Greiman 2013; Klakegg et al. 2008; Merrow 2011; Miller and Hobbs 2005). Megaprojects also have a greater magnitude of aspiration, actor involvement, and impact than smaller projects (Flyvbjerg 2017). Despite the massive investment in

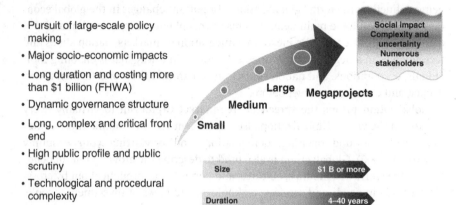

- Pursuit of large-scale policy making
- Major socio-economic impacts
- Long duration and costing more than $1 billion (FHWA)
- Dynamic governance structure
- Long, complex and critical front end
- High public profile and public scrutiny
- Technological and procedural complexity

Figure 1.1 Characteristics of a megaproject. *Source:* Greiman (2013)/John Wiley & Sons.

these projects, many of them fail to meet the expected objectives for which they were built (Dimitriou et al. 2014). Large, complex infrastructure projects often delivered via public–private partnerships entail deep uncertainties, considerable economic and political stakes, and have a significant impact on society. Megaprojects proliferate, despite their problematic relation with sustainability (Lehtonen 2019), and have been defined as large ventures requiring enormous investments and are portrayed as too big to fail and too costly to stop (Denicol et al. 2020). Megaprojects are rarely implemented by a single organization; they are characterized by decentralized decision-making power, rapid resource allocation, and complex stakeholders (Müller 2012).

Megaprojects are not confined to large infrastructure or industrial projects but include research and development, cyber infrastructure, scientific discovery, defense, energy, food security, biotechnology, and many other fields. Given the increasing number of megaprojects and the movement toward the professionalization of the discipline, project management practice has struggled to develop workable solutions and practices to address complexities that exceed the technical concerns of engineering (Clegg et al. 2017; van Marrewijk et al. 2016). Gellert and Lynch (2003, pp. 15–16) consider megaprojects as "displacements" by observing that megaprojects are "projects which transform landscapes rapidly, intentionally, and profoundly in very visible ways, and require coordinated applications of capital and state power." Essentially, looking at society through its megaprojects would reveal its ambitions, problems, as well as its future outlooks. Increasingly, it is evident that the problem areas attached to these projects stretch beyond technical issues: they must be considered as socio-technical endeavors embedded in complex institutional frames. Studying how to deal with institutional differences in the environment of megaprojects has both theoretical and practical implications (Biesenthal et al. 2018).

Though the project management literature is filled with many characterizations of a megaproject, there is no single agreed upon definition. Over the years megaprojects have been defined by different terms. Some of the more common terms or synonyms used to define or characterize megaprojects include: large-scale project, major project, public private partnership, program or portfolio, international development, economic development, global project, disaster relief project, urban megaproject, urban renewal, transportation infrastructure, critical national infrastructure, sustainable development, complex project, systems of systems, large research and development project, transformational project, large engineering project, mega construction project, capital project, giga project, tera project, public works, civil engineering project, systems engineering project, and project finance.

In the United States, the Federal Highway Administration (FHWA) defines megaprojects as:

> Major infrastructure projects that cost more than $1 billion, or projects of a significant cost that attract a high level of public attention or political interest because of substantial direct and indirect impacts on the community, environment, and State budgets. "Mega" also connotes the skill level and attention required to manage the project successfully. (Capka 2004)

Most megaprojects are considered global because of their contribution to trade and economic growth, technology transfer, and regional and international development. Global megaprojects often cross borders and are defined as involving two or more countries.

For example, Levitt and Scott (2017) in their research on institutional and political challenges in projects adopted the formulation developed by Orr and Levitt (2011, p. 17) and defined a "global megaproject" as:

> A temporary endeavor where multiple actors seek to optimize outcomes by combining resources from multiple sites, organizations, cultures, and geographies through a combination of contractual, hierarchical, and network-based modes of organization.

While Flyvbjerg (2014) defines "global megaprojects" as:

> large-scale, complex ventures that typically cost more than 1 billion US dollars, take many years to build, involve multiple public and private stakeholders, are transformational, and impact millions of people.

As described in this chapter, to better capture the changing nature of global megaprojects along with their complex financing and governance structures the following definition is proposed:

> Global megaprojects consist of the creation of a special organizational structure for the purpose of financing a unique multi-billion-dollar investment involving numerous stakeholders and complex interdependencies and interorganizational relations in diverse regions of the world that will provide sustainable and long-term benefits to local communities and the larger eco system of which the megaproject is a part.

Megaprojects: The Global Timeline

> The difficult is what takes a little time, the impossible is what takes a little longer
>
> — *Fridtjof Nansen, Nobel Peace Prize Winner, 1922*

To understand how far we have come in building the great engineering and architectural triumphs of our time, we must look backward to understand the present. Figure 1.2 illustrates the timeline for some of the world's greatest megaprojects. Though the timeline shows only a small representation of these projects, many of them from long ago eras are still in existence showing that megaprojects can be sustainable long beyond the lives of its creators and some are still rendering benefits today.

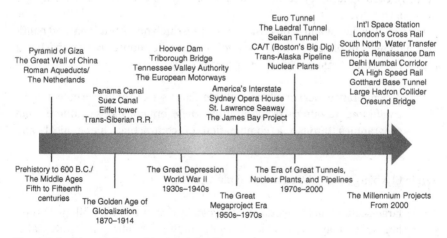

Figure 1.2 Megaprojects: the global timeline.

Prehistory

The Pyramid of Giza (Built Between 2550 and 2490 BCE)

The oldest and last existing Seven Wonders of the Ancient World, the great Pyramid of Giza was built in Egypt. Construction began on the pyramid around 2550 BCE. The Great Pyramid was built over a 60-year period as a tomb, based on markings on an interior chamber referencing fourth dynasty Egyptian Pharaoh Khufu. Originally standing at 481 feet tall, it was the tallest man-made structure in the world for more than 3800 years. Due to erosion, it now stands about 455 feet tall. There is numerous evidence that the Ancient Egyptians were pioneers in geology and geotechnical engineering (Agaby et al. 2014). According to research the workmanship of the Great Pyramid is surprisingly accurate, as the four sides of the base have an average error of only 58 mm. These structures represent the pinnacle of Egyptian pyramid building, built on a foundation of experience from the earlier much smaller pyramids.

Pyramid of Giza. *Source:* Picture Alliance.

The Great Wall of China

Other great medieval projects, include the great wall of China first built in the seventh century as protection against various nomadic groups, and has been proclaimed as one of the most impressive architectural feats in history. Though the

Great Wall, which opened in 220 BCE was built over a 200-year period, it never effectively prevented invaders from entering China, however, it came to function as a powerful symbol of Chinese civilization's enduring strength.

The Aqueducts of Rome

One of the earliest examples of the exploitation of groundwater to sustain human civilization is the aqueduct system of ancient Rome which were being built from 312 BCE to 226 CE. Although some of the aqueducts were fed by surface water, most of them were supplied by springs, usually augmented by tunneling to increase the flow of groundwater (Deming 2020). The Aqueducts were considered a major feat of engineering at that time. Though earlier civilizations in Egypt and India also built aqueducts, the Romans improved on the structure and built an extensive and complex network across their territories. Although there were many more great engineering and architectural achievements during this time demonstrated by the great Cathedrals of the middle-ages, remarkably many of these structures still exist today though some in a different form. For example, the Trevi Fountain is still fed by aqueduct water from the same sources of the ancient Aqua Virgo; however, the Acqua Vergine Nuova is now a pressurized aqueduct.

When we contrast these medieval projects with the megaprojects of today, we wonder if many will even survive the next 20–30 years. The ever-advancing technology we have at present causes us to move more rapidly in building projects, that years ago gave us the luxury of time.

The Middle Ages

The Netherlands

It was during the Middle Ages that some of the most inspiring projects of all time were built. Notably, one of the most impressive engineering feats from the Middle Ages has been "The Netherlands" where the people of this country surmounted near impossible obstacles to reclaim thousands of square kilometers of land from the sea (Butler 1972; Hill 1984). Originating as a fourteenth-century settlement along a small peat river, Rotterdam thanks to the building of The Dam in the Rotte, in 1270, eventually grew into Europe's largest seaport. By 2009, 400 million tons of cargo traveled through the port, but Rotterdam was nearing its capacity. To keep the port competitive, authorities undertook an ambitious project aimed at tripling the port's capacity (Rosenberg 2020). Directly accessible from the North Sea Rotterdam now holds the biggest harbor in Europe that has applied modern, forward-thinking management including serving as a home for the importation of green hydrogen and the leading port for sustainable energy. Green hydrogen can play a valuable role in the energy transition, which in turn is indispensable for meeting the Paris Climate Agreement objectives (Rotterdam 2020).

The Golden Age of Globalization (1870–1914)

During the period 1870–1914 sometimes referred to as the "Golden Age of Globalization" when international trade, international migration flows and international mobility of financial capital reached historical peaks, the world pursued ambitious infrastructure projects, many of which define our world today such as the Panama Canal, the Suez Canal, The Eiffel Tower, and the Trans-Siberian railroad. Not all were intended for transnational use, but they represented a certain confidence and an unprecedented desire to bind the world together. Megaprojects have arisen from a complex set of geographic, economic and, above all, political processes of restructuring occurring throughout the world since that time. Which raises important questions concerning the obstacles that confronted early engineers and how they coped with limited technology and far fewer resources.

The Panama Canal

Connecting the Pacific and Atlantic oceans, the 77 km shipping canal across the Isthmus of Panama, a key international trade channel, was one of the most difficult engineering projects in history. Twenty thousand people lost their lives (mostly due to disease) one of the worst worksite disasters in history. The Canal was opened in 1914 after 34 years of construction. The construction of the Panama Canal early in the twentieth century dramatically changed trade patterns by opening new routes between countries and regions that traditionally could not trade at competitive prices due to the vast distance between them (Sabonge 2014). The Canal was built mainly for military purposes but, over time, became a facilitator for trade by shortening the time and distance between production and consumption markets. From its inauguration until Fiscal Year 2013, more than one million vessels have transited the Canal with more than 9.4B long tons of cargo (Sabonge 2014). Operating around-the-clock, about 40 vessels pass through each day, including tankers, cargo ships, yachts, and cruise ships. The Canal enhances environmental contribution by reducing GHG (green house gas) emissions on the planet with more efficient transport, reducing fuel consumption per cargo unit and fewer emissions than other routes that combine transportation by land.

The Panama Canal. *Source:* Panama Canal Authority.

The Suez Canal

The Suez Canal, a man-made waterway was built in Egypt by ex-French diplomat and developer, Ferdinand de Lesseps and opened in 1869 after 10 years of construction. At 120 miles long, it connects the Mediterranean Sea to the Indian Ocean by way of the Red Sea and separates most of Egypt from the Sinai Peninsula. It was considered an audacious project of engineering at the time and provided a foundation for geographic relationships by connecting the two most densely populated areas in the world – Western Europe and Eastern Asia (Yackee 2018).

The Eiffel Tower

An object of discord, desire, and fascination, the Eiffel Tower was created by Gustave Eiffel and built-in record time – two years, two months, and five days. It was established as a veritable technological feat. Considering the rudimentary means available at that period, this could be considered record speed. The assembly of the Tower, built to be one the main attractions at the Paris World's Fair in 1889, was a marvel of precision, as all chroniclers of the period agree. The construction work began in January 1887 and was finished on 31 March 1889, at a cost of US$1.5 million. On the narrow platform at the top, Gustave Eiffel received his decoration from the Legion of Honor (Eiffel 2021).

The Trans-Siberian Railway. *Source:* Regent Holidays.

The Trans-Siberian Railway

The "Trans-Siberian Main Railroad," is the longest single rail system in Russia, stretching from Moscow to Vladivostok or (beyond Vladivostok) to the port station of Nakhodka spanning 6000 miles and 7 time zones. It had great importance in the economic, military, and imperial history of the Russian Empire and the Soviet Union (Bassin 1991; Wolmar 2014). Conceived by Tsar Alexander III, the construction of the railroad began in 1891 and proceeded simultaneously in several sections – from the west (Moscow) and from the east (Vladivostok) and across intermediate reaches by way of the Mid-Siberian Railway, the Transbaikal Railway, and other lines. Throughout the years, the Trans-Siberian Railway (TSR) has been proven to have the longest history of commercial freight operation between Europe and the Far East (Liliopoulou et al. 2005). Originally, in the east, the Russians secured Chinese permission to build a line directly across Manchuria (the Chinese Eastern Railway) from the Transbaikal region to Vladivostok; this trans-Manchurian line was completed in 1901. After the Russo-Japanese War of 1904–1905, however, Russia feared Japan's possible takeover of Manchuria and proceeded to build a longer and more difficult alternative route, the Amur Railway, through to Vladivostok; this line was completed in 1916. The Trans-Siberian Railroad thus had two completion dates: in 1904 all the sections from Moscow to Vladivostok were linked and completed running through Manchuria. In 1916, there was finally a Trans-Siberian Railroad wholly within Russian territory. "The completion of the railroad marked the turning point in the history of Siberia, opening vast areas to exploitation, settlement, and industrialization" (Britannica 2021).

The Great Depression (1930–1940s)

During the 1930s, in the United States, the Public Works Administration, (later named the Work Projects Administration) and other New Deal agencies funded projects to build and improve the country's infrastructure, including roads, dams, schools, airports and parks. Many of the projects funded by the PWA and WPA remain today as part of the U.S. landscape. The New Deal also created new agencies to fund projects across the country that both improved communities and provided jobs at a time when unemployment was high. While the U.S. was building large projects like the Hoover Dam, Europe was building Motorways.

The Hoover Dam. *Source:* Chris/Adobe stock.

The Hoover Dam

The waters of this great river, instead of being wasted in the sea, will now be brought into use by man. Civilization advances with the practical application of knowledge in such structures as the one being built here in the pathway of one of the great rivers of the continent. The spread of its values in human happiness is beyond computation.

— Herbert Hoover, November 1932

The 1930s brought us the Hoover Dam described as an audacious and courageous undertaking. Built during the Great Depression, the dam would tame the flood-prone Colorado River southeast of Las Vegas – protecting cities and farms, generating cheap electricity to supply power to homes and industry, and providing work for thousands who desperately needed jobs. A consortium called Six Companies Inc., won the right to build the concrete arch dam, at a cost of nearly $49 million – a staggering amount in the early 1930s (roughly equivalent to $860 million today). Skeptics thought it could not be done. Others were convinced that the contractors would go broke. But the workers in the consortium boldly moved forward, drawing on their considerable, collective knowledge and experience, managing huge risks, and pioneering as they went. The dam was built in five years, two years ahead of schedule and under budget, an amazing feat for its time giving company leaders the confidence that they could take on any project, anytime, anywhere (Bechtel 2001). Today, it generates enough hydroelectric power per year to serve 1.3 million people.

Though the Hoover Dam was one of the larger projects of this era, other megaprojects included several major New York projects including the Empire State Building, the Triborough Bridge, as well as the construction or improvement of roughly 800 airports, including LaGuardia Airport in Queens, New York.

The Triborough Bridge

As Robert Caro wrote in his book, The Power Broker, "the man who built the Triborough Bridge would be a man who conferred a great boon on the greatest city in the New World. He would be the man who tied that city together" (Caro 1975). And though the Triborough Bridge was an immense feat of construction, the designer, Robert Moses' vision for the structure did not end with the bridge. Rather, Moses continued to advocate for new roads and parkways that would feed into the bridge, all of which would be part of an interconnected parkway system. To Moses, the Triborough Bridge promised to "slash at a stroke the immense Gordian knot of the East River traffic problem" by creating a direct link between the Bronx and Queens. The parkways connecting to the Triborough were a vital piece of his steel and concrete puzzle (Caro 1975).

The Tennessee Valley Authority Projects

In 1933, the New Deal legislation created a public corporation to improve the Tennessee Valley. One of the biggest projects that this corporation, the Tennessee Valley Authority (TVA), took on was the Chickamauga Dam, located on the

Tennessee River outside of Chattanooga preventing flood damage and a mosquito infestation. As described by Arthur E. Morgan, Chairman of the Tennessee Valley Authority at the time of its implementation, "[t]he proper way to treat the TVA is not as an isolated undertaking, but as an integral part of the whole program of the present administration" (Morgan 1934). FDR's ambitious plan transformed the Tennessee Valley by creating dams and reservoirs for electricity and flood control, controlling soil erosion through forest restoration and better farming techniques, and improving navigation and commerce along the Tennessee River. By 1934, more than 9000 people found employment with the TVA. The agency built 16 hydroelectric dams in the Tennessee Valley between 1933 and 1944 though these projects were not without controversy from the local communities.

The European Motorways

On 26 March 1923, in a formal ceremony, construction of the Milan–Alpine Lakes autostrada officially began, the preliminary step toward what would become the first European motorway. In 1935, the year in which the Genoa–Serravalle Truckway was completed, Italy possessed barely 500 km of motorway. Germany achieved a network of over 3600 km between 1934 and 1941. The last authorizations for construction in Italy were in 1930, that is, three or four years before the German projects got underway (Moraglio 2017). Using the great economic crisis as a dividing line, the Italian motorway projects were mostly realized before the 1929 U.S. stock market crisis made its effects felt in Europe, while in Germany, they were realized after. Italy was not the only nation to propose similar projects. In following the German example, in the second half of the 1930s, France and Holland began construction works on their first motorway trunks, while news of new projects in Denmark, Belgium, Poland, and Czechoslovakia started shortly after.

The Great Megaproject Era (1950–1970s)

In the United States, Altshuler and Luberoff (2003) identify the period between the late 1950s. when the Federal Interstate construction project got underway and the late 1960s when construction peaked, as the Great Megaproject Era in America largely due to the development of the National System of Interstate Highways.

The Interstate Highway System

During the post-war era defined as the period from the late 1940s to the late 1960s growth was defined through several major megaprojects that began the process of connecting cities and states and even countries. In the United States, this was the beginning of the Dwight D. Eisenhower National System of Interstate and Defense Highways, commonly known as the Interstate Highway System called the "Greatest Public Works project in History" (Weingroff 2017).

As these interstates moved from open countryside into dense urban areas requiring the extensive use of eminent domain to take homes and businesses, they triggered a reaction in the form of widespread anti-highway sentiment (Altshuler and Luberoff 2003). The sentiment quickly became politically manifested in the late 1960s and early 1970s. Anti-highway sentiment soon merged with the then nascent but rising environmental consciousness to create the mix of pro transit ideas that fueled the political consensus that sustained the Big Dig and similar efforts across the nation (Greiman and Sclar 2019). The estimated completion cost of the system was $129B (Weingroff 2017). The approval of the Federal-Aid Highway Act of 1956 marked the formal beginning of this project. The completion of the Big Dig in 2007 about one-half century later marked its end.

As he looked back on his two terms in office, former President Dwight D. Eisenhower said of the Interstate System that, "[m]ore than any single action by the government since the end of the war, this one would change the face of America." The impacts of the Interstate System remain controversial, but it did, as President Eisenhower predicted, change the face of America – not simply by altering the landscape during construction, but by supporting changes that transformed our society in the second half of the twentieth century (Weingroff 2017).

The Sydney Opera House

Sydney Opera House. *Source:* Sydney Australia/Wikimedia Commons/CC BY 1.0.

Outside of North America other great projects were under way in the late 1950s. In 1957, when the Sydney Opera House project was awarded by an international jury to Danish architect Jørn Utzon, who left the project almost 10 years later in 1966 and never returned to see his design completed. It marked a radically new approach to construction. Inaugurated in 1973, the Sydney Opera House is a great architectural work of the twentieth century that brings together multiple strands of creativity and innovation in both architectural form and structural design. (Hale and Macdonald 2005).

It cost $102 million and took 14 years instead of the projected 4 years to build. More than 10.9 million people visit the Opera House each year. A great urban sculpture set in a remarkable waterscape, at the tip of a peninsula projecting into Sydney Harbour, the building has had an enduring influence on architecture. The Sydney Opera House comprises three groups of interlocking vaulted "shells" which roof two main performance halls and a restaurant. These shell-structures are set upon a vast platform and are surrounded by terrace areas that function as pedestrian concourses. Described as a daring and visionary experiment on 28 June 2007 the Sydney Opera House was included on the UNESCO World Heritage List under the World Heritage Convention, placing it alongside the Taj Mahal, the ancient Pyramids of Egypt, and the Great Wall of China as one of the most outstanding places on Earth. It is the youngest cultural site to ever be included on the World Heritage list and one of only two cultural sites to be listed during the lifetime of its architect, Jørn Utzon (1918–2008).

St. Lawrence Seaway

The St. Lawrence Seaway is a multi-country project linking the Great Lakes in the United States and the St. Lawrence River in Canada with the Atlantic Ocean via a system of locks, canals, and channels. Recognized as one of the most challenging engineering achievements in history, construction began on the project in 1954 and the project officially opened on 26 June 1959. This famous Canal serves an international need by moving goods from the Great Lakes basin to international destinations.

The James Bay Project

The James Bay Project involved the construction of a series of hydroelectric power stations on the La Grande River in northwestern Quebec. The project represents a series of major construction, each one of which can be considered a megaproject on its own. La Grande River is located between James Bay to the west and Labrador to the east, and its waters flow from the Laurentian Plateau of the Canadian Shield. The project covers an area the size of New York State and is one of the largest hydroelectric systems in the world. It cost more than US$20B to build. The development of the James Bay Project led to an acrimonious conflict with the 5000 Crees and 4000 Inuit of Northern Quebec over land rights, lifestyle and environmental issues. A ruling against the Quebec government in 1973 forced the government to negotiate a far-reaching agreement, known as the James Bay and Northern Quebec Agreement (Peters 1999).

The Era of Great Tunnel, Energy, and Pipeline Projects: (1970s–2000)

The 1970s to 2000 were dominated by mega construction projects such as the Trans-Alaska Pipeline and Boston's Big Dig and the explosion of the nuclear industry. During this time sensitivity developed in the project management industry around sustainability and environmental issues paving the way for the megaprojects of the future. However, sophistication was still lacking in controlling the costs and schedules on these megaprojects as witnessed by the building of Eurotunnel with its private funding and lack of government involvement.

Eurotunnel

Eurotunnel is the largest privately financed infrastructure in history. It is comprised of three tunnels connecting Britain to Continental Europe from terminals in Folkstone in Kent, and Coquelles near Calais in northern France. It had a long history with its first conception in 1802 to its completion in 1994 with lots of stops and starts along the way (Grant 1997). The Tunnel took seven years to build and is still the longest undersea tunnel in the world. It was initially financed by 15 founding shareholders including 5 banks for a 65-year concession. Ultimately growing into a syndicate of 220 banks requiring a massive restructuring of debt and equity due to much higher costs and delays than originally anticipated. The debt will not be fully paid off until 2052 at the earliest.

The Laerdal Tunnel

Opened in 2000, the Laerdal Tunnel is the world's longest road tunnel with a length of 24.5 km (15.2 miles) replacing the Gotthard tunnel as the longest road tunnel in the world. It serves to connect two cities in Norway, Oslo, and Bergen, eliminating the need to drive through narrow mountain passes that are often blanketed in snow. The tunnel cost $153 million to build. To construct it, a total of

3.3 million cubic yards of rock were removed during construction from 1995 to 2000. The tunnel is divided into four sections at 3.7-mile intervals. "The disposal of 2.5 million cubic metres of excavated rock from the tunnel was one of the greatest challenges in planning the tunnel" (RTT 2021). From an environmental perspective, the tunnel was seen as a justifiable investment to avoid destroying large sections of the unspoiled natural landscape.

Seikan Tunnel

Opened in 1988, 27 years after construction commenced, the 53.8 km (32.9 mile) $3.6B Seikan Tunnel in Japan is the world's second deepest and second longest rail tunnel with approximately 23.3 km below the seabed. The Seikan Tunnel is the world's longest tunnel with an undersea segment, the Channel Tunnel, while shorter, has a longer undersea segment). Conceptual planning was during 1939–1940, and construction began in 1971. Construction of the tunnel, which runs 240 m below the sea surface at the deepest part, was extremely difficult, with workers facing numerous problems such as frequent landslides and flooding seawater. Thirty-four workers lost their lives (Japan 2018).

It was built to eliminate the often-challenging crossing across Japan's Tsugaru Strait, which is frequently beset by storms. The tunnel goes to a depth of nearly 800 feet below sea-level and connects the islands Honshu and Hokkaido. Tunneling for this project occurred simultaneously from the northern and southern ends, with the dry land portions using traditional mountain tunneling techniques. For the undersea portion, three bores were used with increasing diameters: a pilot tunnel, a service tunnel, and finally, the main tunnel. Beneath the Tsugaru Strait, the use of a tunnel boring machine had to be abandoned because of the variable nature of the rock. Dynamite blasting and mechanical picking had to be used instead (Japan 2018).

The Central Artery/Tunnel Project (CA/T)

The $14.8B Central Artery/Tunnel Project is famously known as the Big Dig. This project, like most megaprojects, grew from a vision of a small group of people who saw an opportunity for a city in desperate need of revitalization. The Big Dig has been depicted as one of the great projects of the twenty-first century (Tobin 2001). Because of its scale and impacts, the project has been a major issue in national and local politics for more than three decades (Greiman 2013). The project was first conceived in the 1970s, but initial funding was not secured until the early 1980s, and substantial completion did not occur until 2006 more than a 30-year period (CA/T 1990).

The Big Dig was also a record for the United States – it was the first and largest inner-city construction project ever conceived. It was the most complex urban infrastructure project in U.S. history and included unprecedented planning and engineering (EDRG 2006a; EDRG 2006b). Its list of engineering marvels includes the deepest underwater connection and the largest slurry wall application in North America, unprecedented ground freezing and tunnel jacking, extensive deep-soil mixing programs to stabilize Boston's soils, the widest cable-stayed bridge, and the largest tunnel ventilation system in the world (Greiman 2013, p. 39). The Big Dig faced highly unusual challenges, including the necessity of working in one of the most congested urban areas in the country. Coordinating more than 132 major work projects added complexity to the tasks of the project constructors and engineers, and moving 29 miles of gas, electric, telephone, sewer, and water lines maintained by 31 separate companies added extraordinary challenges to the project's utility relocation program (p. 39).

The Trans Alaskan Pipeline. *Source:* Ttstudio / Shutterstock.

The Trans Alaskan Pipeline

By 1920, the American Petroleum Institute estimates there were almost 40 000 miles of pipeline in the country. In the following decade, that number tripled, as welding technology made it easier to build long pipelines. Today, the United States has 2.6 million miles of pipeline crisscrossing the country, more than anywhere else in the world, but that pipeline grid does not work for the shale boom. More than a hundred major pipeline projects are currently planned for the next five years in North America. The scale could easily rival that of the 1950s.

One of the largest pipelines in the world is the Trans Alaskan Pipeline. Between 1968 and 1970, oil exploration efforts resulted in the discovery of a major oil field on Alaska's North Slope. At the end of 1970, the American Petroleum Institute estimated the potential reserves in this Prudhoe Bay field to be approximately 9.6B barrels. Others predicted that when fully developed, the amount of oil ultimately recoverable from the field could be twice as large. The field's size and potential made it one of the single most important discoveries in the history of the domestic crude oil industry (Cichetti 1993). At the same time that oil exploration was occurring in Alaska, the U.S. Congress enacted the National Environmental Policy Act of 1969 (NEPA). The act requires agencies of the federal government undertaking actions that could adversely affect the environment to file an environmental impact statement analyzing and quantifying the expected environmental effects of the proposed action. The act stipulates that when irreversible deleterious effects might be found, the agencies should consider alternatives to a proposed plan. After a bitter debate, construction of the $8B Trans-Alaska Pipeline began in 1973.

As noted by Levitt and Scott (2017) "the Alaska pipeline project became a 'megaproject' due to its regional economic and environmental impact and the resulting complexity of its relational subproject interdependencies and challenges, even though it was neatly divisible into nearly autonomous subprojects in terms of its spatial and technological configuration" (pp. 96–97). Concerted efforts by Congress, the State of Alaska, and other stakeholders have resulted in new momentum to proceed with an Alaska gas pipeline project. Serious reconsideration of the construction of a natural gas pipeline from the Alaska North Slope began around 2000. This reconsideration was prompted, in large part, by tightening gas supplies to the lower-48 states and corresponding increases in natural gas prices and price volatility (Parfomak 2009).

The decision on many megaprojects can be heavily dependent upon the political climate at a particular point in time. For example, in recent years, the Keystone XL Pipeline in the United States has created further controversy over the value of these projects. Some legislators have expressed support for the potential energy security and economic benefits, while others have reservations about its potential environmental impacts. There is also concern over how much crude oil, or petroleum products refined from Keystone XL crude, would be exported overseas (CRS 2015). The Pipeline was officially abandoned in June 2021, however, the future of major pipelines in the United States remains uncertain depending on changes in the political climate and public support.

Nuclear Power Plants. *Source:* Ttstudio / Shutterstock.

Nuclear Power

Nuclear Plants: The Buildup, Downfall and Pending Resurgence

The History of the Nuclear Power in a Nutshell

The science of atomic radiation, atomic change and nuclear fission was developed from 1895 to 1945, much of it in the last six of those years.

From 1939 to 1945, most development was focused on the atomic bomb.

From 1945 attention was given to harnessing this energy in a controlled fashion for naval propulsion and for making electricity.

Since 1956 the prime focus has been on the technological evolution of reliable nuclear power plants.

Source: World Nuclear Association, Outline History of Nuclear Energy, Updated November 2020

As of 2021, the World Nuclear Association reports 440 nuclear reactors are in operation worldwide in 30 countries generating capacity of 390 (GW) which is equivalent to about 10% of the world's electricity. The 10 countries with the largest number of reactors include: the United States (95), France (57), China (47), Russia (38), Japan (33), South Korea (24), India (22), Canada (19), the United Kingdom (15), and Ukraine (15). Figure 1.3 shows the global sources of electricity production as tracked by the World Nuclear Association indicating that coal and gas remain the largest sources of power in the world.

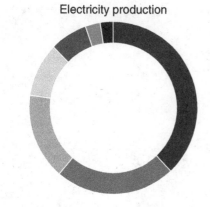

Electricity production

Figure 1.3 Global power sources. *Source:* Adapted from The World Nuclear Association (2021).

- Coal (38%)
- Gas (23%)
- Hydro (16.2%)
- Nuclear (10.1%)
- Solar, wind, geo, tidal (7.2%)
- Oil (2.9%)
- Other (2.5%)

After hydroelectric power, nuclear is the world's second largest source of low-carbon power. According to the World Nuclear Association, 13 countries in 2020 produced at least one-quarter of their electricity from nuclear. France gets around 70.6% of its electricity from nuclear energy, Slovakia and Ukraine get more than half from nuclear, while Hungary, Belgium, Slovenia, Bulgaria, Finland, and Czech Republic get one-third or more (WNA 2021b). South Korea normally gets more than 30% of its electricity from nuclear, while in the United States, the United Kingdom, Spain, Romania, and Russia about one-fifth of electricity is from nuclear. Japan was used to relying on nuclear power for more than one-quarter of its electricity and is expected to return to somewhere near that level (WNA 2021a). Beyond power generation, nuclear technologies have medical applications that will help combat serious viruses such as COVID-19. The International Atomic Energy Agency (IAEA) is providing diagnostic kits, equipment, and training in nuclear-derived detection techniques to countries asking for assistance in tackling the worldwide spread of the novel coronavirus causing COVID-19. Also, small modular reactors (SMRs) are playing a key role in the clean energy transition.

According to the U.S. Energy Information Administration (EIA), most U.S. nuclear power plants were built between 1970 and 1990. Other forms of energy include hydroelectric, coal, natural gas, wind, solar, and petroleum also progressed during that time. As of 2021, the United States has 55 operating nuclear power plants with 93 nuclear power reactors in 28 states across the country. The oldest operating nuclear reactor in the United States was built in 1969. Watts Bar 2, which entered commercial service in 2016, was the first new reactor added since 1996. An additional two are under construction. Of the 99 gigawatts (GW) of total operating nuclear capacity in the country, 95 GW came online between 1970 and

1990. Planned nuclear capacity additions began to slow as early as the late 1970s because of several factors, including slowing electric demand growth, high capital and construction costs, and public opposition. Costs, schedules, and public acceptance were all influenced by the accident at the Three Mile Island plant in 1979 and, more recently, the Fukushima nuclear disaster in 2011. From 1979 through 1988, 67 planned builds were canceled. However, because of the long duration required for permitting and building new nuclear reactors, many plants that had begun the process in the 1970s continued to come online through the early 1990s.

The Millennium Projects

Since 2000 the world has seen an explosion of megaprojects increasing in size, cost, and scope from what has come before and representing multiple industries and initiatives These new megaprojects include areas in transportation, energy, climate change, medical research, education, water supply, innovation and scientific projects, roads, bridges, tunnels, dams, and high-speed rail. The projects stand at various stages of completion and continue to drive economic, social, and institutional development. It is during the present era that we see a focus on ecological impact, sustainability, and international project expansion in the developing world through the establishment and growth of international development bank investments. Table 1.1 shows some of the larger and more global megaprojects impacting in some cases multiple countries and regions of the world.

Gotthard Base Tunnel. *Source:* Gavin van Marle.

Table 1.1 The millennium megaprojects.

Project	Description	Cost/Budget	Funding source
International Space Station 1988–2024	An international partnership representing 15 countries to develop a modular space station in low Earth orbit to study the Earth's environment and the universe. As of April 2021, 244 individuals from 19 countries have visited the International Space Station. Top participating countries include the United States (153 people) and Russia (50 people) (NASA 2021)	$150B to develop; $48B to operate $3.6B over budget	Multinational collaborative involving the United States (NASA) (largest contributor), Russia, Europe, Japan, Canada
London's Crossrail 2009–2022	Europe's largest megaproject, Crossrail Limited is delivering the Elizabeth line – a new railway for London and the South-East to increase capacity and improve connectivity running from Reading and Heathrow in the west, through 42 km of new tunnels under central London to Shenfield and Abbey Wood in the east. From end to end the Elizabeth line will stretch over 60 miles (Crossrail 2021)	Cost: £18.9B Budget: £14.8B	UK Parliament DfT and TfL and £600M developer
South North Water Diversion/Transfer Project, China 2002–	A $79B multi-decade infrastructure megaproject, the objective of which is to divert approximately 44.8B cubic meters of fresh water annually from four rivers in the Southern region of China to the more arid and industrialized northern region. Pollution and resettlement of 333,000 people are major challenges. Society, ecology, and the economy are central to the success of this project (Rogers et al. 2020; Zhuang 2016)	Cost: $79B Budget: $62B	Government of China
The Grand Ethiopian Renaissance Dam (GERD) Benishangul-Gumuz Region 2011–2020	The $5B cost of the GERD is about 7% of the 2016 Ethiopian gross national product. The primary purpose of the dam is electricity production to relieve Ethiopia's acute energy shortage and for electricity export to neighboring countries. With a planned installed capacity of 6.45 GW, the dam is the largest hydroelectric power plant in Africa, as well as the seventh largest in the world (Abtew and Dessu 2019)	Cost: $5B Budget: $4.8B	Ethiopian Government via crowdfunding, bonds, and income tax

Project	Description	Budget/Cost	Funding
The Delhi Mumbai Corridor 2006–2037	A planned industrial development project between India's capital, Delhi, and its financial hub, Mumbai, spread across six Indian States along the 1500 km long western dedicated Freight Corridor will serve as the corridor's transportation backbone. In addition to infrastructure and airports, it also gives the country a unique opportunity to plan, develop, and build new cities that are economically, socially, and environmentally sustainable (Macomber and Muthuram 2014)	Budget: $90B	Governments of India and Japan
California High Speed Rail (CHSR) 2015–2033	A publicly funded high-speed rail system under construction in the U.S. State of California. Despite serious funding concerns, the project is projected to connect the Anaheim Regional Transportation Intermodal Center in Anaheim and Union Station in Downtown Los Angeles with the Salesforce Transit Center in San Francisco via the Central Valley, in 2 hours and 40 minutes, which is 380 miles (612 km). The intent of the project is to divert traffic from the congested roads and airports (CHSR 2021). Costs have risen dramatically from an initial plan of 33B to a $100B and there is uncertainty as to when it will complete due to funding shortfalls	$33B Budget (2008) $100B Budget (2021)	Government of California
Gotthard Base Tunnel 1998–2016	Switzerland's largest construction project and the world's longest rail tunnel, with two parallel lines and a total length of 57 km. One of the deepest railway tunnels constructed to date; in some parts, 2300 m separate the tunnel from the earth's surface. The tunnel has a capacity of 50 passenger trains and 160 commodity trains per day. Due to the complexity and size of the project, the Swiss parliament passed a project-related legal framework, which formed the bases for the construction of this corridor and following a 17-year construction period the project was inaugurated on 1 June 2016 on schedule and budget (Drouin and Müller 2021). Sadly, eight workers died in accidents while the tunnel was under construction	(2015 USD) Budget: $13.2B	The Swiss Government
Hadron Collider, France–Switzerland Border 1998–2008	Designed to study the structure of the subatomic world and the laws of nature governing it, the Collider lies in a tunnel 27 km (17 miles) in circumference and as deep as 175 m (574 ft.) (CERN 2021)	$4.7B Budget: $5B	U.S. and EU via CERN (the European Organization for Nuclear Research)
Øresund Bridge 1984–2000 1984 – Planning 1993 – Construction 2000 – Completion	A 10-mile combined rail and road bridge-tunnel spanning the Oresund Strait that links the cities of Copenhagen, Denmark, and Malmo, Sweden. Largest European bi-national project since the Channel Tunnel and the longest concrete tunnel in the world (UCL 2014)	(2010 USD) Cost: $4.10B Budget (1991): $2.96B	Loans guaranteed by Sweden and Denmark and User Fees

Why Study Megaprojects?

Megaprojects have transformed the way we live in the world and have solved many of the world's great problems, yet we still do not understand much about how they are conceived, implemented, and achieve success, or why some megaprojects fail.

Scholars have expressed the need to know how these megaprojects are initiated, financed, developed, and delivered over a period of many decades. Recent literature on megaprojects searches for answers to the following questions (i) How are megaprojects organized and structured and does it matter? (ii) How do they really perform and is it possible that not all megaprojects are over budget and behind schedule as described in the literature or are there other factors that drive this perception; and (iii) what are the major contributions to society that are often not known for years after project completion (Soderlund et al. 2017).

In this section we review five of the following major reasons to study megaprojects: (i) improving economic prosperity for all; (ii) meeting sustainable development goals; (iii) creating opportunities for economic revitalization, and social, and technological innovation; (iv) understanding how megaprojects impact globalization and contribute to the wider ecosystem; and (v) learning from megaproject theory and practice how all projects can deliver better outcomes.

1) Improving Global Economic Prosperity

> Ending poverty and promoting decent work are two sides of the same coin. Decent work is both the major instrument to make development happen and also in effect, the central objective of sustainable development.
> — *Guy Ryder, Director-General, ILO*

One of the major reasons to study megaprojects is set forth in The World Bank's 2020 Flagship Report, *Poverty and Shared Prosperity: Reversals of Fortune* (WB, 2020). This study found that extreme poverty increased globally for the first time in two decades – and that COVID-19, combined with the ongoing effects of climate change and conflict, would impede progress toward ending poverty. Extreme poverty, defined as living on less than $1.90 a day, is likely to affect between 9.1% and 9.4% of the world's population.

The report provides recommendations for a complementary two-track approach: responding effectively to the urgent crisis in the short run while continuing to focus on foundational development problems, including conflict and climate change. Responding to this crisis is complex and challenging and involves the need for large-scale solutions that involve multiple stakeholders, local, regional, and national governments, technological innovation, and capital market solutions. Chapters 2 and 3 will provide a few specific examples of how megaprojects are addressing the impact of poverty and aiding in the growth of global economic prosperity.

2) Meeting Sustainable Development Goals Through Infrastructure Development

One of the critical ingredients in meeting the new continental and global sustainable development goals, namely the African Union (AU)'s Agenda 2063 and the 2030 Agenda for Sustainable Development Goals (SDGs) is infrastructure (UN 2021). As an example, rural infrastructure investments can lead to higher farm and non-farm productivity, employment and income opportunities, and increased availability of wage goods, thereby reducing poverty by raising mean income and consumption.

According to the McKinsey Global Institute (MGI) the world needs to invest around 3.8% of its GDP, or an average of $3.3 trillion a year, in economic infrastructure just to support expected rates of growth from 2016 through 2030 (Woetzel et al. 2016). Emerging economies account for some 60% of that need. But if the current trajectory of underinvestment continues, the world will fall short by roughly 11%, or $350B a year.

Price Waterhouse Coopers (PwC) (2020) Report on Infrastructure Trends, reveals that the infrastructure sector sits at a collision point of global disruptions, including shifts in capital availability, evolving social and environmental priorities, and rapid urbanization. Successful infrastructure delivery demands close alignment and collaboration between a wide range of participants, each with its own agenda and interest.

> Oxford Economics (2017) in its detailed analysis of global infrastructure found that the need for infrastructure investment, is forecast to reach $94 trillion by 2040, and a further $3.5 trillion will be required to meet the United Nations' Sustainable Development Goals for electricity and water (UN 2021).

The Oxford Economics Outlook (2017) reveals where investment is most likely to fall short, and therefore where the needs are greatest, across 50 countries and seven sectors. It considers what investment is needed and what is likely to occur based on a range of factors, such as a country's historic infrastructure spending levels and how its population and economy is changing, hence identifying investment gaps. The findings are compelling. For instance, Asia has the largest overall need, requiring just over 50% of global investment in infrastructure, however the region is forecast to have a relatively small investment gap. The picture is quite different in other regions where investment gaps are more prominent. The Americas and Africa, by contrast, are forecast to have proportionally much larger infrastructure investment gaps. In these regions the investment gap is 32% and 28%, respectively of investment need. Africa's investment gap is forecast to widen further to 43% if investment need includes SDGs. Quantifying country-level needs is a powerful and positive step. These insights will help governments identify and respond to infrastructure needs, and guide opportunities for private sector investors. Many countries are increasingly focused on the role of infrastructure to improve economic growth and community wellbeing.

Electricity and roads are the two most important sectors – together they account for more than two-thirds of global investment needs. The investment gap between the two scenarios is greatest in the roads sector, where investment needs are 31% higher than would be delivered under current trends. The gap is also relatively large for ports and airports (Oxford 2017).

3) Creation of Opportunities for Economic Revitalization and Social, and Technological Innovation

Megaprojects can not only make huge impacts at local, regional, and even international scale but can also be significantly affected by the complex project environment. It is inevitable that megaprojects can create opportunities for economic revitalization and technological innovation for a sustainable future. In this book, we look at the past and present of megaprojects, but we also consider the future of megaprojects and its many emerging issues:

- The social, economic, and institutional value of megaprojects
- The global market for infrastructure investment
- Emerging challenges in implementation of international megaprojects
- Megaproject finance and value creation
- Resilience of mega infrastructure
- Megaproject Innovation – digitization, artificial intelligence, and machine learning
- Corporate social responsibility and human rights

"Grand challenges," related, for instance, to environmental and health issues, have become increasingly pervasive in both policy discourse and in the science, technology, and innovation (STI) policy literature (Ludwig et al. 2021). The appropriate policy responses to these societal challenges differ from mission-oriented policy interventions that relied on large research and development (R&D) programs such as the Manhattan and Apollo projects. These new initiatives will require greater focus on the user perspective to determine whether what is procured is an innovation or not. Today's equivalent might be the Mars Exploration Program which would call for public procurement of innovation (PPI). Grand challenges call for systemwide transformations where a single instrument is not sufficient (Kuhlman and Rip 2018).

4) The Megaproject Ecosystem: Asphalt, Silk, and the Digital Roads

Infrastructure connectivity has received a significant amount of attention in recent years. Based on country case studies, infrastructure development played a significant role in the fast economic development of East Asia, including the PRC (Peoples Republic of China), Japan, and the Republic of Korea (Pascha 2020).

The goals of the One Belt Road, the Interstate Highway System, the English Channel, and the Trans-Siberian Rail, all highlight the significant role that megaprojects play in connecting cities, regions, nations, and the world. In this context, the links between megaprojects and development could not be clearer and more dependent on carefully planned national strategies to promote growth and competitiveness.

In addition to regional and urban opportunities for expansion, megaprojects can also create ecosystems that can last for decades even centuries as evidenced by the U.S. Interstate Highway System and High-Speed rail in Europe and Asia (Greiman and Sclar 2019; Nunno 2018; Weingroff 2017).

Øresund Bridge. *Source:* reduper.com

All megaprojects have the potential to develop an ecosystem of its own or become part of an existing ecosystem. They also can have impacts on an entire Region. For example, the completion and opening of the Oresund Fixed Link in 2000 marked an upturn in mobility at an international, national, regional, and local level for one of the busiest and most important traffic routes between the Scandinavian peninsula and the European continent. The Øresund Fixed Link is a combined bridge and tunnel link across the Øresund Sound between Denmark and Sweden. The fast link to the center of Copenhagen has also had a significant impact on the potential of Copenhagen's Kastrup Airport to attract more international flights. Further, it triggered the formation of a common labor and housing market, which lies at the heart of the political vision of the Øresund Region.

Sustainability normally refers to environmental practices. In megaprojects, a broader definition, including concepts of economic, social, and institutional sustainability, is appropriate (Dimitriou et al. 2012).

The Asphalt Road

One of the most expensive road projects in history and the largest road network in the world is the United States Interstate Highway System that has cost the federal government roughly $528B, adjusted for inflation. That is almost three times the cost of the International Space Station, and almost five times the cost of the Kashagan oil field in the Caspian Sea. Interstates have transformed the way we move goods and people in the United States. In 1919, then Lt. Colonel Eisenhower traveled in an 80-vehicle military convoy from Washington, DC to San Francisco. The trip took 62 days, inspiring him to create the system. Today that drive could be completed in about three days. Referred to as the greatest public works project in history, the Interstate encompasses 47000 miles of roadway, and now runs through all 50 states, the District of Columbia and Puerto Rico (Weingroff 2017). But large road projects are not unique to the United States, the estimated $4.2B Trans-African Highway network comprises transcontinental road projects in Africa being developed by the United Nations Economic Commission for Africa (UNECA), the African Development Bank (ADB), and the African Union in conjunction with regional international communities. The goal of the highway network is to promote trade and alleviate poverty in Africa through highway infrastructure development and the management of road-based trade corridors. The total length of the nine highways in the network is 56683 km (35221 miles). India's road project network is estimated as the second-largest in the world with about 6 million kilometres of roads, while China has the world's third-biggest road network, exceeding 4.24 million kilometres.

All of these road projects worldwide have several important goals in common. Advancing trade, accommodating growing populations, and connecting cities and countries are just a few of the reasons why road construction projects are so important, and they all equate to help improve the quality of life at the local, national, and regional level, and permit industries to continue to move their economies forward. Though important, alleviating congestion, providing jobs, and reducing safety risk seem to take a back seat to the holistic social and economic gains of these amazing projects for the entire ecosystem.

China's Maritime and Silk Road: One Belt One Road

Two thousand years ago, the states of Kazakhstan, Kyrgyzstan, Tajikistan, Turkmenistan, Uzbekistan, and the four western Chinese provinces of Gansu, Ningxia, Shaanxi, and Xinjiang were at the heart of the ancient Silk Road

(UNCTAD 2014). Over time empires were destroyed, cities fell into ruins, and new forms of transport emerged, while the trade route of antiquity gradually fell into disuse for centuries (p. 1). It was not until after the fall of the Soviet Union in the 1990s, that the Central Asian Countries began their transformation to market economies and became reintegrated into the world economy (p. 1). Importantly, the ancient Maritime Silk Road is becoming a renewed focus for both economic development and for interactions with China and its many neighbors. In recent years, "there has been a new spirit of energy and the Silk Road region has experienced strong economic growth and is emerging as an important foreign investment destination" (UNCTAD 2014). A much-discussed aspect of the Silk Road is the "opening up" of opportunity for China – through its One Belt, One Road Initiative. This initiative consists of two segments – one is centered on the Asian land mass called the Silk Road Economic Belt; the other looks to the South China Sea, the South Pacific, and Indian Ocean and is known as the twenty-first century Maritime Silk Road (Economist 2015). This initiative has many important advantages for sustainable development not least of which is the building of a cohesive policy among different nations, economies, and cultures to facilitate trade and investment, increase financial cooperation, and promote economic development. The plan aims to restore the country's old maritime and overland trade routes and lift the value of trade with more than 40 countries to $2.5 trillion within a decade (Economist 2015).

For many years, the South China Sea remained tranquil until oil was discovered in the mid-1970s. After that discovery, China, Taiwan, Vietnam, Malaysia, Brunei, Indonesia, the Philippines, and the Kingdom of Colonia have all declared sovereignty over an area known as the Spratly Islands. Despite recent efforts by international organizations including the Association of Southeast Asian Nations (ASEAN) to calm the waters, the South China Sea continues to cause considerable turmoil among the eight claimants and other interested nations (Greiman 2014). Without a willingness to set aside at least temporarily claims of sovereignty, and to focus on a paradigm shift from sovereign rights to advance the political, economic, and social goals of the Region, an opportunity to change the quality of life for people living in the region forever is sadly missed. Global megaprojects can do great things if we find a common ground to let them happen as they pave the way for a better future for all of society (Greiman 2014).

The Digital Road: The Industrial Revolution 4.0

In a world that is increasingly characterized by enhanced connectivity and where data is as pervasive as it is valuable, Africa has a unique opportunity to leverage new digital technologies to drive large-scale transformation and competitiveness. Africa cannot and should not be left behind (Ndung'u and Signé 2020). The Fourth Industrial Revolution (4IR) – characterized by the fusion of the digital, biological, and physical worlds, as well as the growing utilization of new technologies such as artificial intelligence, cloud computing, robotics, the Internet of Things, and advanced high-speed wireless technologies, among others – has ushered in a new era of economic disruption with uncertain socioeconomic consequences for our global society (Ndung'u and Signé 2020).

This revolution is powered by electric cars and the ultra-fast train (Prisecaru 2016). In 2020, the European Commission and the Inter-American Development Bank (IDB) made significant efforts to jointly assist Central America and Caribbean countries on issues including the COVID-19 response, sanitation, energy, and digitalization (IDB 2021). Poor bandwidth infrastructure is like traveling on a dirt road. According to the IDB the technological disruption the world is experiencing is something unprecedented as compared to the industrial revolution and the enormous impact it has on per capita gross domestic product (GDP). This explosion of technology occurs in the context of the world of data, data that can only be moved with the necessary infrastructure of digital networks. Thus, just like in any traditional highway infrastructure megaproject, without routes to permit the movement of trucks, buses, and automobiles, we could not connect destinations, trades, and people (Cabañas 2020). In the world of technology, data are the means of mobility such as automobiles, and networks would be the digital highways or routes available in a country, which is directly determined by the country's regulatory framework and capital investments in the country made by mobile or satellite service operators.

Digitization is also critical to the African economy. As reported by the World Bank in its Africa Plan 2019–2023, from mobile money to drones, the digital economy in Africa is driving growth and innovation, bringing more people into the formal economy, and connecting people to each other and to markets. Together, investments in digital infrastructure, skills, and platforms could help Africa accelerate growth while tapping in to the US$11.5 trillion global digital economy. The World Bank is supporting Africa's vision to connect every African individual, business, and government by 2030 – a vision that, if realized, can boost growth by up to 2% points per year and reduce poverty by 1% point per year in sub-Saharan Africa. In Benin, the World Bank is expanding digital connectivity to around 1.9 million smallholder farmers and increasing use of digital financial services in rural communities (WB 2019).

In order for governments and companies such as Uber, Amazon, Alibaba, and Airbnb to deliver its services we must build and improve digital highways so that this exponential quantity of data can be processed. To ensure the positive effects of the twenty-first century's digital economy, constructing and investing in those digital highways must be the priority for economic development and social inclusion in Africa, Latin America, and the Caribbean.

> IDB Invest has invested and will continue to invest in the region in projects that help to expand the digital highway, because we know that 10% penetration by broadband has an average economic effect of 2% to 3% on GDP and 2.6% on productivity. (Cabañas 2020)

As we discuss in this book, innovation and the use of digital technologies including the Internet, AI, machine learning, the Cloud, the Internet of Things, Blockchain, and other digital transformations will depend upon megaprojects to provide the infrastructure necessary for the twenty-first century. For example, recent research shows that the use of artificial intelligence deployment in megaprojects is rapidly advancing and includes AI analytics in the defense industry, occupational safety and health incident tracking in construction projects, and intelligence systems in risk analysis (Greiman 2020). Researchers are recognizing that awareness of the architecture of the 4th Industrial Revolution (4IR) digital tools will be essential to meet the needs of the megaprojects of the future (Whitmore et al. 2020). Megaprojects are becoming the digital platforms of the future. As this development continues, the digital road will become by far the road most traveled.

5) Learning from megaproject theory and practice how all projects can deliver better outcomes.

We conclude this section and this first chapter with a brief focus on megaproject theory and practice. As you will learn throughout this book from reviewing the literature and case studies megaproject theory assists policy makers, industry, project stakeholders including contractors, designers, engineers, and government in understanding better the challenges we face in selecting the right projects, implementing these large and complex endeavors, and then ensuring that we deliver the benefits promised. Megaprojects can provide frameworks for structural decision making, strategy alignment, and investment choices that can be beneficial to all projects (Esty 2004). There is now a growing body of research on megaprojects that provides valuable insights that can be applied not only on megaprojects but provides valuable lessons for projects of all sizes large and small. The megaproject literature is interdisciplinary and crosses the fields of engineering, management, law, sociology, psychology, physics, finance, and economics to name a few. The lessons learned will assist all those involved in the future challenges of megaproject management.

Summary

This chapter explores how megaprojects can be used to advance society in unimagined and innovative ways. They require risk taking, meeting impossible challenges, and the willingness to invest in and manage the power and politics of the time. Lessons from the past will assist in meeting the grand challenges of the future. This may include megaprojects that develop cures for the world's pandemics or infrastructure projects that connect cities, countries, and regions. It also includes technological advancement that helps the human connect with machine learning to create greater efficiencies and to solve problems that the human mind cannot resolve alone. The impact of these projects on world trade and development is central to the future of all countries. Continued research will assist in helping overcome some of the myths and perceptions about these massive undertakings and will encourage others to take on these challenges and create sustainable and resilient projects for the future.

In Chapter 2, we will explore how global megaprojects are financed. A general overview of the basic project finance concepts, the sources of financing, and the advantages and limitations of project finance, along with the motivations for using project finance will be discussed and global models of project finance will be presented.

Lessons and Best Practices

1) Global megaprojects can be transformative and provide insights into opportunities for the wider ecosystem.
2) Megaprojects can provide frameworks for structural decision making, strategy alignment, and investment choices that can be beneficial to all projects.
3) Partnerships and alliances should be forged to build bonds between countries, companies, and cultures and create better opportunities for investment.
4) Global megaprojects create an opportunity to share knowledge across countries and cultures that will make future projects more sustainable and affordable.
5) Data collection and statistical analysis should be the primary focus of the upfront planning of projects to better measure performance and outcomes.
6) Perceptions of megaprojects should be studied to determine the truth behind the success and failure of megaprojects.
7) Megaprojects are shaped over time and cannot be evaluated by a snapshot picture of cost and schedule at a specific point in the project's long life.
8) Megaprojects are critical to economic growth and prosperity in both developed and developing countries and can be used to advance societal interests.

Discussion Questions

1 To meet the growing demand for megaprojects what are the major challenges we must overcome?

2 What are the lessons from the great megaprojects, built over the centuries, that apply to our megaprojects today?

3 How can infrastructure development be used to advance and improve societal interests?

4 As a project manager in a developing country where half the population lives on a few dollars a day, what strategies would you implement to address poverty alleviation and social improvement?

5 What is the meaning of sustainability in today's megaproject practice?

6 Why does Paul Schulman an authority in policy making argue that large-scale project policy represents the pursuit of objectives that cannot be fulfilled by a series of individualized, partial, and disaggregated steps?

7 How can data collection and statistical analysis be used to transform megaprojects and better connect across cities, countries, and regions of the world?

8 How can megaprojects be used to meet the United Nations Sustainable Goals and alleviate poverty the world over?

9 What are the types of frameworks that we will need to finance trillions of dollars of infrastructure in the future, based on the OECD's estimates, particularly since private financing depends on a reasonable return on investment?

10 Why will the digital road be the one most traveled in the future in infrastructure development?

References

Abtew, W. and Dessu, S.B. (2019). Financing the Grand Ethiopian Renaissance Dam. In: *The Grand Ethiopian Renaissance Dam on the Blue Nile* (ed. W. Abtew and S.B. Dessu). Springer Geography.

Agaby, S., El-Ghamrawy, M.K., and Ahmed, S.M. (2014). Learning from the Past: The Ancient Egyptians and Geotechnical Engineering. *4th International Seminar on Forensic Geotechnical Engineering*, Bengaluru, India https://doi.org/10.13140/RG.2.1.2398.4164.

Altshuler, A. and Luberoff, D. (2003). *Mega Projects: The Changing Politics of Urban Public Investment*. Washington, DC, Cambridge, MA: Brookings Institution Press, Lincoln Institute of Land Policy.

Bassin, M. (1991). Inventing Siberia: Visions of the Russian East in the Early Nineteenth Century. *The American Historical Review* 96 (3): 763–794. https://doi.org/10.2307/2162430.

Bechtel (2001). *Hoover Dam: A Project for the Ages*. Reston, VA: Bechtel https://www.bechtel.com/projects/hoover-dam/.

Biesenthal, C., Clegg, S., Mahalingam, A., and Sankaran, S. (2018). Applying Institutional Theories to Managing Megaprojects. *International Journal of Project Management* 36 (1): 43–54.

Britannica, The Editors of Encyclopedia (2021). Trans-Siberian Railroad. *Encyclopedia Britannica*, 22 Jan. 2021, https://www.britannica.com/topic/Trans-Siberian-Railroad.

Butler, M. (1972). Netherlands: Dutch Continue to Reclaim Land from the Sea. *Science* 176 (4038): 1002–1004. http://www.jstor.org/stable/1734091.

CA/T (1990). (FEIS/R) Final Supplement Environmental Impact Report, Central Artery (I-93)/Tunnel (I-90) Project. Boston: Commonwealth of Massachusetts, Department of Public Works. November.

Cabañas, E. (2020). *Digital Highways and Their Similarity to Transportation Highways*. Washington, DC: IDB Invest, Inter-American Development Bank Group.

California High Speed Rail (CHSR) (2021). *CEO October 2021 Report*. Sacramento, CA: CHSR Authority https://hsr.ca.gov/about/board-of-directors/ceo-report/2021-ceo-reports/ceo-report-october-2021/.

Capka, J.R. (2004). Megaprojects: They Are a Different Breed. *Public Roads* 68 (1): 2–9. Federal Highway Administration, Washington, DC.

Caro, R. (1975). *The Power Broker: Robert Moses and the Fall of New York*. New York, NY: Vintage Books, Random House.

CERN (2021). *CERN Annual Report 2020*. Geneva: CERN (European Council for Nuclear Research).

Cichetti, C.J. (1993). The Route not Taken: The Decision to Build the Trans-Alaska Pipeline and the Aftermath. *American Enterprise, Washington, D.C.* 4 (5), n. p. web. ISSN 1047-3572.

Clegg, S., Shankar, S., Biesenthal, C., and Pollack, J. (2017). Power and Sensemaking in Megaprojects. In: *The Oxford Handbook of Megaproject Management* (ed. B. Flyvbjerg), 238–258. Oxford: Oxford University Press.

Congressional Research Service (CRS) (2015). *Keystone XL Pipeline: Overview and Recent Developments*. Updated April 1, 2015. Washington, DC: CRS.

Crossrail (2021). *Crossrail Project Update November 22, 2021*. Crossrail Ltd. https://www.crossrail.co.uk/news/articles/crossrail-project-update.

Deming, D. (2020). The Aqueducts and Water Supply of Ancient Rome. *Ground Water* 58 (1): 152–161. https://doi.org/10.1111/gwat.12958.

Denicol, J., Davies, A., and Krystallis, I. (2020). What Are the Causes and Cures of Poor Megaproject Performance? A Systematic Literature Review and Research Agenda. *Project Management Journal* 51 (3): 328–345.

Dimitriou, H.T., Ward, E.J., and Wright, P.G. (2012). *Mega Projects Executive Summary – Lessons for Decision-Makers: An Analysis of Selected International*

Large-Scale Transport Infrastructure Projects, OMEGA Project 2, OMEGA Centre, and VREF, Bartlett School of Planning, University College London, December.

Dimitriou, H.T., Low, N., Sturup, S. et al. (2014). What Constitutes a "Successful" Mega Transport Project? *Planning Theory & Practice* 15 (3): 389–430.

Drouin, N. and Müller, R. (2021). The Gotthard Base Tunnel: The Work of a Century. In: *Megaproject Leaders: Reflections on Personal Life Stories* (ed. N. Drouin, S. Sankaran, A. van Marrewijk and R. Müller). Cheltenham, UK: Elgar.

Economic Development Research Group, Inc. (EDRG) (2006a). *Economic Impact of the Massachusetts Turnpike Authority and Related Projects, Volume I: The Turnpike Authority as a Transportation Provider*. Boston, MA: EDRG.

Economic Development Research Group, Inc. (EDRG) (2006b). *Economic Impact of the Massachusetts Turnpike Authority and Related Projects, Volume II: Real Estate Impacts of the Massachusetts Turnpike Authority and the Central Artery/Third Harbor Tunnel Project*. Boston, MA: EDRG.

Economist (2015, September 12). *The New Silk Road*. https://www.economist.com/special-report/2015-09-12

Eiffel Tower (2021). *The Origins and Construction of the Eiffel Tower*. https://www.toureiffel.paris/en/the-monument/history.

Esty, B. (2004). *Modern Project Finance*. Hoboken, NJ: John Wiley & Sons.

Flyvbjerg, B. (2014). What You Should Know about Megaprojects and Why: An Overview. *Project Management Journal* 45 (2): 6–19.

Flyvbjerg, B. (2017). Introduction: The Iron Law of Megaproject Management. In: *The Oxford Handbook of Megaproject Management* (ed. B. Flyvbjerg), 1–18. Oxford: Oxford University Press.

Friedman, T.L. (2000). *The Lexus and the Olive Tree: Understanding Globalization*. New York, NY: Random House.

Gellert, P.K. and Lynch, B.D. (2003). Mega-projects as Displacements. *International Social Science Journal* 55 (175): 15–25. https://doi.org/10.1111/1468-2451.5501009_1. Wiley Online Library.

Grant, M. (1997). Financing Eurotunnel. *Japan Railway & Transport Review* 11: 46–52. http://www.ejrcf.or.jp/jrtr/jrtr11/pdf/f46_gra.pdf.

Greiman, V.A. (2013). *Megaproject Management: Lessons on Risk and Project Management from the Big Dig*. Hoboken, N.J.: John Wiley & Sons, Inc.

Greiman, V.A. (2014). A Model for Joint Collaboration in the South China Sea. *The Journal of Asian Finance, Economics and Business (JAFEB)* 1 (1): 31–40. Online ISSN 2288-4645, https://doi.org/10.13106/jafeb.

Greiman, V.A. (2020). Artificial Intelligence in Megaprojects: The Next Frontier. *19th European Conference on Cyber Warfare and Security*, Chester, UK. https://doi.org/10.34190/EWS.20.123.

Greiman, V.A. and Sclar, E. (2019). Mega-Infrastructure as a Dynamic Ecosystem: Lessons from America's Interstate System and Boston's Big Dig. *Journal of Mega*

Infrastructure and Sustainable Development 1 (2): 188–200. https://doi.org/10.1080/24724718.2020.1742624.

Hale, P. and Macdonald, S. (2005). The Sydney Opera House: An Evolving Icon. *Journal of Architectural Conservation* 11 (2): 7–22. https://doi.org/10.1080/13556207.2005.10784942.

Hill, D. (1984). *A History of Engineering in Classical and Medieval Times*. New York: Routledge.

Inter-American Development Bank (IDB) (2021). *Annual Report: The Year in Review 2020*. IDB Group https://publications.iadb.org/en/inter-american-development-bank-annual-report-2020-year-review.

Japan Times (2018, March 14) *30 Years on, World's longest Undersea Tunnel Faces Challenges*. https://www.japantimes.co.jp/news/2018/03/14/national/30-years-worlds-longest-undersea-tunnel-faces-challenges-japan-balances-bullet-trains-freight/.

Kardes, I., Oxturk, A., Cavusgil, S.T., and Cavusgil, E. (2013). Managing Global Megaprojects: Complexity and Risk Management. *International Business Review* 22 (6): 905–917.

Klakegg, O.J., Williams, T., Magnussen, O.M., and Glasspool, H. (2008). Governance Frameworks for Public Project Development and Estimation. *Project Management Journal* 39 (1): 27–42.

Kuhlman, S. and Rip, A. (2018). Next-Generation Innovation Policy and Grand Challenges. *Science and Public Policy* 45 (4): 448–454. Oxford University Press.

Lehtonen, M. (2019). Ecological Economics and Opening up of Megaproject Appraisal: Lessons from Megaproject Scholarship and Topics for a Research Programme. *Ecological Economics* 159 (C): 148–156.

Levitt, R.E. and Scott, W.R. (2017). Institutional Challenges and Solutions for Megaprojects. In: *The Oxford Handbook of Megaproject Management* (ed. B. Flyvbjerg), 96–116. Oxford: Oxford University Press.

Liliopoulou, A., Roe, M., and Pasukeviciute, I. (2005). Trans-Siberian Railway: From Inception to Transition. *European Transport* 29: 46–56.

Ludwig, D., Blok, V., Garnier, M. et al. (2021). What's Wrong with Global Challenges? *Journal of Responsible Innovation*, Routledge 9 (1): 6–27. https://doi.org/10.1080/23299460.2021.2000130.

Macomber, J. and Muthuram, V. (2014). *Delhi-Mumbai Industrial Corridor: India's Road to Prosperity?* Harvard Business School Case 214-077. Brighton, MA: Harvard Business School Publishing https://www.hbs.edu/faculty/Pages/item.aspx?num=46077.

van Marrewijk, A., Ybema, S., Smits, K. et al. (2016). Clash of the Titans: Temporal Organizing and Collaborative Dynamics in the Panama Canal Megaproject. *Organizational Studies* 37 (12): 1745–1769.

Merrow, E.W. (2011). *Industrial Megaprojects: Concepts, Strategies and Practices for Business*. Hoboken, NJ: John Wiley & Sons, Inc.

Miller, R. and Hobbs, B. (2005). Governance Regimes for Large Complex Projects. *Project Management Journal* 36 (3): 42–50.

Mirabile, M., Marchal, V., and Baron, R. (2017). *Technical Notes on Estimates of Infrastructure Investment Needs: Background Note to the Report. Investing in Climate, Investing in Growth*. Paris, France: Office of the Secretary General, Organization for Economic Cooperation and Development (OECD).

Moraglio, M. (2017). The 1930s: The European Utopia and the Nationalist Fulfillment. In: *Driving Modernity: Technology, Experts, Politics, and Fascist Motorways, 1922–1943* (ed. G. Mom, M. Sheller and G. Clarsen). Oxford, New York: Berghahn Books.

Morgan, A.E. (1934). The Tennessee Valley Authority. *The Scientific Monthly* 38 (1): 64–72. American Association for the Advancement of Science.

Müller, R. (2012). *Project Governance*. Aldershot, UK: Gower Publishing, Ltd.

National Aeronautic Space Administration (NASA) (2021). International Space Station Facts and Figures. https://www.nasa.gov/feature/facts-and-figures.

Ndung'u, N. and Landry Signé, L. (2020). *Top Priorities for the Continent 2020-2030, Capturing the 4th Industrial Revolution: A Regional and National Agenda*. Washington,DC:Brookingshttps://www.brookings.edu/research/the-fourth-industrial-revolution-and-digitization-will-transform-africa-into-a-global-powerhouse/.

Nunno, R. (2018). Environmental and Energy Study Institute (EESI) Fact Sheet: High Speed Rail Development Worldwide, July 19. https://www.eesi.org/papers/view/fact-sheet-high-speed-rail-development-worldwide.

Orr, R.J. and Levitt, R.E. (2011). Local Embeddedness of Firms and Strategies for Dealing with Uncertainty in Global Projects. In: *Global Projects: Institutional and Political Challenges* (ed. W.R. Scott, R.E. Levitt and R.J. Orr), 183–246. Cambridge, UK: Cambridge University Press.

Oxford Economics 2017 Global Infrastructure Outlook (2017). https://oxfordeconomics.com/recent-releases/GlobalInfrastructure-Outlook.

Parfomak, P.W. (2009). *The Alaska Natural Gas Pipeline: Background, Status, and Issues for Congress*. Washington, DC: Congressional Research Service (CRS).

Pascha, W. (2020). *Belts, Roads, and Regions: The Dynamics of Chinese and Japanese Infrastructure Connectivity Initiatives and Europe's Responses*. Asian Development Bank Institute http://hdl.handle.net/11540/11781.

Peters, E.J. (1999). Native People and the Environmental Regime in the James Bay and Northern Quebec Agreement. *ARCTIC Journal* 52 (4): 395–410.

Peterson Institute for International Economics (PIIE) (2021). *What Is Globalization? And How Has the Global Economy Shaped the U.S.* Washington, DC: PIIE.

Pitsis, A., Clegg, S., Freeder, D. et al. (2018). Megaprojects Redefined – Complexity vs Cost and Social Imperatives. *International Journal of Managing Projects in Business* 11 (1): 7–34.

Price Waterhouse Cooper (PwC) (2020S). *Global Infrastructure Trends: The Global Forces Shaping the Future of Infrastructure*. London, UK: PwC https://www.pwc.

com/gx/en/industries/capital-projects-infrastructure/publications/infrastructure-trends.html.

Prisecaru, P. (2016). Challenges of the Fourth Industrial Revolution. *Knowledge Horizons – Economics* 8 (1): 57–62.

Road Traffic Technology (RTT) (2021). *The World's Longest Road Tunnel – Laerdal Tunnel, Norway (1992–2000)*. London, UK: Global Data https://doi.org/10.1108/IJMPB-07-2017-0080.

Rogers, S., Chen, D., Jiang, H. et al. (2020). An integrated assessment of China's South—North Water Transfer Project. *Geographical Research* 58 (1): 49–63.

Rosenberg, M. (2020). *How the Netherlands Reclaimed Land from the Sea*. ThoughtCo https://www.thoughtco.com/polders-and-dikes-of-the-netherlands-1435535.

Rotterdam (2020, May 19). *Port of the Future 7 Building Blocks* [White Paper]. Port of Rotterdam. https://magazines.portofrotterdam.com/magazine/portofthefuture/lead-on-pipelines/.

Sabonge, R. (2014). *The Panama Canal Expansion: A Driver of Change for Global Trade Flows*. Economic Commission for Latin America and the Caribbean, the United Nations.

Söderlund, J., Sankaran, S., and Biesenthal, C. (2017). The Past and Present of Megaprojects. *Project Management Journal* 48 (6): 5–16.

Tobin (2001). *Great Projects: The Epic Story of the Building of America, from the Taming of the Mississippi to the Invention of the Internet*. New York, London: The Free Press.

United Nations (UN) (2021). *The Sustainable Development Goals Report 2021*. New York, NY: The United Nations https://unstats.un.org/sdgs/report/2021/.

United Nations Conference on Trade and Development (UNCTAD) (2014). *Investment Guide to the Silk Road*. Geneva: The United Nations https://unctad.org/webflyer/investment-guide-silk-road.

University College London (UCL) (2014). *Project Profile: Sweden, The Oresund Link, Bartlett School of Planning*. London: The Omega Centre, UCL.

Weingroff, R.F. (2017). *The Greatest Decade 1956–1966: Celebrating the 50th Anniversary of the Eisenhower Interstate System, Part 1 Essential to the National Interest*. Washington, DC: U.S. Department of Transportation, Federal Highway Administration, (FHWA).

Whitmore, D, Papadonikolaki, E, and Krystallis, I. (2020). Are Megaprojects Ready for the Fourth Industrial Revolution? *Proceedings of the Institution of Civil Engineers: Management, Procurement and Law*, The University of Leeds, Woodhouse, Leeds, UK. ISSN 1751-4304.

Woetzel, J., Garemo, N., Mischke, J. et al. (2016). *Bridging Global Infrastructure Gaps*. In collaboration with Mckinsey's capital projects and infrastructure practice. Mckinsey Global Institute (MGI).

Wolmar, C. (2014). *To the Edge of the World: The Story of the Trans-Siberian Express, the World's Greatest Railroad*. New York, NY: Public Affairs.

World Bank (WB) (2019). *Africa Strategy for 2019–2023: Supporting Africa's Transformation*. Washington, DC: World Bank Group.

World Bank (WB) (2020). *2020 Flagship Report, Poverty and Shared Prosperity: Reversals of Fortune*. Washington, DC: World Bank Group.

World Nuclear Association (WNA) (2021a). *Press Pocket Guide 2020/21 Edition*. WNA.

World Nuclear Association (WNA) (2021b). *Nuclear Power in the World Today*. London: WNA.

Yackee, J.A. (2018). The First Investor-State Arbitration? The Suez Canal Dispute of 1864 and some Reflection on the Historiography of International Investment Law. In: *International Investment Law and History Series: Frankfurt Investment and Economic Law* (ed. S. Schill, C.J. Tams and R. Hofmann), 70–101. Elgaronline.

Zhuang, W. (2016). Eco-Environmental Impact of Inter-Basin Water Transfer Projects: A Review. *Environmental Science and Pollution Research* 23: 12867–12879. https://doi.org/10.1007/s11356-016-6854-3.

2

Megaproject Finance: Innovation and Value Driven Megaproject Management

With public sentiment nothing can fail. Without it nothing can succeed
— Abraham Lincoln

Introduction: Financing the World's Infrastructure

Project finance has been used for centuries to build great and monumental projects which we still experience today. As expressed in the above quote, public sentiment matters particularly in the building of the great projects of the time. Imagine the world without the great engineering projects or the symbols of beauty that stand large and proud in every country. Imagine the world without its great Cathedrals or Rome without the Colosseum, Paris without the Eiffel Tower or Athens without the Acropolis and its long and storied history with many other remnants still existing today. The list is way too long to mention all the insights we have gained about our history and civilization through these lasting monuments.

Project finance is a type of financing used to finance large capital-intensive projects. The financing mechanism has been used for many decades across multiple sectors including energy, infrastructure, research and development, agriculture, telecommunications, transport, and aerospace. Project financing is one of the most important aspects of project management, yet surprisingly little research has been conducted on the subject (Müllner 2017). This chapter will provide a general overview of the basic project finance concepts, the sources of financing, and the advantages and limitations of project finance, along with the motivations for using project finance. The research compares the project finance mechanism with corporate financing and the use of public private partnerships as a financing device. Sustainable development and the goals of the leading multinational

Global Megaprojects: Lessons, Case Studies, and Expert Advice on International Megaproject Management, First Edition. Virginia A. Greiman.
© 2023 John Wiley & Sons, Inc. Published 2023 by John Wiley & Sons, Inc.

development banks will be reviewed along with the challenges faced by these banks and the legal and regulatory environment of multinational project finance.

Project Finance: A Brief History

Tracing the history of project finance is not easy without historical records. The first documented project finance transaction dates to 1299 when the English Crown used a loan from the Italian merchant bank Frescobaldi to finance the exploration and development of the Devon silver mines (Claughton 2010; Guyol 2016; Kessinger and Martin 1988). Project finance was also used to finance maritime operations and infrastructure developments in ancient Greece and Rome where the loans would be repaid through the sale of shipped cargo (Temin 2004). In the nineteenth century project finance was used to develop America's railroads by private companies using land grants, state and federal government subsidies, and mortgage bonds. In more recent history, project finance has been used in the development of the Panama Canal, North Sea oil field infrastructure, the Trans-Alaskan pipeline, and a range of other projects. Project finance was used to develop oil and gas fields in the 1970s, and to realize transport projects such as bridges and tunnels from the 1980s on (Yescombe 2014, p. 9).

As recognized by the Japan Bank for International Cooperation (JBIC) (2021) following steady progress into the mid-1990s, project finance entered a temporary period of stagnation due to the emergence of political risks such as the Asian currency crisis that occurred in the late 1990s, balance sheet recession among Japanese companies, and the added impact of revisions in developing country portfolios by European and U.S. power companies. However, against the backdrop of the subsequent strong recovery and expansion of developing economies, which became markets with solid potential for return on investment, and the widening acceptance of the concept of public–private partnerships (PPP) among developing countries, the demand for financing has increased as demand for infrastructure development increased. In turn, the number of project finance loans has steadily grown, despite the temporary stagnation stemming from the Lehman Shock in 2008 (JBIC 2021).

Even after a few difficult years in response to the worldwide pandemic, the cumulative value of recovery funds intended for long-term investment worldwide is approaching \$3.5T (about \$11 000 per person in the United States). Considering the potential to use these funds to mobilize additional private funds, the total investment could exceed \$10T (UNCTAD 2021). Three sectors have accounted in major part for infrastructure expenditures and include power generation, oil and gas, and transport infrastructure.

A recent development is the increasing use of project finance for renewable energy projects such as solar and onshore wind, and even small modular reactors

---

Box 2.1   Types of Renewable Energy

1) *Hydropower* – Electricity produced from flowing water
2) *Geothermal Energy* – Heat from the hot interior of the earth or near the earth's surface
3) *Wind Energy* – Wind turbines use blades to collect the wind's kinetic energy
4) *Solar Energy* – Radiation from the sun to produce heat and electricity through solar thermal systems, solar energy power plants, and solar Photovoltaic (PV) systems
5) *Biomass Energy* – Produced from non-fossilized plant materials. Wood and wood waste are the largest sources of biomass energy in the United States, followed by biofuels and municipal solid waste.

Source: Adapted from U.S. Energy Information Administration (2022).

(SMRs) many of which are smaller in scale and less complex than conventional power plants that traditionally used project finance. (Offshore wind projects, in contrast, resemble more conventional plants concerning size and complexity). Based on available equity and debt data from BloombergNEF (BNEF), 2015 marked the first year in which project finance represented more than half of total asset finance in renewable electricity and biofuels at US$104B (Moslener et al. 2016). Box 2.1 breaks down the types of renewable energy which in 2021 attracted $366B for new projects and small-scale systems (up 6.5% from 2020) (BNEF 2022).

The Japan Bank for International Cooperation (JBIC) has been active in financing a number of sustainable renewable energy projects. As shown in Box 2.2 the Muara Laboh Geothermal Power Project in Indonesia is an example of a megaproject which features many of the positive sustainable characteristics of a project finance including economic, social, and environmental benefits to a developing world country.

Project Finance Definition and Characteristics

The financing of megaprojects is one of the most important aspects of project management, yet, as mentioned earlier, surprisingly little research has been conducted on the subject. The breadth of the types of projects included under the general rubric of "project finance" is staggering since it is broadly defined to include the "financing of long-term assets through cash flows." Complex projects like the Hoover Dam, the English Channel Tunnel (the Chunnel) and Boston's Big Dig often require multiple sources of finance from both the public and private sector and multiple entities may be involved.

Box 2.2   The Muara Laboh Geothermal Power Project in Indonesia

- Contributes to the construction and operation of an environmentally friendly geothermal power plant in line with the policies of the Indonesian government for the development of renewable energy.
- Fulfills Indonesia's surging electricity demand which has increased due to steady economic growth,
- Contributes to achieving a sustainable electricity supply.
- Co-financed with the Asian Development Bank (ADB), an international organization.
- Revenues are supported by a 30-year power purchase agreement (PPA) with PLN (the off-taker) and a 20-year support from the Ministry of Finance, as stipulated under a business viability guarantee letter (BVGL).

Project Description:

Construct, own, and operate a geothermal power plant with a capacity of 80 MW
Project Company: PT. Supreme Energy Muara Laboh
Location: West Sumatra Province, Indonesia
Loan Signing: 26 January 2017 (Overseas investment loan)
Sponsors: Sumitomo Corporation, Electrabel S.A., PT. Supreme Energy Related
Japanese Companies: Sumitomo Corporation, Mizuho Bank, Ltd., Sumitomo Mitsui Banking Corporation, MUFG Bank, Ltd.
Total Cofinancing Amount: US$439M JBIC Loan Amount.

Source: Adapted from JBIC (2021)

There are differences in the definition of project finance in the literature and practice though these definitions have some common characteristics. Some examples include:

- Project finance involves the creation of a legally independent project company financed with nonrecourse debt (and equity from one or more corporate entities known as sponsoring firms) for the purpose of financing investment in a single-purpose capital asset usually with a limited life (Esty 2004).
- The financing of long-term infrastructure, industrial projects and public services based upon a nonrecourse or limited recourse financial structure where project debt and equity used to finance the project are paid back from the cash flow generated by the project. (The International Project Finance Association)
- Project finance is the structured long-term financing of infrastructure, industrial projects and public services with limited recourse to the sponsors, where project debt is repaid from future cash flow generated by the project once operational (The World Bank)

These characteristics are reflected in the definition that is part of the Bank for International Settlements, Basel III framework (an international regulatory accord), defining project finance as:

> [A] method of funding in which the lender looks primarily to the revenues generated by a single project, both as the source of repayment and as security for the exposure. This type of financing is usually for large, complex, and expensive installations that might include, for example, power plants, chemical processing plants, mines, transportation infrastructure, environment, and telecommunications infrastructure. Project finance may take the form of financing of the construction of a new capital installation, or refinancing of an existing installation, with or without improvements (BIS 2021, Art. 30.10). In such transactions, the lender is usually paid solely or almost exclusively out of the money generated by the contracts for the facility's output, such as the electricity sold by a power plant. The borrower is usually a Special-Purpose Vehicle (SPV) that is not permitted to perform any function other than developing, owning, and operating the installation. The consequence is that repayment depends primarily on the project's cash flow and on the collateral value of the project's assets. In contrast, if repayment of the exposure depends primarily on a well-established, diversified, credit-worthy, contractually obligated end user for repayment, it is considered a secured exposure to that end-user. *(BIS 2021, Art. 30.11)*

Why Project Finance?

Under-investment in infrastructure can have a serious impact on lives and quality of life for everyone by prioritizing cost reduction over safety. Since the start of the COVID-19 crisis, the World Bank Group has committed over $157B to fight the impacts of the pandemic that includes over $50B of IDA resources on grant and highly concessional terms (WBG 2022). As expressed by David Malpass World Bank Group President, "[We] are working to save lives, protect the poor and vulnerable, support business growth and job creation, and rebuild in better ways toward a green, resilient, and inclusive recovery" (WBG 2022).

Despite this unprecedented global effort, the pandemic has reversed gains in global poverty reduction for the first time in a generation, pushing nearly 100M people into extreme poverty in 2020 creating more hardships for the already fragile states. The Organization for Economic Cooperation and Development (OECD) estimates global infrastructure investment needs of US$6.3T per year over the period 2016–2030 to support growth and development, without considering further climate action (Mirabile et al. 2017). That is more than the total estimated value of the infrastructure already on the ground today. Without project finance many lives around the world would be impacted economically, socially, and environmentally. Whether you are a lender, a borrower, an equity sponsor, or a beneficiary, project finance will be critical to achieving the goals and objectives of a more sustainable society.

Megaprojects create value to society by making possible critical infrastructure to society while mobilizing large amounts of cash and expertise. A few of the most important benefits of these massive endeavors are to:

- Provide the potential for economic, social, and institutional benefits that would not otherwise be available.
- Develop financial markets in the poorer countries and to improve quality of life.
- Provide resources for long term financing for a project that may take 10 years or more to build.
- Provide connectivity and integration among cities, states, countries, and the larger ecosystem.
- Open competitive markets for oil, gas, mining, and other natural resources.
- Lower the cost of a product or project through highly leveraged debt financing.
- Attract private investment through public private partnerships for public projects. Share the risk with those most capable of controlling the risk such as engineers and contractors.
- Provide greater expertise in control and management of large-scale projects.

Project Finance v. Corporate Finance

As illustrated in Figure 2.1, project finance and corporate finance have quite different financing structures. Since a project does not have assets until the project is complete, a project finance statement includes the sources of cash and the use of cash. To

Off-balance sheet financing project finance		Balance sheet financing corporate finance	
Debt and equity (sources of cash)		**Assets**	
• Loans	$50M	• Cash	$100M
• Bonds	$50M	• Real property	$50M
• Grants	$50M	• Equipment	$50M
• Equity	$50M	**Total assets**	$200M
Total cash sources	**$200M**		
		Liabilities	
Costs (uses of cash)		• Bank debt	$50M
• Oil drilling costs $100M		• Other liabilities	$50M
• Export costs $100M		**Total liabilities**	$100M
Total uses of cash			
	$200M	**Capital (owner's equity)**	$100M
Revenues			
• Cash flow from operations	$900M	**Total liabilities and capital**	$200M

Figure 2.1 Corporate v. project finance comparison.

the extent revenue is generated during the project life a separate cash flow statement is used to show the revenues earned by the project. Once the project is complete the cash flow or revenue stream is transferred to the operator's financial statements.

Some projects use both corporate and project finance depending on the project's risk analysis and the economics of the transaction. For example, the Chad Cameroon Pipeline, a $3.7B project to extract oil and build a distribution pipeline in Africa, used a combination of corporate and project finance. The sponsors financed the $1.5B upstream project (the oil field development) using corporate finance and the $2.2B downstream project (the export pipeline) using project finance. The $1.4B of project debt would come from three sources: the International Finance Corporation (IFC), two export credit agencies (ECAs), and the capital markets (Esty 2004). By creating legally independent entities, the project sponsors as the borrowers can protect their balance sheets. In other words, the risk that a sponsor might normally face is transferred to the new entity without placing the corporation at risk for absorbing the project risk, particularly if the project should fail. Project finance also includes a form of risk sharing where the sponsors may share the risks with other project participants such as the lenders or other equity sponsors.

Investment in the Developing Countries

Foreign Direct Investment (FDI) has grown substantially since the 1970s as governments have liberalized their markets and introduced measures to facilitate foreign investment. The growth in foreign investment was especially marked from the mid-2000s onward as countries sought to encourage cross-border investment, improve competitiveness, and foster growth. The upward trend in FDI came to a sharp but temporary end with the financial crisis of 2008 and has since progressed on an upward trend. For example, in the last 20 years, the amount of FDI in the United States has more than doubled. In 2000, FDI was US$1.26T and in 2020 it had risen to US$4.63T (Statista 2021).

In developing countries, as reflected in the projects financed by the world's leading development banks, an essential requirement for sustainable economic and social growth is the provision of infrastructure services, such as secure energy, resilient water and sanitation systems, efficient transport, and reliable communications systems. Shakirat (2018) emphasizes that government expenditure on infrastructure serves as a catalyst for public development in the entire government agenda, such as healthcare delivery, poverty, transportation, education, and food security. Infrastructure level affects the developmental ratings of a nation. The availability of infrastructure promotes investment because it reduces operational costs (Chukwuebuka et al. 2020).

According to the African Development Bank's 2021 Outlook for Africa, the creditor base for Africa's debt continues to shift away from traditional multilateral and bilateral Paris Club sources (creditor nations) toward commercial creditors and non-Paris Club official lenders. The share of commercial creditors in Africa's external debt stock has more than doubled in the last two decades, from 17% in 2000 to 40% by the end of 2019. At least 21 African countries accessed international capital markets between 2000 and 2020. Non-frontier-market economies and low-income countries which do not have access to international capital markets have continued to rely on bilateral and multilateral concessional credit.

There are three major reasons that governments and private industry use project finance over the option of corporate finance: (i) To share the risks inherent in projects from political, environmental, economic, technical, financial, and operational uncertainty; (ii) to increase the likelihood of finding additional financing resources not available through the sponsor corporation; (iii) and internationalizing the project through international participation including developers, designers, contractors, inter-governmental organizations, suppliers, purchasers, and users.

The Economic Case for Nature

A major focus of multilateral lending recently has been on ecosystem services. A new World Bank report estimates that the collapse of select ecosystem services provided by nature – such as wild pollination, provision of food from marine fisheries and timber from native forests – could result in a decline in global GDP of $2.7T annually by 2030 (Johnson et al. 2021). *The Economic Case for Nature* is part of a series of papers by the World Bank that lays out the economic rationale for investing in nature and recognizes how economies rely on nature for services that are largely underpriced (Johnson et al. 2021). The analysis looks at two key approaches to mobilizing private finance for biodiversity. First, it assesses opportunities for "financing green," that is, the financing of projects that contribute – or intend to contribute – to the conservation, restoration, and sustainable use of biodiversity and its services to people. Second, it looks at "greening finance," that is, directing financial flows away from projects with negative impact on biodiversity and ecosystems to projects that mitigate negative impact, or pursue positive environmental impact as a co-benefit. Despite growing innovation in both categories, significant challenges to scaling up private finance remain. These include policies that exacerbate the underpricing of biodiversity; lack of data, measurement, and reporting standards; and issues with biodiversity investment opportunities, which tend to be small scale and noncommercial – making private sector financing a challenge (Johnson et al. 2021).

Global Project Finance Loans 2021 First Quarter Investments

Sustainability-themed investment in the global capital markets experienced strong growth despite volatile markets in 2020, according to the United Nations Conference on Trade and Development (UNCTAD's) World Investment Report 2021. With a growing need to mobilize the vast sums of capital needed to meet the Sustainable Development Goals (SDGs) by 2030, the report analyzed the latest trends in sustainable finance and the global capital market, or the upstream segment of the investment chain.

> UNCTAD estimates that the value of sustainability-themed investment products amounted to \$3.2T in 2020, up more than 80% from 2019, showing that the capital market is increasingly aligning itself with sustainable development outcomes, including the SDGs. These products include sustainable funds (over \$1.7T), green bonds (over \$1T), social bonds (\$212B), and mixed-sustainability bonds (\$218B) (UNCTAD 2021).

With the passage of a \$1T infrastructure bill in the United States known as the *Infrastructure Investment and Jobs Act*, project financing will be used to build new infrastructure and make much needed repairs in all 50 states with the largest allocations going to California, Texas, and New York to fund new roads and bridges, passenger and freight rail, and public transit. A smaller allocation will be used for climate change initiatives (U.S. Congress 2021).

Infrastructure Financing in the United States

Every year, the United States spends over \$400B on public infrastructure (Deloitte 2017). This figure appears high, but annual spending routinely falls short of major maintenance requirements and results in a continuing deterioration of the country's infrastructure assets. Infrastructure supports every aspect of life from airports, energy, railroads, shipping ports, transport of goods, roads, high speed internet, transmission and distribution lines, drinking water, dams, and levees to name a few. To combat the infrastructure gap, state and local governments are moving toward a more coordinated infrastructure planning approach in collaboration with key government agencies and other stakeholders.

In its *2021 Infrastructure Report Card*, the American Society of Civil Engineers (ASCE) reveals that for the first time in 20 years America's Infrastructure GPA is a C− up from a D+ in 2017 (ASCE 2021, p. 2). The only category out of the D to C range is rail with a GPA of B. Categories like ports, drinking water, and inland waterways have been the beneficiaries of increased federal funding. According to ASCE (2021), the United States is still just paying half of its infrastructure bill – and the total investment gap has gone from $2.1T over 10 years to nearly $2.59T over 10 years. As stated in the ASCE Report (p. 6), *by 2039, America's overdue infrastructure bill will cost the average American household $3300 a year, or $63 a week.*

Three steps are essential to fill the infrastructure gap (ASCE 2021, pp. 8–9):

1) Promote sustainability, or the "triple bottom line" in infrastructure decisions, by considering the long-term economic, social, and environmental benefits of a project. This will occur through strong leadership, decisive action, and a clear vision.
2) Increase investment from all levels of government and the private sector from 2.5% to 3.5% of U.S. (GDP) by 2025.
3) Advancement in resilience across all infrastructure sectors through enabling communities to institute their own resilience pathway, incentivizing and enforcing the use of codes and standards, prioritizing projects that improve safety and security, improving land use, and enhancing natural or "green" infrastructure. Although demand for transit solutions is increasing nationwide, US public transit systems face a $90B rehabilitation backlog created by a lack of adequate funding and maintenance. The push to develop and upgrade public transit systems has been strong but uneven, and millions of Americans still do not have proper access to public transit solutions.

Infrastructure Financing in the European Union

In the European Union, EU financial assistance programs provide a wide array of grants, loans, loan guarantees, and co-financing for feasibility studies and projects in several key sectors (e.g. environmental, transportation, energy, telecommunications, tourism, and public health). Several centralized financing programs are also generating procurement and other opportunities directly with EU institutions. The EU supports economic development projects within its Member States, as well as EU-wide "economic integration" projects that cross both internal and external EU borders. In addition, the EU assists candidate and neighbor countries (European Union (EU) 2021).

The EU provides project financing through grants from the EU budget and loans from the European Investment Bank. Grants from the EU Structural and Investment Funds program are distributed through the Member States' national and regional authorities. Projects in non-EU countries are managed through the Directorate-Generals Enlargement, Development and Cooperation (EuropeAid), Humanitarian Aid and Civil Protection (ECHO).

The Cohesion Fund is another instrument of the EU's regional policy. It finances in two areas which are co-financed by national authorities, the European Investment Bank, and the private sector. The Cohesion Fund supports investments in the field of environment and trans-European networks in the area of transport infrastructure (TEN-T). For the 2021–2027 period, the Cohesion Fund concerns Bulgaria, Czechia, Estonia, Greece, Croatia, Cyprus, Latvia, Lithuania, Hungary, Malta, Poland, Portugal, Romania, Slovakia, and Slovenia. Thirty-seven percent of the overall financial allocation of the Cohesion Fund are expected to contribute to climate objectives.

The European Regional Development Fund (ERDF) finances programs in shared responsibility between the European Commission and national and regional authorities in Member States. The Member States' administrations choose which projects to finance and take responsibility for day-to-day management. Other sets of sector-specific grants such as Horizon 2020 aids EU Member States in the fields of science, technology, communications, energy, security, environmental protection, education, training, and research. In 2021–2027, the fund will enable investments to make Europe and its regions:

> *More competitive* and smarter, through innovation and support to small and medium-sized businesses, as well as digitization and digital connectivity.
> *Greener,* low-carbon and resilient
> *More connected* by enhancing mobility
> *More social,* supporting effective and inclusive employment, education, skills, social inclusion, and equal access to healthcare, as well as enhancing the role of culture and sustainable tourism.
> *Closer to citizens*, supporting locally led development and sustainable urban development across the EU.

Loans from the European Investment Bank

Headquartered in Luxembourg, the European Investment Bank (EIB) is the financing arm of the European Union. Since its creation in 1958, the EIB has been a key player in building Europe. As a non-profit banking institution, the EIB assesses, reviews and monitors projects, and offers cost-competitive, long-term lending. Best known for its project financial and economic analysis, the EIB makes loans to both private and public borrowers for projects supporting four key areas: (i) innovation and skills, (ii) access to finance for smaller businesses, (iii) climate and environment, and (iv) infrastructure.

While the EIB mostly funds projects within the European Union, it lends outside the European Union as well (e.g. in Southeastern Europe, Africa, Latin

America, and Pacific and Caribbean states). In 2020, the EIB loaned €66B for projects. The EIB also plays a key role in supporting EU enlargement with loans used to finance improvements in infrastructure, research, and industrial manufacturing to help those non-EU countries prepare for eventual EU membership.

Sources of Funding for Project Financing

Project financing has been used in various ways for many years, but in the 1970s and 1980s, it emerged as a leading way of financing large infrastructure projects that might otherwise be too expensive or speculative for any one individual investor to carry on its corporate balance sheet. Project financing has been particularly important to project development in emerging markets, with participants often relying on guarantees, long-term offtake or purchase agreements, or other contractual relationships with the host sovereign or its commercial appendages to ensure the long-term viability of individual projects. These were typically backstopped by multilateral lending agencies that mitigated some of the "political" risks to which the project lenders (and, sometimes, equity investors) were exposed.

With the rise of the independent power industry in the United States starting in the 1980s, project finance became essential to the domestic market, as the new entrants to the sector did not have the same rate-based revenue streams and capital-raising abilities of the incumbent utilities. This same pattern continues to hold true with new entrants in the hydrogen renewable power sector; project finance remains essential for the realization of their business plans. As shown in Figure 2.2, the major actors in a project finance include the Project sponsor(s), owner(s), the grantor if a government entity is hosting or sponsoring the project, the project company sometimes referred to as the delivery team and the suppliers, offtake purchaser, the contractor, operator, and labor. Although a single megaproject may have thousands of stakeholders, certain key stakeholders will play a more dominant role.

With the implementation of Basel III brought on by the global financial crisis, the ways in which project finance deals are structured has begun to change because the new regulatory framework requires banks to hold significantly more liquid assets and reduce their reliance on short-term funding (Ma 2016). As reliance on loans decreases bonds have become more attractive to institutional investors. Many public projects in Africa, Europe, and the United States have been funded by a combination of project bonds and government funding sources.

As provided in Table 2.1, there are many sources of project financing and in recent years, creative financing has been critical in meeting shortfalls in project finance, particularly in the developing world. Both debt and equity instruments continue to evolve as governments and private industry work to find solutions to fill the desperate need for project financing to address poverty, improve quality

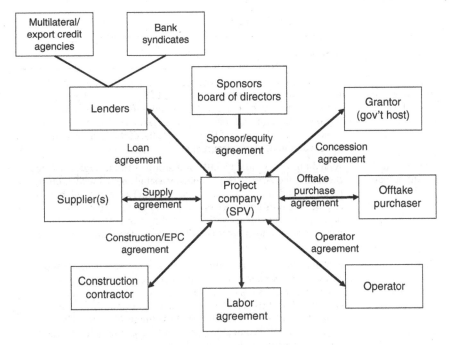

Figure 2.2 Project company structure and legal agreements.

of life, provide essential transport, fulfill energy needs, create greater mobility, and meet food security and sustainability requirements throughout the world.

Innovative Finance can mean different things to different people. At the World Bank Group, innovative finance includes:

- Generating additional development funds by tapping new funding sources (that is, by looking beyond conventional mechanisms such as budget outlays from established donors and bonds from traditional international financial institutions).
- Engaging new partners (such as emerging donors and actors in the private sector).
- Enhancing the efficiency of financial flows, by reducing delivery time and/or costs, especially for emergency needs and in crisis situations.
- Making financial flows more results-oriented, by explicitly linking funding flows to measurable performance on the ground.

Innovative financing is also increasingly used in the United States. Over the last two decades, as revenues have lagged behind investment requirements, Congress and States have sought ways to expand the capacity of the *Federal-aid program* to deliver projects. Today, States and other project sponsors have at their disposal an

Table 2.1 Global sources of project finance.

Source	Description
Debt (Commercial Bank Loans)	Commercial bank debt has historically been the main source of project finance. The simplest meaning of debt financing is borrowing money on credit with a promise to repay the amount borrowed, plus interest. Local commercial banks are important in project finance because of their knowledge of the local regulatory system and political environment
Syndicated credit or loan	A credit or loan granted jointly by a group of banks, typically when the loan is too large to be provided by a single bank (such as euro credits). Syndicated loans are managed by a lead bank which assesses the borrower's needs and tries to get other banks to participate in the loan. For the developing world, the multilateral banks serve as a loan organizer through the development of bank syndicates
Global Institutional Investors	Global institutional investors encompass a heterogeneous group of investors that seek to maximize financial returns on their assets. Institutional investors are entities that invest in different asset classes and pool risks on behalf of their members, while being bound to meet a minimum threshold of financial return. They include insurance companies, pension funds, mutual funds, and sovereign wealth funds among others
Public Private Partnerships (PPP or P3)	P3 agreements usually involve a government agency contracting with a private company to renovate, construct, operate, maintain, and/or manage a facility or system. While the public sector usually retains ownership in the facility or system, the private party will be given additional decision rights in determining how the project or task will be completed
Reserve-Based Lending: Natural Resource Projects	A type of asset-based lending (ABL) commonly used in the oil and gas sector, reserve-based loans are made against, and secured by, an oil and gas field or a portfolio of undeveloped or developed and producing oil and gas assets. The amount of the loan facility available to the borrower is based on the value of the borrower's oil and gas reserves, as adjusted from time to time. The loan facility is repaid using the proceeds from sales in the field or portfolio
Subordinated Debt/ Mezzanine Financing	A combination of financing instruments, with characteristics of both debt and equity, providing further debt contributions through higher-risk, higher return instruments, sometimes treated as equity
Equity	The part of a company's capital belonging to its shareholders. In accounting terms, equity is what is left when all the company's liabilities have been deducted from its assets, except for those liabilities due to shareholders. The value of equity of a quoted company corresponds to the price per share multiplied by the number of shares

Table 2.1 (Continued)

Source	Description
Bonds	A financial security, bearing a fixed interest rate, issued by private businesses or governments as a means of raising money and long-term funds (i.e. borrowing). When an investor buys bonds, he is lending money to the issuer. Bonds are repaid by the issuer at maturity
General Obligation Bond	A general obligation bond is a common type of municipal bond in the United States that is secured by a state or local government's pledge to use legally available resources, including tax revenues, to repay bond holders. The Bond is usually fully backed by the assets of the municipality or government entity
Green Bonds	A green bond is a type of fixed-income instrument that is specifically earmarked to raise money for climate and environmental projects. These bonds are typically asset-linked and backed by the issuing entity's balance sheet, so they usually carry the same credit rating as their issuers' other debt obligations
Zero Bonds/ Zero Coupon Bonds	A Bond on which no periodic interest (coupons) is paid during its life, and where both the principal and the interest are paid at the bond's maturity date. The bond is sold at a discount and its maturity value equals the par value. Its yield interest rate is determined by the rise in the bond's value over time. This kind of instrument is widespread in the United States and on the Euromarkets
Grants	A government grant is a financial award given by a government authority for a beneficial project. A grant does not include technical assistance or other financial assistance, such as a loan or loan guarantee, or an interest rate subsidy. The grantee is not expected to repay the money but is expected to use the funds from the grant for their stated purpose, which typically serves some larger good. It is used by governments to fill gaps in project finance and sometimes can be substantial
Grant Anticipation Notes	Grant Anticipation Notes is a short-term obligation of an issuer to repay a specified principal amount on a certain date, together with interest at a stated rate, usually payable from a defined source of anticipated revenues. Notes usually mature in one year or less, although notes of longer maturities are also issued. These notes have been used in the United States to fund shortfalls in financing
National and Local Governments	Government financing includes revenues from operations, taxes, bonds, notes, and guarantees
Multilateral/ Bilateral Development Banks	*Bilateral* – Created by a single country and is solely funded by the Government. *Multilateral* – Agency or Institution created by international agreement among multiple countries funded by its members. The major global multilateral, the World Bank Group consists of the IBRD and the IDA, the primary lenders to the developing world. The IFC encourages the investment of private capital

(*Continued*)

Table 2.1 (Continued)

Source	Description
Transnational Financing	The European Investment Bank, the European Regional Development and the European Social Fund, and the EU Structural and Investment Funds program funds projects across the European Union
Export Credit Agencies	Government departments or financial institutions that benefit from government guarantees or direct funding, which provide financing as a means of supporting exports from their countries
Guarantees	A contractual engagement protecting the holder of the guarantee against default, bankruptcy or failure of a third party. The guarantor will pay the debt to the obligee if the third party (principal debtor or obligor) fails to perform. It also refers to the commitment provided by an Export Credit Agency to reimburse the lender (frequently a commercial bank) if the exporter fails to repay a loan. In return for this coverage, the lender pays a guaranteed fee. Although guarantees may be unconditional, they frequently have conditions attached to them
Take or Pay Contract	A contractual term whereby the buyer is unconditionally obligated to take any product or service that he is offered (and pay the corresponding purchase price), or to pay a specified amount if he refuses to take the product or service
Industrial Development Funds	Established government run financial institutions that will provide support to a particular group or sector.
Islamic Finance	Finance that is structured to be compliant with the principles of Islamic law (known as Shariah law in Arabic). Key principles require financing without interest and uncertainty and the sharing of profits and losses. While originating from Muslim majority countries has now spread globally and is undertaken in more than 75 countries (Khan et al. 2021). Indonesia, Malaysia, and Gulf Cooperation Council (GCC) countries have been the geographical regions that have contributed most to growth in the Islamic Finance industry in recent years (S&P 2022)
Value Capture	Value capture mechanisms are a type of public financing where increases in the private land values generated by public transportation investments are "captured" to repay the cost of the public investment. Using value capture mechanisms to finance new or existing transportation infrastructure connects the benefit of the infrastructure investment with the cost to provide it

array of project finance tools to facilitate the delivery of projects. One example in Riverside, California is, the $1.3B SR 91 Corridor Improvement Project an 8-mile extension of the Orange County SR 91 Express Lanes east into Riverside County through conversion of existing HOV lanes completed in 2017. Two general purpose lanes will also be added, along with improvements made to interchanges and bridges. The project is supported by a TIFIA loan and is almost exclusively paid for with local funding through a combination of toll revenue and voter-approved county-level sales tax proceeds.

The 22 sources of project finance included in Table 2.1 highlight the variety of sources that are being used globally to fund mega infrastructure projects.

The Legal Framework: The Major Project Participants and Agreements

Project Finance Agreements provide the legal framework for all project financings. Several of the major agreements are shown in Figure 2.2. These agreements provide the governance structure, the applicable regulations, and the rights and obligations of all the parties to the project finance. The central focus of most project financings is the concession agreement between the government host or the grantor and the project company (grantee) that is granted the right to build the project. The project company is usually established as a special purpose vehicle (SPV) such as a private company or a joint venture between two or more parties. The concession agreement sets the framework for the financing as it includes (i) the purpose of the project, (ii) the grantee's and grantor's rights and obligations including the right to develop the project site, (iii) the requirements to achieve financial closing, and (iv) assurance against nationalization, expropriation, and discrimination.

The Lender and Project Finance

"The investment gap in infrastructure is not the result of a shortage of capital. Real long-term interest rates are low, there is ample supply of long-term finance, interest by the private sector is high, and the benefits are obvious." However, a number of factors hold back investment in terms of financing and funding. "The main challenge is to find bankable and investment-ready projects."

Source: Adapted from B20 Policy Paper (2017).

The lenders in a project finance may include commercial banks, multilateral development banks (MDBs), export credit agencies, national government agencies such as the International Development Finance Corporation (DFC) in the United States, and other financial institutions established for the purpose of financing large scale development.

The basic premise of project finance is that the future cash flows from the development of a project serve as collateral to the lender and as a future equity return on investment to the sponsors. The lenders to a project have either no recourse or only limited recourse to the project company assets. Unlike corporate finance there are no assets in the project to attach until the project is completed. "Non-recourse" refers to the lenders' inability to access the capital or assets of the Sponsor to repay the debt incurred by the special purpose entity that owns the project (the "Project Company"). In cases where project financings are limited recourse as opposed to truly nonrecourse, the Sponsor's capital or assets may be at risk only for specific purposes and in specific (limited) amounts set forth in the project financing documentation. This means they might have limited recourse to the equipment or accounts receivable but only prior to project completion. Lenders will also take an interest in how the legal regime of the relevant jurisdiction treats foreign sponsors, because should they need to enforce their security and sell the project company assets, they may eventually need to replace the original sponsors.

The major benefits to the lender in a project finance include the following:

1) To finance large infrastructure projects that are too expensive and too risky for a corporate investment to carry on the corporate balance sheet
2) To provide lenders with a source of funding to support their loans through equity financing
3) To provide the expertise that is necessary to build large infrastructure
4) To provide equity sponsors with an ability to maximize equity returns, take the liabilities off their balance sheet, and protect key assets.

In a project finance, lenders have lots of risk including political, technical, construction, and economic that can be mitigated in the early stages of the project development process. Euro tunnel being fully funded by private lenders and investors rather than the traditional public financing of public infrastructure, is an example where the lenders waived many of their rights just to see the project through to completion. The main risk turned out to be the cost of private finance, which escalated by 220% by the start of operations and would have grown unsustainable without suspension of interest payments in 1995 (Goldsmith and Boeuf 2019). As of 2016, Eurotunnel still had outstanding debts of around €4B, and according to recent financial reports, debt will not be paid off until at least 2050 (GETLINK SE 2022), even though average interest rates through restructuring have been reduced to 4%. Moreover, Eurostar the passenger train operator continues to receive new funds due to an unanticipated drop in ridership.

Megaprojects not only have a long life but also have debt payments that can also last for decades well beyond project completion. Although Euro tunnel delivered many long-term benefits including urban and regional regeneration, the lenders and the investors will not fully recover their debt for 20–30 years. One of the major concerns of lenders to a project finance is the debt service coverage ratio (DSCR). In the context of a project finance, the debt-service coverage ratio is a measurement of a project company's available cash flow to pay current debt obligations as well as the obligations due throughout the life of the loan. Key questions for lenders are highlighted in Table 2.2.

Borrower Credit Conditions

There are many factors in determining whether credit can be extended by a lender that may even be beyond the lender's control based on regulatory requirements, expectations of stakeholders, investors willingness to commit, government agencies overseeing the project and other internal and external factors. Figure 2.3 highlights eight factors that impact the global economic landscape of borrowing by countries seeking loans and equity from potential sponsors including governments, MDBs, export credit agencies, commercial banks, private equity sponsors, private industry, contractors, and other sources. Countries with high debt, a lack of policy development, limited law and regulatory development, and high inflation, make it difficult for sponsors to agree to a project finance. On the other hand, these factors

Table 2.2   Key questions for the lender.

Who will perform the Benefit Cost Analysis and ensure its viability?
How was the project selection process conducted and what were the options that were rejected and why?
Is there opposition to the project and has it been resolved?
Is the project Bankable?
What controls, if any, will the lender have over the Project Company? Will the government sign a direct agreement with the lenders so the lender can step-in if the project company fails to do its job and hire new management?
What is the expected cash flow or revenue stream of the project company on an annual basis?
Based on the cash flow is it realistic that the project debt can be paid back in the time allotted?
What is the Debt Service Cover Ratio, and is it sufficient?
Will there be any assets available for security by the lender?
Who are the equity sponsors and how stable are their balance sheets?
What commitments have the equity sponsors made to provide funding for cost overruns?
What is the debt equity ratio?
What is the amount of funded equity overrun commitment by the equity sponsors?
What are the risks in this transaction that are known and where is the uncertainty in the project?

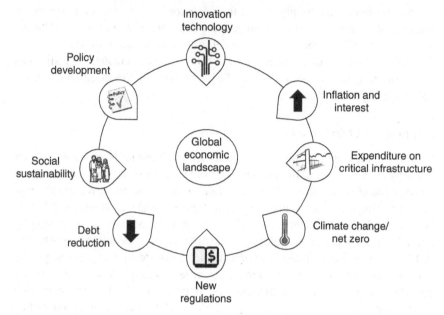

Figure 2.3 Borrower credit conditions.

are weighed against a country's desperate need for twenty-first century innovation and technology infrastructure, climate change and environmental improvements, critical infrastructure, and social sustainability. The struggle to meet the goals of a project finance to achieve the conditions that impact quality of life and a country's future are a continuous tension in the project finance life cycle.

Equity Investment

Sponsors and investors should conduct an internal evaluation of whether an equity investment would best serve their respective strategic objectives. Considerations for this decision are shown in Table 2.3. Governments have always turned to private investors to develop public infrastructure. Until the late twentieth century, the typical model divorced public infrastructure construction and operation from private finance via bond markets (Clifton et al. 2011). In the mid-1980s, that model began to change. A new model, the public–private partnership (P3), dissolved the separation and effectively reversed the partners' roles. The public partner was no longer the "managing" partner and the private one was no longer a "passive" investor. Role reversal affords the private partner new opportunities to secure investment return. Under the older investment regime, investors obtained a safe but low interest rate on the government-guaranteed bonds that financed the project. The new finance regime permits investors to potentially claim revenues from three separate sources:

Table 2.3 Key questions for the equity sponsors.

What was the process for the selection of the project, by the project owner or sponsoring organization, and what were the alternatives that were not selected and why?

How is the return on investment calculated?

Is the project benefit cost analysis, the environmental assessment, and the feasibility study based on accepted standards?

Has an audit been conducted to ensure the assumptions made in these studies are realistic?

How much capital is required for the project? What are the sources of equity financing and how large is the investor pool?

Who are the owners of this project and what is their level of experience?

Who are the lenders and is the project highly leveraged?

What is the Lender's right to control cash flow?

Who will verify the cash flows, and are there sufficient funds to cover debt/service ratios?

What is the experience of the project delivery team and the government owners in projects of this type?

If this is a public private partnership what are the risks and the mitigation plan? Have any of the partners worked together before and what is the success rate of these partnerships?

returns on invested capital; equity appreciation; and fee income as project service manager. The public partner's role reduces to that of service purchaser (Sclar 2015).

The equity sponsors in a project finance provide the essential funding to make a project bankable for the lenders. If there is too much equity, the project will suffer from low leverage. On the other hand, if there is high leverage (60%) or more investment from the commercial and multilateral banks, the project tends to be more stable due to the discipline of having to pay off lenders short term. The debt/equity ratio will change according to the lenders' perceived risk profile of the project in question and may also require the sponsors to make "standby" or "contingent" equity available in the event of construction delays or cost overruns, or in response to specific "in-country" risks.

Projects in early-stage emerging markets will typically require greater equity commitments as a percentage of overall project costs from the project sponsors. Even where debt is guaranteed by a government sponsor or through completion agreements, lenders usually require that a substantial proportion of the project budget be funded by equity investment. The amount is determined by a variety of factors including the DSCR. The debt service cover ratio (DSCR) is a key measure of a project company's ability to repay its loans, take on new financing and make dividend payments. Typically, equity sponsors may include the project owner (typically a public entity), the project developer or main contractor, and often institutional investors that do not play an active role in the project and are known as passive investors. These institutional investors include sovereign wealth funds, pension funds, hedge funds, and insurance companies. Equity can be contributed by capital investment in the project company or by shareholder loans to the project company. The decision to invest in a project company depends on many of the factors raised in the questions for equity sponsors in Table 2.3.

Project Bonds

Project bonds open up an alternative debt funding avenue to source financing for infrastructure related projects. Bonds have been successfully used for infrastructure in Europe and America and some projects in the developing world. As reported by Deloitte (2022) in Europe, corporate bond markets continue to grow in spite of the increase in market volatility and it is anticipated that the use of corporate bonds to fund infrastructure projects in Europe will play a significant role in boosting the economy. The United States has served as a primary source for project bonds that have been issued to the U.S. market for U.S. Projects, Bonds have also been used successfully in Kenya and Nigeria and bonds are desperately needed to fund infrastructure in South Africa.

> Government and the banks alone cannot fund South Africa's infrastructure program. The use of bonds allows project developers to tap into R3 trillion worth of assets under management by South African institutional investors. In addition, Sovereign Wealth Funds are beginning to invest directly into infrastructure projects, so this may also provide an additional source of funding for capital projects into the future.
>
> (Deloitte 2022)

Though bonds can be more difficult than loans to execute due to the regulatory requirements and the involvement of the SEC in the United States, the investor base is usually quite broad and includes insurance companies, bank treasuries, pension funds and asset managers. Credit ratings are also important to bond issuers as they usually will be required to obtain a credit rating of BBB+ or better. In the United States, the federal Grant Anticipation Note Program known as The GARVEE program enables States and other public authorities to issue debt-financing instruments, such as bonds, to pay for current expenditures on transportation construction projects and repay the debt using future federal apportionments. The benefit of the GARVEE financing mechanism is that it generates upfront capital to keep major highway projects moving forward at tax-exempt rates and enables a State to construct a project earlier than is possible with traditional pay-as-you-go financing. With projects completed sooner, costs are lower because of inflation savings, and the public realizes safety and economic benefits. By paying with future federal highway reimbursements, the cost of the infrastructure is spread over its useful life rather than just over the construction period.

In 2021, Fitch assigned an "A+" rating on North Carolina's $245M of Series 2021 grant anticipation bonds. Ratings for standalone GARVEE bonds are derived in large part from the nature of the federal surface transportation funding program. The program has proven to be an essential investment for the federal government with funding disseminated in a formulaic nature across states (Fitch 2021a).

Another form of Bond used to finance megaprojects in the United States are General Obligation Bonds (GOBs) issued as debt instruments by state and local governments to fund highway and infrastructure projects. In the case of Boston's Big Dig project, General Obligation Bonds totaling $1.7B and Commonwealth General Obligation Bonds totaling $675M were issued in November and December 2000. In accordance with the Commonwealth's financial reports the debt service will not be paid off until 2039 (HC 2011).

Export Credit Agencies (ECAs)

ECAs as shown in Table 2.4, make an important contribution to financing projects around the world. ECAs are established by nation states to support overseas opportunities for domestic exporters. Most G20 countries have at least one ECA. In the United States, the Export-Import Bank provides funding through loans, credit, insurance, and guarantees. Many of these Export Banks work with developing countries to provide financing for trade and investments including large energy and infrastructure projects. For example, in 2020 the Export-Import Bank of the United States (EXIM) signed a memorandum of understanding (MOU) with the Republic of Indonesia's Ministry of Finance and Ministry of Planning aimed at enhancing trade and economic opportunities between the two countries. In 2020, the top 10 Export Import Agencies in the world based on total assets included:

The conduct of ECAs is often governed both directly and indirectly by certain international legal obligations because their conduct may be attributed to the nation state. For example, ECAs are held accountable by the World Trade Organization (WTO)'s mandates and regulations. The regulations were implemented closely after ECA's rapid growth toward becoming a major player in international trade. Regulations for one country may vary from others depending on their economic policies and beliefs. When policies vary between

Table 2.4 Export import (EXIM) banks.

EXIM banks	Total assets	Selected project financings
China Exim Bank	$609.7B	Belgrade Bypass opened for operation (Serbia) 2021
		West Africa's first double track standard gauge Modern Railway opened for operation (West Africa) 2021
Japan Bank for International Cooperation	$159B	Suez Canal Onshore Wind Project (Egypt) 2017
		The offshore gas field in Cabo Delgado Province (Mozambique) 2020
Korea Eximbank	$87.3B	Railroad Project ($300M) (Egypt) 2021
		Digital Infrastructure Project ($600M) (Tunisia) 2021
Export Development Canada	$52.9B	Baja Mining (Mexico) 2010
		Kachi Lithium Project (Argentina) 2020
		Xewbec Adsorption (Canada) 2018
Nippon Export and Investment Insurance	$16.1B	Optical Submarine Cable Export Project (Palau) 2021
		FPSO Project for Buzios and Marlim Oil Fields (Brazil) 2020
Export-Import Bank of the United States	$15.8B	Air Products LNG (Mozambique) 2020
		Rural Electrification Project (Senegal) 2020
		Energy and infrastructure sectors. $7B largest in history (Romania)
Finnvera (Finland)	$13.8B	Horizonte's Araguaia Project (Brazil) 2021
		Mantoverde mine expansion project (Chile) 2021
Korea Trade Insurance Corp.	$4.4B	Jawa 9 and 10 Power Plants (Indonesia) 2023
		Ultra Large Container Vessels (South Korea) 2021
Swiss Export Risk Insurance	$3.8B	Porto de Sergipe, I Power Project (Brazil) 2020
		Istanbul Waste to Energy Generation Project (Turkey) 2021

Source: Adapted from Annual Reports and EXIM Bank Documents.

countries, the WTO's guidelines will come into play, making for fair overseas imports and exports. ECAs (themselves and through the appointment of independent advisers) will undertake diligence including legal, technical, market, insurance, and environmental and social diligence, before committing to finance a project. The detailed level of diligence required by an ECA can enhance the perceived viability of a project and assist an owner in attracting potential co-investors or other forms of financing.

Multilateral Development Banks

The Department of Treasury leads the Administration's engagement in the MDBs, which include the World Bank, Inter-American Development Bank, Asian Development Bank, the African Development Bank, and the European Bank for Reconstruction and Development. Emerging markets have launched two new MDBs: the Chinese-led Asian Infrastructure Investment Bank (AIIB) and the New Development Bank (NDB). Brazil, Russia, India, China, and South Africa (the BRICS countries) signed an agreement in July 2014 to establish the New Development Bank (NDB), often referred to as the "BRICS Bank." Launched in Shanghai in 2015, it committed to an initial capitalization of $100B. As the first major MDBs created in decades, questions have been raised how they will fit in with existing MDBs (Nelson 2019).

The MDBs provide financial and technical support to developing countries to help them strengthen economic management and reduce poverty. Together, the MDBs provide support to the world's poorest in every corner of the globe, strengthening institutions, rebuilding states, addressing the effects of climate change, and fostering economic growth and entrepreneurship. At a time when few institutions were lending during the global financial crisis, the MDBs provided $222B in financing, which was critical to global stabilization efforts. In 2018, the United States increased its funding to the IBRD the Bank that lends to middle income countries at market rates, and in 2020 increased the funding request to $1.69B for the MDBs (Nelson 2019).

The MDBs are known for their powerful financial model. Member governments invest a relatively small amount of shareholder capital to establish the financial foundations of an MDB, and the MDB in turn raises the majority of resources for development projects by borrowing on international capital markets. For example, as reflected in the 2020 Sustainable Development Bonds & Green Bonds Impact Report (IBRD 2021). "[A] key priority for the World Bank's capital markets engagement is building strategic partnerships with investors and other market participants and working groups to promote the importance of private sector financing in its sustainable development" (p. 26).

Providing aid through multilateral organizations offers different benefits for donor countries. By pooling the resources of several donors, multilateral organizations allow donors to share the cost of development projects (often called burden-sharing). America's leadership in these institutions ensures that the United States can help shape the global development agenda, leveraging its investments to ensure effectiveness and on-the-ground impact. For example, due largely to U.S. pressure, a significant component of MDB lending is now in the form of grants, helping to break the lend-and-forgive cycle that previously crippled development in many countries. U.S. leadership has also led to the adoption of new lending policies that focus on results, instead of volumes, and reward the strongest performers (Nelson 2019).

Return on Investment and the Revenue Stream: Should Public Projects Be Profitable?

Every project is analyzed from a return-on-investment (ROI) perspective with profitability often a key consideration. In a discussion of the attempt by the United Kingdom to turn operation of rail infrastructure into a private and profitable activity, a Japanese rail executive observed:

> Operation of infrastructure resulting in profits is unthinkable in Japanese railway operations. If profits by Railtrack (the UK's former, now bankrupted, private rail infrastructure provider) have resulted from high levels of usage rates, the UK government should reduce subsidies to operating companies and insist on cuts in track usage fees, as would certainly be the case if this happened in Japan. If the profits of Railtrack have come about by cutting costs required for upkeep and maintenance of infrastructure, there would be problems regarding operational safety. (Quote as cited in Wolmar 2005, p. 82)

The important point here is that infrastructure is a public good. It should not, be, as was attempted with the Channel Tunnel, a profit center in itself. It creates value through its connectivity contribution to the services that the larger economy attains because it is in place. Paying too much for its use, lowers the amount of social value it can create through more widespread use. Trying to make a profit by under maintaining or under creating the public benefits it generates could be a safety hazard (Greiman and Sclar 2019).

Infrastructure can be viewed in one of two ways, as a series of public works or as an interconnected and evolving whole. Each approach makes sense in terms of underlying analytic concern. Much of the literature on project management and cost benefit analysis takes the former approach, focusing on efforts to assess the value added for any given infrastructure project. Presumably, if the benefits exceed the costs, it is worthy of consideration, otherwise not. However, if one is concerned with the overall social and economic vitality of evolving geographic places then every infrastructure item needs to be considered in terms of its role in that evolution. "In the case of the standard evaluation of individual projects, this larger dynamic is treated as methodological background, its impacts for good or ill are described as 'externalities,' sometimes approximately estimated but usually beyond the scope of analysis. In the case of the geographic whole as interconnected networks, the externalities are internalized as foreground and the isolated and project-based value-added estimations are background" (Greiman and Sclar 2019).

Public–Private Partnership Structures

Public–private partnerships (P3s) in transportation are contractual relationships typically between a state or local government, which are the owners of most transportation infrastructure, and a private company. P3s provide a mechanism for greater private sector participation in all phases of the development, operation and financing of transportation projects.

P3s have emerged, in part, because of the growing demands on the transportation system and constraints on public resources. As reported by the Congressional Research Service (CRS) (2021) to date, the number of transportation P3s in the United States is relatively small, as is the amount of long-term private financing provided. Among the reasons for this are the availability to state and local governments of tax-preferred municipal bonds; the need for some kind of revenue stream, such as a toll, fare, or tax, to provide funding; and the fact that many states have limited experience with P3s.

Unlike conventional methods of contracting for a project, in which discrete functions are divided and procured through separate solicitations, P3s contemplate a single private entity being responsible and financially liable for performing all or a significant number of functions in connection with a project. The "private partner" is typically a consortium of private companies with expertise in the different functions to be performed (design, construction, financing, operation, and/or maintenance). In transferring responsibility and risk for multiple project elements to the private partner, the procuring agency shifts certain risks to the private partner and focuses on desired outcomes instead of detailed project specifications.

The first and most common reason to form a P3 is to obtain private financing when public financing is unavailable. The second reason is to transfer risk from the government or public owner to parties better able to manage the risk. However, in recent years, a third and arguably the most important rationale for partnerships has emerged to promote and enhance technological innovation and to improve the lives of citizens in the local communities where these projects are constructed.

A 2018 study by the World Bank Group of public private partnerships in 135 countries revealed that many countries have established government departments to manage public private partnerships (IBRD 2018). While the role and functions of the PPP units vary, most PPP units have a common set of core tasks: PPP regulation policy and guidance (in 85% of the economies with a PPP unit); capacity building for other government entities (in 88%); promotion of the PPP program (in 88%); technical support in implementing PPP projects (in 80%); and oversight of PPP implementation (in 75%) (IBRD 2018). These functions are consistent with the PPP unit performing mainly an advisory role supporting the actual procuring authorities (usually the relevant line ministry). Successful international PPPs include the Australia-Sydney Harbor Tunnel, the M5 Toll Motorway in Hungary, and the Hong Kong County Park Motorway. Further examples of the challenges of implementing P3 structures will be discussed in later chapters.

Evaluation Criteria for Project Finance

In this section, we review evaluation criteria for project finance by looking at the three emerging drivers for increasing project finance opportunities. These include: (i) the role of sustainable development; (ii) the role of value creation; and (iii) the role of value capture.

1) Sustainable Development
According to the United Nations Commission on Trade and Development (UNCTAD) (2021) World Investment Report, capital markets can have a decisive impact on the level and direction of sustainable investment and can contribute toward filling the financing gap for the SDGs (Sustainable Development Goals). Increasingly, financial institutions, such as stock exchanges and derivatives exchanges, have been integrating sustainability values and performance criteria in their activities. There has been a proliferation of sustainability-themed financial products in recent years, including sustainability-themed funds, bonds, and derivative products. Institutional asset owners, such as pension and sovereign wealth funds, are having an impact on companies and markets through asset allocation decisions and active ownership practices. Global efforts to fight the pandemic have also helped accelerate a transition toward sustainable investment.

Addressing these challenges requires three fundamental transitions in the sustainable investment market which would take it from where it is today to where it needs to be in the future (UNCTAD 2021):

Three Fundamental Transitions in the Sustainable Investment Market
1) Growing sustainable investment from "market niche" to "market norm," by making sustainability integration universal rather than a strategy of a sub-set of the larger market.
2) Transforming the sustainable investment market from a developed-country phenomenon to a global market, which benefits all countries, in particular developing economies.
3) Strengthening the credibility of sustainability ratings and reporting with more robust and regulated standards and taxonomies.

2) Megaprojects as Value Creators

Many of the benefits of megaprojects may not be realized for years after the project is complete. For example, emission reductions help in the long term to avoid disastrous climate change, but its effects will not be felt immediately. Governments can win and maintain public support for the benefit of future generations if they are able to develop a broader understanding of what these projects can mean and integrate stakeholders into active participation in their ongoing creation (Greiman and Sclar 2019). In this section, we explore how megaprojects create value and how this value creates opportunity and resources for project finance.

The Project Management Institute in its *2021 edition of, A Guide to the Project Management Body of Knowledge (PMBOK)* has proscribed a system for value delivery as an essential part of an organization's internal environment. There are various mechanisms, such as portfolios, programs, projects, products, and operations, which can be used individually and collectively to create value. *PMI PMBOK* (2021) describes ways that projects provide value to include but are not limited to:

Value Creation Examples
• Creating a new product, service or result that meets the needs of customers or end users
• Creating positive social or environmental contributions
• Improving efficiency, productivity, effectiveness, or responsiveness
• Enabling the changes needed to facilitate organizational transition to its desired future state, and
• Sustaining benefits enabled by previous programs, projects, or business operations.

Value creation in megaprojects has been approached from outcome-based and system lifecycle-based perspectives. From the outcome-based perspective, a megaproject creates value after the project's completion for the organizations participating in it, when it achieves the desired outcomes set initially in the strategic front-end phase (Edkins et al. 2013). The outcome-based view emphasizes the sponsor's role with a notion that the project must create value for the project's sponsor (Eweje et al. 2012). Systems development life cycle phases include planning, system analysis, system design, development, implementation, integration and testing, and operations (Artto et al. 2016). Opportunities for creating value should be looked at not only during the project systems life cycle, but also after the project is completed in the operations phase.

Value Co-Creation

Working closely with stakeholders and local communities as co-creators of a project is an often-underutilized strategy. Globally, when local communities have been involved in project planning and development the outcomes have delivered well-received and long-lasting benefits. In private industry when customers are involved in co-creation of a product or service it has reduced costs of innovation, built stronger market relationships and increased brand loyalty. The studied examples of BMW Group Co-Creation lab, Pixar Innovation Partnerships, and Volkswagen's People's Car Project clearly showed the advantages of engaging customers in the co-creation process.

In the project management literature, projects and programs are being re-conceptualized as value creation processes for multiple actors, and there have been numerous calls for more research on value creation at the front-end (Liu et al. 2019; Martinsuo et al. 2019; Smyth et al. 2018). In recent years, the concept of value co-creation has been gaining some traction in the construction management literature (Candel et al. 2021; Eriksson et al. 2016; Fuentes 2019). One example of value co-creation is through understanding how housing developers and municipalities attempt to resolve conflicts between each other in order to reach mutually beneficial agreements and thereby co-create value propositions (Candel et al. 2021).

3) Value Capture

The growth in demand for new urban rail lines in the twenty-first century has been dramatic in all parts of the globe, especially in China and India but also in most developed cities (Newman et al. 2021). The patronage of existing urban rail systems has seen a significant rise in this period suggesting there is now a major market for urban rail. While traditionally these projects have been predominantly government-funded across the globe they are now struggling to meet the required finances to cater for transit demand. The conventional loan and subsidy based public investments have been unable to meet the demand. The relevance of knowledge about the development phase specific character of value capture policies becomes apparent because value capture policies might not be easily transferable between the divergent institutional contexts of nations at different development stages (Kresse et al. 2020).

Opportunities to pay for infrastructure by leveraging property values may range in scale, but value capture remains consistently viable. Boston's "Big Dig" project, for example, created value to the landowners in the area and transformed the downtown and Seaport District and created greater air rights throughout the metropolitan area after the project was completed (EDRG 2006a, 2006b). Perhaps even greater investment would have resulted if both private and public interests shared risk and potential benefits. Value capture can help give local, state, and federal governing bodies the financial means to improve infrastructure and accommodate growing populations. Many members of the development community support leveraged financing like this, and growing numbers of both developers and public leaders are beginning to better understand the potential of value capture.

Some examples of value capture are shown in Table 2.5.

Value Capture: Case Studies

Because well-planned transportation investments increase people's access to desirable destinations, locations near these investments command higher land prices, benefiting landowners and developers. Studies of the Chicago region show a 10–20% increase in land values near transit stations (Nichols 2012). Value capture mechanisms are a type of public financing where increases in the multi-party computation (MPC) private land values generated by public transportation investments are "captured" to repay the cost of the public investment. Using value capture mechanisms to finance new or existing transportation infrastructure connects the benefit of the infrastructure investment with the cost to provide it (Nichols 2012). Examples of Value capture are shown in Figure 2.4 and described in the following section.

Value Capture at the MTRC Hong Kong

The Hong Kong Mass Transit Railway Corporation (MTRC) is one of the best examples for the use of value capture to fund public transport (Aveline-Dubach and Blandeau 2019). The MTR system in Hong Kong is fully constructed, operated, and maintained without a financial subsidy from the government. In fact, the MTRC is a publicly traded corporation that earns profits for its shareholders, chief among them the government of Hong Kong. This railway serves

Table 2.5 Value capture.

Value capture	Description
Pricing appropriately	A highway that charges high tolls will not be able to capture value if drivers use a less expensive alternative route
Impact fees	One-time charges to developers on new development. Revenues are used to pay for infrastructure improvements – such as schools, hospitals, and roads – to support growth generated by development
Privatization	On the spectrum of P3s, San Juan Airport is representative of a situation where the public sector may want to privatize an unprofitable endeavor and private industry is in fact well suited to make important changes that recover profitability
Special assessments	These are levied on special property taxing districts, self-imposed by residents and/or business owners to support infrastructure needs. The cost of infrastructure is paid for by the owners of properties that are deemed to benefit from the improvements
Tax increments	Captures a portion of the increase in property value as a result of redevelopment that infrastructure improvements may facilitate. Tax-increment financing is typically structured through property taxes
Development	In addition to impact fees, development contributions (also known as exactions) can take the form of contributions, land donations or in-kind donations, such as construction of public infrastructure, parks, or the provision of public services. Development exactions are negotiated and agreed on as part of the permitting process of development
Joint development	A formal arrangement between a public entity and a private developer for the development of a specific asset on publicly owned or controlled property

approximately 20% of transit trips in Hong Kong – a city where the transit mode share is close to 90% of all trips (Zhang 2007).

Cities throughout Asia are currently studying the MTRC business model with the hope of adapting it for use elsewhere. Hong Kong's MTRC model is known as Rail + Property (R+P) whereby the corporation concurrently develops property and the MTR system. (Cervero and Murakami 2009). To enable this, the Hong Kong government does provide an indirect subsidy to the MTRC in the form of land provision at favorable rates. The MTRC obtains land from the government at pre-MTR rates and sells or leases those lands at post-MTR rates. Because Hong Kong is such a dense and public transport-rich city, new rail stations add substantial value to surrounding land, especially for commercial uses.

Crossrail Project in London	Bus Rapid Transit (BRT) in Bogotá, Columbia and Sao Paulo, Brazil
Direct levies were assessed on property owners in the areas benefitting from the additional transport capacity	Local government has recouped public funds used to finance the system through increased value of government-owned properties along the line

Value capture

Hong Kong Mass Transit Railway Corp.	Seattle Washington, Alaskan Way Viaduct
Funded with no government subsidies through land sales, joint development, long term leasing of development rights and commercial space in and around stations	Benefits from property owner assessments ($2.84M), Port Revenues ($267.7M), private funding through joint ventures and over $100M in philanthropic investment

Figure 2.4 Value capture projects.

Over the last decade, almost half of MTRC operating revenue came from property development activities, rather than the corporation's core business of providing rail service (Tang and Lo 2008).

Value Capture in Latin America

Latin American cities have been leaders in the implementation of bus rapid transit (BRT) systems – a transportation mode often characterized by infrastructure improvements that prioritize transit over other vehicles, provide off-vehicle fare payment, and allow quick vehicle access (Rodriguez and Vergel-Tovar 2013). In Latin America, value capture has been utilized to help fund Bus Rapid Transit (BRT) in cities such as Bogota, Columbia, and Sao Paulo, Brazil. Property values have increased dramatically along BRT corridors due to improved transit, and the local government has been able to recoup public funds used to finance the system through increased value of government-owned properties along the line. Both Bogota and Sao Paulo helped pay for new transit lines through the increase in property values along those corridors (Toma 2011).

Porto Alegre in southern Brazil employed a far more aggressive model of value capture. As population in the city grew during the 1990s, speculators hoarded undeveloped land in the city center – which in turn created a housing shortage in the city and forced residents to balloon out into suburban slums. In response,

Porto Alegre created an incentive for landowners to develop vacant properties and implemented a land value increment tax to help capture increased values from the city's investment (Toma 2011).

Value Capture at Crossrail

At the outset of the establishment of the Crossrail project (Europe's largest megaproject) to deliver the new Elizabeth line east–west railway across London, the UK government's preferred delivery model was a privately financed concession. When this proved to be unachievable the Treasury capped the contribution from the central exchequer at around a third of the overall cost, requiring the remaining funding to be generated from the "beneficiaries" of the project. The Department for Transport (DfT) and Transport for London (TfL) – together the "sponsors" – collaborated with the business and property development communities in London to assemble a finance and funding package (Buck 2017). When GVA the independent commercial property consultant for Crossrail updated their original impact study in 2017 they concluded that by 2026 with the construction of $180 000 new homes it would add £20B to property value. The Montague Review (2004) identified that there was strong resistance to upfront contributions from landowners. However, they were less dismissive of increased post opening levies. While Community Infrastructure Levy (CIL) contributions for Crossrail have been an important means of capturing some value, the gains accruing to property owners in the vicinity of the Elizabeth line suggest that the public sector could have captured more of the value upside and that value capture would be explored in all future projects including Crossrail 2.

Value Capture at Seattle Alaskan Way Viaduct Replacement Project

The $3.3B Alaskan Way Viaduct, an elevated section of State Route 99 in Seattle, Washington was built in the 1950s, and decades of daily wear and tear took a toll on the structure. Because of the viaduct's age and vulnerability to earthquakes, replacing it was critical to public safety. The program included 30 projects led by the Washington State Department of Transportation. The decision to replace the viaduct arose from a combination of physical infrastructure concerns and shifting political and public opinion about the best use of public space along Seattle's waterfront. In 2001, the 6.8-magnitude Nisqually Earthquake struck the Puget Sound region, resulting in damage and settlement on sections of the viaduct. Crews stabilized the structure, but engineers agreed that if the earthquake had lasted a few moments longer, the viaduct would have collapsed. Damage from the earthquake, along with cumulative wear and tear from decades of use, made replacing the viaduct critical to public safety. WSDOT and its partner agencies used a variety of funding and financing strategies to generate the estimated $3.3B needed for the Seattle Alaskan Way Project in addition to federal and state funds. These additional sources included: Local Improvement District property owner assessments and other local funds – $28.4M. Port of Seattle – $267.7M. In addition to public dollars, the waterfront park and other public amenity projects connected with the demolition of the viaduct and the tunneling project received private funding through a variety of joint ventures and over $100M in philanthropic investment (WSDOT 2021).

Megaproject Evaluation: Beyond the Iron Triangle

> Building an evaluation culture is not only about strengthening the supply side but, also about strengthening the demand side. . .by building a common understanding of the role of evaluative evidence in learning and account-ability. (WBG 2019)

The above quote reflects the position of the World Bank in the development of its principles of evaluation of all its projects across the entire World Bank Group. To ensure that the goals of a project finance are being met, multilateral and commercial lending institutions have project evaluation principles and guidelines on which they rely to evaluate the project finance at various stages of the life cycle. Though the criteria may vary across these organizations and across projects the goals are similar. The common principles include core principles for evaluation, and principles for selecting, conducting, and using evaluations. These aim to strengthen accountability and learning for evidence-based decision-making and program improvement to enhance development results. They are designed to: (i) align the World Bank Group's evaluative efforts with global challenges and the World Bank Group's strategic focus; (ii) clarify the roles and responsibilities of key actors and encourage synergy among them throughout the evaluation process; and (iii) ensure that all World Bank Group evaluations are robust, of high quality, and credible (WBG 2019, p. 2).

As discussed in the prior section, sustainability, value creation, and value capture will play a major role in evaluating our project financings in the future. How infrastructure systems are selected, designed, financed, and managed today will have a major role in how those systems affect society and the environment now and in future years (Weber et al. 2016). The extensive literature on sustainability and infrastructure address three sustainable development dimensions – namely, social, environmental, and financial. Social sustainability refers to the social benefits and equity derived from infrastructure; environmental sustainability refers to the impact of infrastructure and its use on the health and welfare of the population and the environment; and financial sustainability refers to the governments capacity to meet the financial obligations resulting from infrastructure investments, both in the short- and long-term (Garrido and Vassallo 2021).

The literature is filled with examples of how projects have achieved successful outcomes through cost and schedule adherence, but emerging theories focus on the institutional, the environmental and the ecological approach to evaluating megaprojects. Lehtonen (2019) in his analysis of ecological economics: "institutionalist deliberationism" and the environmental justice approach found that both can help to "open up" megaproject appraisal (i) beyond the "iron triangle" of project appraisal criteria employed by the mainstream megaproject scholarship; (ii) to a broader range of appraisal perspectives; and (iii) toward "informal" appraisal, for example, the wider sociopolitical discourse including media interventions and NGO initiatives. Lehtonen suggests four areas of further work: (i) integrating formal and informal appraisal; (ii) the role of appraisal and the appraiser in megaproject governance; (iii) the role of deliberation in the face of power asymmetries; and (iv) the possibility of sustainable megaprojects (Lehtonen 2019).

While megaprojects in the past typically served the objectives of industrialization and development – as they still do in the developing countries – today's "resurgence of high-modernist reliance on big infrastructure" in the developed world is driven also by climate concerns, as well as global competition between cities for investments, knowledge workers, tourists, and prestige (Perry and Praskievicz 2017).

Looking Back and Looking Forward: Ex-ante and Ex-post Evaluations

Ex-ante Evaluations

Transcribed from Latin, ex-ante is the prediction of an event before it happens, or before the participants become aware of the event and it is the opposite of the Latin word ex-post. In the case of project finance, ex-ante means assessment of a project during the early conception and planning phases to test its viability and feasibility. The predicted outcome serves as a basis for comparing the prediction to the actual results (ex-post).

For example, in the early conception stages of a project, predictions of a project's performance will be assessed through various approaches including benefit cost analysis, feasibility studies, credit rating services, economic and sustainability evaluations, and environmental assessments. The approach is often based on an economic analysis of future benefits based on a series of assumptions including prior experience. Since it is based on a future event it is usually uncertain what will be the actual economic performance.

During the long life of a megaproject, unexpected events can occur such as a global pandemic realized in the COVID-19 experience, or an economic downturn that was not foreseen when the project commenced. If the events turn out to be better than expected the decision to move forward with the project will appear to be the right decision, but if events push the project into insolvency or cause serious cost overruns it may mean that the project selected was a poor decision in hindsight.

Ex-post Evaluation

Unlike ex-ante, which is based on an estimation of project performance sometimes 10–20 years in advance of project completion, ex-post represents the actual results attained by the project, which is the return on investment that the project yielded.

Ex-post evaluations are used throughout the European Commission to assess whether a specific intervention was justified and whether it worked (or is working) as expected in achieving its objectives and why. Ex-post evaluations also look for unintended effects (i.e. those which were not anticipated at the time of the initial project feasibility study).

Ex-post evaluation analysis is an approach often used by the European Commission, the European Investment Bank (EIB), or the World Bank to evaluate the impact of the support they provide to PPP projects, and whether they have reached their sustainable development goals. However, these evaluations usually focus on specific impacts of interest for the institution providing the support, and the role played by that institution during the implementation of the project. Thus, for example, the evaluations carried out at the EU level focus on the financial value added to the projects and whether the projects meet cost and schedule commitments. In the United States, all federally funded projects are required to have both ex-ante and ex-post evaluations; however, these evaluations are not always carried out in a way that provides sufficient data or input for future projects. For instance, the main variable analyzed by the World Bank support is the business performance of the projects.

Most project-development organizations or project sponsors are reluctant to spend a significant amount of money on early-stage engineering and design for three reasons. First, they often lack the funds in the early stage of a project to spend significantly on design and engineering. Second, they are eager to break ground and start construction. Finally, they worry that the design will be modified once construction is under way and thus make the expenditure on up-front design pointless.

Evaluation of a Project Finance: The Critical Front End: Ex-ante Evaluation

In this section, we look at a variety of ex-ante evaluations and ex-post evaluations to determine whether the project delivered what it promised from a financial perspective. If the front end of a project finance (the cost/benefit analysis) is compared with the back end of the project (the results) ideally, they should be similar. However, this is often not the case except for smaller, well controlled projects without complex technology, systems, and diverse stakeholders.

Globally, projects are evaluated by different mechanisms and groups. These include evaluations by government agencies, public auditors, intergovernmental organizations, legacy reports, credit rating agencies and commercial banks, and private equity groups. These reports are readily available both before a project commences and after a project completes. Too often these reports are not consulted, or they are reviewed too late after the project has serious cost overruns or delays, or long after a project has been abandoned or completed.

Ex-post evaluations that occur within five years of project completion are often not extended over a long enough period to be able to see a more holistic view of the project success or failure. Some benefits may not surface immediately and may take decades to develop particularly in urban renewal and transportation projects.

The scholarly research suggests that the early appraisal of an investment case or a project should apply the same evaluation criteria that will be used in ex-post evaluation, and thus increase the likelihood of a successful project outcome. However, the initial plan might be altered because of subsequent analysis, assessment, negotiation, positioning, and the exercise of power (Samset and Christensen 2017).

Megaprojects are subject to tremendous change over their long life due to unexpected events, technological and scientific evolution, stakeholder demands, political influences and a host of other reasons thus making ex-ante and ex-post evaluations harder to compare. However, understanding these unanticipated events can make it easier for megaprojects in the future.

Too often these evaluations are not used by organizations, particularly if it is a private financing where government regulations may not apply or may not even exist. As noted by Samset and Christensen (2017) in their research on ex-ante and ex-post evaluations it is a paradox that systematic ex-ante and ex-post evaluations are rarely used for project evaluations.

In today's modern societies, it is common that either evaluations are clearly managed, or that results are interpreted, slanted, politicized, and oversold. This analysis may contribute to more control and greater breakthrough but also may undermine the credibility of and support of projects (Samset and Christensen 2017).

Evaluating Projects Through the Life Cycle

As illustrated in Figure 2.5, projects typically pass through four phases in a project life cycle that include: (i) Project Conception and Selection; (ii) Project Implementation: (iii) Project Closure and Transition; and (iv) Project Evaluation and Operations. During each of these phases the project is continually evaluated with the most important phases being during conception when an ex-ante evaluation is conducted to see if the project is worth doing and determining the feasibility and sustainability of the project, and during operations after project closure to determine if the benefits are being realized and if the benefits are sustainable throughout the life cycle of the operational phase. In Figure 2.5, some common evaluation techniques for each phase represented are included and are described in more detail in this section. Recognizing that a full evaluation of global techniques would be a book unto itself the goal here is to identify some of the more common approaches to project evaluations both looking forward and looking back.

Ex-ante Evaluation Criteria

Cost Benefit Analysis

An important evaluation method frequently used by business and government officials to support the ex-ante evaluation is cost-benefit analysis (CBA), which provides a monetary value quantifying the benefits and costs and thus allows a

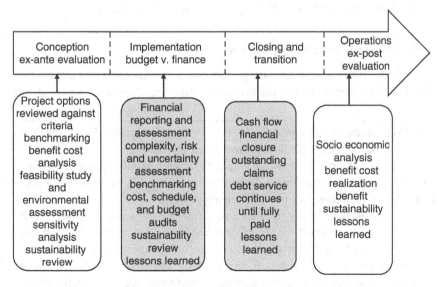

Figure 2.5 Project evaluation phases.

comparison of a broader range of scenarios. For the purpose of evaluating an investment an important question is:

> How are investment appraisal tools such as cost-benefit analysis used to shape investment decision outcomes rather than merely providing objective criteria for selection between alternatives?

A CBA requires an investigation of a project's net impact on economic, and social welfare. Benefits and costs are typically evaluated for a period that includes the construction period and an operations period ranging from 20 to 50 years after the initial project investments are completed. Given the permanence and extended design life of high-speed rail investments, longer operating and thus, evaluation periods are applicable. A rail project may last a 100 years, but a bridge project 60 years (Unless it is the Oakland Bridge in SF) a suspension bridge built to last 150 years, whereas a nuclear plant may only last 40 years.

Importantly, the approach taken to the analysis of benefits by the multilaterals is different than the approach taken by a privately financed project. For example, since a societal perspective is taken in CBA at the Inter-American Development Bank for Bank funded projects, the benefits should include all societal benefits.

As recognized by the World Bank's Evaluation Group, cost-benefit analysis entails measuring results, valuing results, and comparing results with costs, and hence is highly relevant to the results agenda. Cost-benefit analysis can provide a comprehensive picture of the net impact of projects and help direct funds to where their development effectiveness is highest (WBG 2010).

Project Finance Credit Rating Agencies

The credit rating agencies are another important ex-ante source of information about project finance. In the United States, three major agencies – Standard &

A (2010) study by the World Bank's Independent Evaluation Group on Cost Benefit Analysis (WBG 2010) drew two broad conclusions:

First, the bank needs to revisit its policy for cost-benefit analysis in a way that recognizes the legitimate difficulties in quantifying benefits while preserving a high degree of rigor in justifying projects.

Second, the bank needs to ensure that cost-benefit analysis is done with quality, rigor, and objectivity: poor data and analysis misinform, and do not improve, results. Reforms are required to project-appraisal procedures to ensure objectivity, improve both the analysis and the use of evidence at appraisal, and ensure effective use of cost benefit analysis in decision making.

Poor's, Moody's Investor Service, and Fitch Ratings – play a significant role in evaluating the likelihood of success of a public financing undertaking. For example, Standard & Poor's (S&P) project debt ratings address default probability – or, put differently, the level of certainty with which lenders can expect to receive timely and full payment of principal and interest according to the terms of the financing documents. Unlike corporate debt, project-finance debt is usually the only debt in the capital structure, and typically amortizes to a schedule based on the project's useful life. Importantly, also unlike corporate ratings, which reflect risk over three-to-five years, project debt ratings are assigned to reflect the risk through the debt's tenor (S&P 2007).

Although there is tremendous demand for private financing (through project finance or other techniques) of basic public infrastructure in developing markets, emerging-market (EM) sovereign ratings must be carefully reviewed to ensure that the ratings have stabilized particularly after major events. Since May 2021 Fitch had downgraded only two EM countries (Columbia and Tunisia), matched by two upgrades (Cote d'Ivoire and Gabon) after a net 30 EM country downgrades in 2020 (Fitch Ratings 2021b).

In addition to garnering public support, it might also be assumed that a vital role for the government is to conduct a benefit-cost analysis that justifies the project as advancing the welfare of the public. This analysis should include the impact of negative externalities of the project. For instance, a government is peculiarly situated to evaluate the effects of the projects on all that will be affected by it, whether it is an explicit stakeholder in the project. Rating agencies look at various aspects of the project financing to ensure that the project is viable. Figure 2.6 shows a typical analysis of various factors in the rating of a project finance to include economic, technical, financial, and social viability.

Ex-post Evaluation Criteria

Ex-post evaluations can serve multiple important purposes. They can be beneficial for managing authorities in order to build internal capacity and to improve the project selection process (de Jong et al. 2019), and they contribute to accountability and/or control and learning and improvement (Worsley 2014). Zidane et al. (2015) develop a holistic model for project evaluation on strategic, tactical and operational levels (PESTOL) by reviewing different definitions of project success and/or failure. As discussed in this chapter, whatever, the methodology, all megaprojects should have ex-post evaluation criteria that should be conducted over the operational life of a completed project. For some projects such as oil and gas that could be a 40- to 60-year period and for infrastructure that will depend on the expected life of the infrastructure.

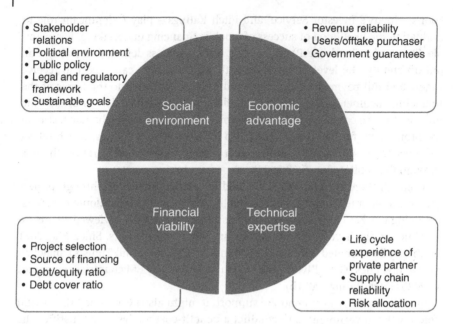

- Stakeholder relations
- Political environment
- Public policy
- Legal and regulatory framework
- Sustainable goals

- Revenue reliability
- Users/offtake purchaser
- Government guarantees

- Project selection
- Source of financing
- Debt/equity ratio
- Debt cover ratio

- Life cycle experience of private partner
- Supply chain reliability
- Risk allocation

Social environment

Economic advantage

Financial viability

Technical expertise

Figure 2.6 Project finance evaluation ratings. *Sources:* Adapted from Standard and Poor's (2007) *Updated Project Finance Summary Debt Rating Criteria for Infrastructure and Project Finance: Master Criteria*; and Fitch Group India Ratings & Research (2019) *Rating Criteria for Infrastructure and Project Finance: Master Criteria*.

Equator Principles

One of the important ex-post evaluation criteria used by the major international multilateral banks and import/export agencies are the Equator Principles (EPA 2022). Traditionally, lenders have required, at a minimum, that the project company undertakes to comply with all applicable environmental and social laws and regulations; however, in recent years, lenders (especially ECAs and DFIs) have typically required the project company to adhere to a set of guidelines known as the "Equator Principles," which are a financial industry's benchmarks for determining, assessing, and managing social and environmental risk in project financing (EPA 2022). The Equator Principles incorporate the International Finance Corporation and World Bank environmental performance standards and guidelines. Thus, the Equator Principles extend these international project-based environmental and social standards into the realm of private financings. Amongst other things, adherence to the Equator Principles requires the project company to develop and comply with an agreed environmental and social management plan focusing on areas such as: labor and working conditions; pollution prevention and abatement; community health, safety, and security; biodiversity, conservation, and sustainable natural resource management; and protection of Indigenous people and their cultural heritage.

Looking Back and Looking Forward and Qualitative Scoring

Many countries regulate the early stages of a project through project selection criteria, project feasibility studies, and project benefit cost analysis, however, the ex-post evaluation process is more problematic. Once a project is completed there is less pressure on the project sponsors and often there are no clear requirements or a regulatory process that mandate ex-post evaluations. Ex-post evaluation is essential in showing whether projects really delivered the benefits expected and to *learn which* projects do better and which do worse than expected, and why (de Jong et al. 2019). An ex-post evaluation of 10 major transport infrastructure projects supported by the European Regional Development Fund (ERDF) and Cohesion Fund (CF) between 2000 and 2013 were evaluated ex-post using a common methodology. The methodology includes a CBA (looking backward and forward), as well as a more qualitative scoring of project outcomes. The project performance was also related to its determining factors. The assessment showed that 9 out of the 10 projects had a benefit-cost ratio above 1 ex-post, though the benefits for about half of the projects were clearly lower than had been anticipated ex-ante. Even the worst performing projects according to the research could not be called failed projects, only underachieving projects.

Evaluation over the Entire Operational Life of a Megaproject

As shown in Figure 2.6, evaluation of projects is essential to their viability and sustainability at various phases including (i) an ex-ante evaluation before the project is selected, (ii) an evaluation during the implementation phase, (iii) an evaluation during transition of the project to operations, (iv) and regular ex-post evaluations over the operational phase. For some projects this may mean evaluations over a 20-year period or longer. The purpose of evaluation is to determine whether the project meets the expectations for which it was created. Those expectations are no longer limited to financial profitability but include societal benefits and sustainability that require a more qualitative analysis than that achieved using traditional evaluation methodologies. The literature and practice suggest that the early appraisal of an investment case or a project should apply the same evaluation criteria that will be used in ex-post evaluation, and thus increase the likelihood of a successful project outcome.

Summary

For most of the last century, infrastructure development and management have focused primarily upon capital construction. In addition, a substantial amount of the financing for infrastructure, particularly within the United States, has come from federal sources, as evidenced by the Interstate Highway System Program and the Environmental Protection Agency Construction Grants Program (Garvin 2007). Globally, megaprojects require long up-front planning to clarify political, community, and governmental concerns. Projects must demonstrate benefits that are clear to the decision makers, and all impacts – economic, social, and financial must be addressed in the planning process. Project finance should provide the appropriate balance and incentives between the roles of the public sector and the private sector in terms of both management and funding of the project. Projects should be evaluated for feasibility from the lender's, the equity sponsors, and the owner's perspectives. Project financings in the future will need to go beyond the traditional sources of government and multilateral organizational sponsorship. Creativity will be required to capture the real value of the project as well as those who will benefit the most. Sustainable public private partnerships will need to be created to tap private sector financial, technical and management resources to achieve public objectives such as greater cost and schedule certainty, innovative technology applications, specialized expertise, and access to private capital.

Projects reliant on multilateral support and export credit agencies depend upon a complete up-front analysis of the project selection process, feasibility analysis, benefit cost determination and sustainability measurements including compliance with the Equator Principles, the UN Sustainable Goals and the World Bank's Legal and Social Framework. Unfortunately, too often the up-front analysis is limited and is controlled by political expediency and urgency rather than a complete analysis of the benefits and sustainability of the project. In this chapter, we reviewed the structural aspects of a project finance, the sources of project finance including innovative techniques such as value creation and value capture and the means of evaluating projects, both ex-ante and ex-post to determine if projects are delivering what they promised but also to learn what achievements are possible when innovation and sustainability are central to decision making.

Lessons and Best Practices

1) Project finance is not a new concept and existed in various forms since the Middle Ages.

2) Project finance provides the impetus for building projects that otherwise would not happen.

3) It takes many forms including debt, equity, bonds, guarantees, offtake purchasers, creative financing, and requires solutions to complex and uncertain problems.

4) In addition to relevant environmental and related laws, lenders now require compliance with the Equator Principles which incorporate the International Finance Corporation and World Bank environmental performance standards and are a financial industry benchmark for determining, assessing, and managing social and environmental risk in project financing.

5) The real cost of a megaproject includes not just the cost for the project, but the financing required to pay for the project which may be substantial and take decades to pay off if not close to a century as was the case in Eurotunnel.

6) Many megaprojects are so large and the need for them is so critical that pay as you go through tolls or tax revenues is not a viable option. Instead, states are stepping up with higher contributions and using innovative financing techniques including federal loans and state bonding initiatives (Greiman 2013, p. 71).

7) Regardless of how they are financially structured, major infrastructure projects (for example, new roads, ports, dams, power plants, and airports) typically have significant environmental and social impacts that are causing project costs to continue to rise dramatically (Greiman 2013, p. 71).

8) Ex-post and ex-ante evaluations should be conducted on all megaproject financings so that lessons can be learned and shared to benefit all future projects.

Discussion Questions

1 What are the common characteristics of a project finance?

2 How does project finance differ from corporate finance? What are the benefits of project finance and the challenges that must be overcome to ensure successful outcomes?

3 What are the important questions that lenders and equity sponsors must ask before embarking on a project finance?

4 What role can bonds play in a project finance?

5 What are the best opportunities to fill financing gaps?

6 Why is project finance important to economic and social development of a city, nation or a region of the world?

7 How can private financing be used to fund major infrastructure projects globally?

8 What is meant by innovative financing? How might you use this type of financing to fill gaps in the project financing?

9 Distinguish between ex-ante evaluations and ex-post evaluations. What are some techniques that you can use on megaprojects to evaluate the project finance in both the early and late stages of the project?

10 How would you go about gathering data about a megaproject so that your forecasts will be realistic and in line with standard practice?

11 How do you mobilize capital for projects in the developing world with high technical, financial, economic, social, and political risk?

12 What are the greatest challenges to project finance and how would you increase financing opportunities?

References

American Society of Civil Engineers (ASCE) (2021). *A Comprehensive Assessment of America's Infrastructure: 2021 Report Card for America's Infrastructure*. Reston, Virginia: ASCE.

Artto, K., Ahola, T., and Vartiainen, V. (2016). From the front end of projects to the back end of operations: Managing projects for value creation throughout the system lifecycle. *International Journal of Project Management* 34 (2): 258–270. http://dx.doi.org/10.1016/j.ijproman.2015.05.003.

Aveline-Dubach, N. and Blandeau, G. (2019). The political economy of transit value capture: The changing business model of the MTRC in Hong Kong. *Urban Studies* 56 (16): 3415–3431.

B20 Task Force (2017). *Investing in Resilient, Future-Oriented Growth Boosting Infrastructure Investment and Balancing Financial Regulation*. B20 Taskforce Financing Growth & Infrastructure Policy.

Bank for International Settlements (BIS) (2021). *Basel Committee on Banking Supervision CRE Calculation of RWA for Credit Risk CRE30 IRB Approach: Overview and Asset Class Definitions.* Basel III Publication.

BloombergNEF (BNEF) (2022). *Energy Transition Investment Trends 2022.* London, England: Bloomberg.

Buck, M. (2017). Crossrail project: Finance, funding and value capture for London's Elizabeth line. *Civil Engineering* 170 (6): 15–22. http://dx.doi.org/10.1680/jcien. 17.00005.

Candel, M., Karrbom, T.G., and Eriksson, P.-E. (2021). Frontend value co-creation in housing development projects. *Construction Management and Economics* 39 (3): 245–260. http://dx.doi.org/10.1080/01446193.2020.18510.

Cervero, R. and Murakami, J. (2009). Rail and property development in Hong Kong: Experiences and extensions. *Urban Studies* 46 (10): 2019–2043.

Chukwuebuka, N., Azolibe, B., Okonkwo, J.J., and Adigwel, P.K. (2020). Government infrastructure expenditure and investment drive in an emerging market economy: Evidence from Nigeria. *Emerging Economy Studies* 6 (1): 61–85. http://dx.doi. org/10.1177/239490152090772.

Claughton, P. (2010). The crown silver mines and the historic landscape in Devon (England). The work of the Bere Ferrers project. *Les mines d'argent du roi anglais et le paysage historique du Devon: le Projet Bere Ferrers* 299–308. https://doi. org/10.4000/archeosciences.2862.

Clifton, J., Lanthier, P., and Schröter, H. (2011). Regulating and deregulating the public utilities 1830–2010. *Business History, Taylor & Francis Journals* 53 (5): 659–672.

Congressional Research Services (CRS) (2021). *Public-Private Partnerships (P3s) in Transportation.* Updated March 26, 2021. Washington, DC: CRS.

Deloitte (2017). *Investing in Infrastructure. Leading Practices in Planning, Funding and Financing.* Deloitte Development LLC https://www2.deloitte.com/content/ dam/Deloitte/us/Documents/risk/us-risk-infrastructure-investment-funding.pdf.

Deloitte (2022). *Project Bonds.* Johannesburg, South Africa: Deloitte https://www2. deloitte.com/za/en/pages/finance/articles/project-bonds-an-alternative-to-financing-infrastructure-projects.html.

Edkins, A., Geraldi, J., Morris, P., and Smith, A. (2013). Exploring the front-end of project management. *Engineering Project Organization Journal* 3 (2): 71–85. http:// dx.doi.org/10.1080/21573727.2013.775942.

Economic Development Research Group, Inc. (EDRG) (2006a). *Economic Impact of the Massachusetts Turnpike Authority and Related Projects, Volume I: The Turnpike Authority as a Transportation Provider.* Boston, MA: EDRG.

Economic Development Research Group, Inc. (EDRG) (2006b). *Economic Impact of the Massachusetts Turnpike Authority and Related Projects, Volume II: Real Estate Impacts of the Massachusetts Turnpike Authority and the Central Artery/Third Harbor Tunnel Project.* Boston, MA: EDRG.

Equator Principles Association (EPA) (2022) *Equator Principles 2022*. https:// equator-principles.com/.

Eriksson, P.E., Leiringer, R., and Szentes, H. (2016). The role of co-creation in enhancing explorative and exploitative learning in project-based settings. *Project Management Journal* 48 (4): 22–38.

Esty, B. (2004). *Modern Project Finance*. Hoboken, New York: John Wiley & Sons.

European Union (EU) (2021). *EU – Country Commercial Guide*. Brussels: European Union, International Trade Administration.

Eweje, J., Turner, R., and Müller, R. (2012). Maximizing strategic value from megaprojects: The influence of information-feed on decision-making by the project manager. *International Journal of Project Management* 30 (6): 639–651. http://dx.doi.org/10.1016/j.ijproman.2012.01.004.

Fitch Action Commentary (2021a). *Fitch Assigns North Carolina's GARVEE Bonds at 'A+'; Outlook Stable*. Fitch Ratings, Inc https://www.fitchratings.com/research/ us-public-finance/fitch-assigns-north-carolina-garvee-bonds-at-a-outlook-stable-16-08-2021.

Fitch Group India Ratings & Research (2019). *Rating Criteria for Infrastructure and Project Finance: Master Criteria*. https://www.indiaratings.co.in/Uploads/CriteriaReport/ Rating%20Criteria%20for%20Infrastructure%20and%20Project%20Finance.pdf.

Fitch Ratings (2021b). *Emerging-Market Ratings Broadly Stabilize, But Risks Remain*. New York: Fitch, Inc., June 12. https://www.fitchratings.com/research/sovereigns/ emerging-market-ratings-broadly-stabilise-risks-remain-14-10-2021.

Fuentes, M. (2019). Co-creation and co-destruction of experiential value: A service perspective in projects. *Built Environment Project and Asset Management* 9 (1): 100–117.

Garrido, L. and Vassallo, J.M. (2021). Is ex-post fiscal support to PPPs sustainable? Analysis of government loans granted to shadow-toll roads in Spain: A case study. *Sustainability* 13: 219. https://doi.org/10.3390/su1301021.

Garvin, M.J. (2007). Rethinking mega-project development strategies: A case study of the Central Artery/Tunnel Project. *Structure and Infrastructure Engineering* 3 (2): 147–157. http://dx.doi.org/10.1080/15732470600590796.

GETLINK SE Universal Registration Document (2022). *GETLINK GROUP, Paris, France*. https://www.businesswire.com/news/home/20210318005772/en/ Getlink%E2%80%99s-2020-Universal-Registration-Document-Made-Available

Goldsmith, H. and Boeuf, P. (2019). Digging beneath the iron triangle: The Chunnel with 2020 hindsight. *Journal of Mega Infrastructure and Sustainable Development* 1 (1): 79–93. http://dx.doi.org/10.1080/24724718.2019.1597407.

Greiman, V.A. (2013). *Megaproject Management: Lessons on Risk and Project Management from the Big Dig*. Hoboken, New York, London: John Wiley & Sons.

Greiman, V.A. and Sclar, E. (2019). Meg-infrastructure as a dynamic ecosystem: Lessons from America's interstate highway system and Boston's Big Dig. *Journal of Mega Infrastructure and Sustainable Development* 1 (2): 188–200. Published online 2020.

Guyol, C. (2016). The Frescobaldi of Florence and the English Crown. In: *Kings, Knights and Bankers: The Collected Articles of Richard W. Kaeuper (Later Medieval Europe) 13* (ed. R.W. Kaeuper and C. Guyol), 43–92. Leiden, The Netherlands: Koninklijki Brill NV https://doi.org/10.1163/9789004302655_006.

House Committee on Bonding (HC) (2011). *Capital Expenditures and State Assets, Report on Capital Planning and Spending in the Commonwealth of Massachusetts (2011–2012)*. Representative Antionio F.D. Cabral Chairman, Boston, Massachusetts State Legislature.

International Bank for Reconstruction and Development (IBRD) (2018). *Procuring Infrastructure Public-Private Partnerships Report: Assessing Government Capability to Prepare, Procure, and Manage PPPs*. Washington, DC: World Bank Group https://openknowledge.worldbank.org/bitstream/handle/10986/29605/PIPPP_ 2018.pdf?sequence=1&isAllowed=y.

International Bank for Reconstruction and Development (IBRD) (2021). *Sustainable Development Bonds & Green Bonds Impact Report*. Washington, DC: World Bank Group https://issuu.com/jlim5/docs/world-bank-ibrd-impact-report-2020?mode= window.

Japan Bank for International Cooperation (JBIC) (2021). *JBIC Project Finance Initiatives: Japan Business and Global Prosperity*. Japan: Tokyo https://www.jbic. go.jp/ja/information/contents/project_finance_en.pdf.

Johnson, J.A., Ruta, G., Baldos, U. et al. (2021). *The Economic Case for Nature: A Global Earth Economy Model to Assess Development Policy Pathways*. Washington, DC: World Bank Group.

de Jong, G., Vignetti, S., and Pancotti, C. (2019). Ex-post evaluation of major infrastructure projects. *Transportation Research Procedia* 42: 75–84.

Kessinger, J.W. and Martin, J.D. (1988). Project finance: Raising money the old-fashioned way. *Journal of Applied Corporate Finance* 1 (3): 69–81.

Khan, A., Rizvi, S.A.R., Ali, M., and Haroon, O. (2021). A survey of Islamic finance research – Influences and influencers. *Pacific-Basin Finance Journal* 69: 101437.

Kresse, K., Kangb, M., Kimc, S., and van der Krabben, E. (2020). Value capture ideals and practice – Development stages and the evolution of value capture policies. *Cities* 106: 102861.

Lehtonen, M. (2019). Ecological economics and opening up of megaproject appraisal: Lessons from megaproject scholarship and topics for a research programme. *Ecological Economics* 159: 148–156.

Lui, Y., van Marrewijk, A., Houwing, E.-J., and Hertogh, M. (2019). The co-creation of values-in-use at the front end of infrastructure development programs. *International Journal of Project Management* 37 (5): 684–695. https://doi.org/ 10.1016/j.ijproman.2019.01.013.

Ma, T. (2016). Basel III and the future of project finance funding. *Michigan Business and Entrepreneurial Law Review* 6 (1): 109–126.

Martinsuo, M.M., Vuorinen, L., and Killen, C. (2019). Lifecycle-oriented framing of value at the front end of infrastructure projects. *International Journal of Managing Projects in Business* 12 (3): 617–643.

Mirabile, M., Marchal, V., and Baron, R. (2017). *Technical note on estimates of infrastructure investment needs: Background document to the report Investing in Climate, Investing in Growth*. Paris, France: Organization for Economic Development (OECD).

Montague, A. (2004). *Crossrail Review*. London, UK: The Stationery Office (TSO).

Moslener, U., McCrone, A., d'Estais, F. et al. (2016). *Global Trends in Renewable Energy Investment 2016*. Frankfurt School of Finance & Management – UNEP Collaborating Centre and Bloomberg New Energy Finance.

Müllner, J. (2017). International project finance: Review and implications for international finance and international business. *Management Review Quarterly* 67: 97–133.

Nelson, R.M. (2019). *Multilateral Development Banks: Overview and Issues for Congress*. Updated February 11, 2020. Congressional Research Service (CRS). https://crsreports.congress.gov R41170.

Newman, P., Davies-Slate, S., Conley, D. et al. (2021). From TOD to TAC: Why and how transport and urban policy needs to shift to regenerating main road corridors with new transit systems. *Urban Science* 5 (3): 52. https://doi.org/10.3390/urbansci5030052.

Nichols, C.M. (2012). *Value Capture Case Studies: What Is Value Capture?* Chicago, IL: Metropolitan Planning Council https://www.metroplanning.org/news/6311/Value-capture-case-studies-What-is-value-capture.

Perry, D.M. and Praskievicz, S.J. (2017). A new era of big infrastructure? (Re) developing water storage in the U.S. west in the context of climate change and environmental regulation. *Water Alternatives* 10 (2): 437–454.

Project Management Institute (2021). *Project Management Body of Knowledge (PMBOK)*. Newtown Square, PA: PMI.

Rodriguez, D. and Vergel-Tovar, C.E. (2013). Bus rapid transit and urban development in Latin America. *Land Lines* 25: 14–20.

Samset, K. and Christensen, T. (2017). Ex ante project evaluation and the complexity of early decision-making. *Public Organization Review* 17: 1–17.

Sclar, E. (2015). The political economics of investment Utopia: Public–private partnerships for urban infrastructure finance. *Journal of Economic Policy Reform* 18 (1): 1–15. https://doi.org/10.1080/17487870.2014.950857.

Shakirat, A.B. (2018). Government spending on infrastructure and economic growth in Nigeria. *Economic Research-Ekonomska Istraživanja* 31 (1): 997–1014.

Smyth, H., Lecoeuvre, L., and Vaesken, P. (2018). Co-creation of value and the project context: Towards application on the case of Hinkley Point C Nuclear Power. *International Journal of Project Management* 36 (1): 170–183.

Standard and Poor's (S&P) (2007). *Ratings Direct: Updated Project Finance Summary Debt Rating Criteria for Infrastructure and Project Finance: Master Criteria.* New York, NY: Standard & Poor's, McGraw-Hill Companies, Inc. https://www.maalot.co.il/publications/MT20120529105458.pdf.

Standard and Poor's (S&P) (2022). *Islamic Finance Outlook. S&P Global Ratings 2022 Edition.* https://www.spglobal.com/ratings/en/research/pdf-articles/islamic-finance-outlook-2022-28102022v1.pdf.

Statista (2021). *Foreign Direct Investment Flows.* www.statista.com/statistics/188870/foreign-direct-investment-in-the-united-states-since-1990/

Tang, B.S. and Lo, H.K. (2008). The impact of public transport policy on the viability and sustainability of mass railway transit: The Hong Kong experience. *Transportation Research Part A: Policy and Practice* 42 (4): 563–576.

Temin, P. (2004). Financial intermediation in the early Roman empire. *The Journal of Economic History* 64 (3): 705–733. Cambridge University Press, Cambridge, UK. https://www.jstor.org/stable/3874817.

Toma, W. (2011). Value capture: An innovative strategy to fund public transportation projects, Smart Growth America. *Transportation Research* 42: 75–84. https://doi.org/10.1016/j.trpro.2019.12.008.

U.S. Congress (2021) H.R.3684 - *Infrastructure Investment and Jobs Act*,117th Congress (2021–2022).

U.S. Energy Information Administration (EIA) (2022, February) *Monthly Energy Review.* Washington, DC. https://www.eia.gov/totalenergy/data/monthly/pdf/mer.pdf.

United Nations Conference on Trade and Development (UNCTAD) (2021). *World Investment Report.* Geneva: UNCTAD.

Washington State Department of Transportation (WSDOT) (2021). *Alaskan Way Viaduct Replacement Program.* Seattle, Washington: Washington DOT.

Weber, B., Staub-Bisang, M., and Alfen, H.W. (2016). *Infrastructure as an Asset Class: Investment Strategy, Sustainability, Project Finance and PPP*, 2nde. Chichester, UK: John Wiley & Sons.

Wolmar, C. (2005). *On the Wrong Line: How Ideology and Incompetence Wrecked Britain's Railways.* London, UK: Aurum Press, Ltd.

World Bank Group (WBG) (2010). *Cost-Benefit Analysis in World Bank Projects.* World Bank Independent Evaluation Group https://openknowledge.worldbank.org/handle/10986/2561.

World Bank Group (WBG) (2019). *Evaluation Principles.* World Bank, International Finance Corporation, Multilateral International Guarantee Association https://ieg.worldbankgroup.org/sites/default/files/Data/reports/WorldBankEvaluationPrinciples.pdf.

World Bank Group (WBG) (2022). *World Bank Group's Operational Response to COVID-19 (Coronavirus) – Projects List.* Washington, DC: International

Development Agency (IDA), the World Bank https://www.worldbank.org/en/news/press-release/2021/07/19/world-bank-group-s-157-billion-pandemic-surge-is-largest-crisis-response-in-its-history.

Worsley, T. (2014). *Ex-post Assessment of Transport Investments and Policy Interventions*. [Discussion Paper] 2014 International Transport Forum, OECD, Paris, France.

Yescombe, E.R. (2014). *Principles of Project Finance*, 2nde. London, UK: YCL Consulting Ltd.

Zhang, M. (2007). The Chinese edition of transit-oriented development. *Transportation Research Record* 2038: 120–127.

Zidane, Y.J.-T., Johansen, A., and Ekambaram, A. (2015). Project evaluation holistic framework – Application on megaproject case. *Procedia Computer Science* 64: 409–416.

3

The Multilaterals and World Development

Everything we do during and after this crisis [COVID-19] must be with a strong focus on building more equal, inclusive, and sustainable economies and societies that are more resilient in the face of pandemics, climate change, and the many other global challenges we face.
— António Guterres, *Secretary-General, United Nations*

Introduction

The United Nations 2030 Agenda for sustainable development goals (SDGs) is among the most ambitious of global initiatives to adopt a results-based approach toward sustainable development. In 2000, the Millennium goals became a guidepost for what is essential in improving the quality of life and reducing poverty in the developing nations. Fifteen years later, the United Nations General Assembly officially adopted 17 SDGs as shown in Figure 3.1 which replaced the Millennium Development Goals, and which will guide global development efforts for the next decade or longer. There are an increasing number of international initiatives including those proposed by the multilateral banks focused on reform and, above all, results. The SDGs are central to the mission of all multilateral development and to the countries that host these mega commitments. With a holistic approach to sustainable development, we should not have to choose between zero hunger, clean water, good health, or energy efficient societies.

In this chapter, we cover a wide range of topics to demonstrate not only the need for mega infrastructure investment in the developing world, but also what is being done to fulfill that need. We look at developing and transitional economies to learn the challenges they have faced and how they have fared over the past 30 years since the fall of Communism in many countries the world over. Project finance

Global Megaprojects: Lessons, Case Studies, and Expert Advice on International Megaproject Management, First Edition. Virginia A. Greiman.
© 2023 John Wiley & Sons, Inc. Published 2023 by John Wiley & Sons, Inc.

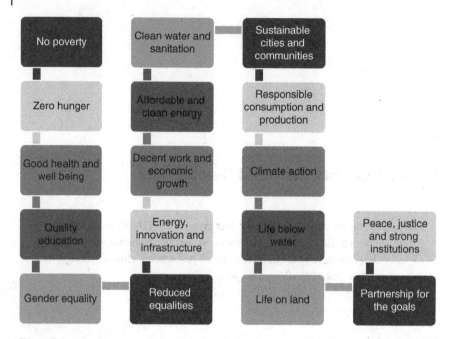

Figure 3.1 The 2030 agenda for sustainable development goals. *Source:* Adapted from The United Nations (2021).

sources are explored with a focus on structures for mobilization of capital. The world's largest multilateral banks are evaluated in terms of their contributions and the focus of their mission, and selected current projects are presented. Finally, several case studies from various regions of the world are described in terms of the investment, and the successes and challenges these projects have experienced.

Growing Demand for Global Infrastructure

Developing and emerging market economies have been and will be faced with an enormous demand for infrastructure. The G20 Initiative, Global Infrastructure Outlook, estimates that global infrastructure investment needs to reach US$94 trillion by 2040 and forecasts the investment gap of about US$15 trillion – equal to a 16% infrastructure investment deficit by that year. The Outlook also predicts that meeting the SDGs (Figure 3.1) increases the need by a further US$3.5 trillion, growing the gap to about US$18 trillion (GIH 2017).

Developing and some emerging countries continue to have relatively large infrastructure needs and investment gaps (GIH 2017). According to the Global

Infrastructure Hub, the private investment gap between low- and high-income countries persisted in 2020 (GIH 2021). The World Bank, reports that new infrastructure could cost low- and middle-income countries anywhere between 2 and 8% of gross domestic product (GDP) per year to 2030 and that investments of 4.5% of GDP would enable them to achieve the infrastructure-related SDGs (Fay et al. 2019). Furthermore, the worldwide spread of COVID-19 pandemic in 2020 creates much larger challenges (WB 2020). The International Monetary Fund (IMF) staff estimates that "on average the public and private sectors will together have to spend some 14 percent of GDP additionally every year between now and 2030 to meet the SDGs in the five sectors, 21 percent more—than before the pandemic" (Benedek et al. 2021, p. 5). In accordance with the growing demand for infrastructure, the "private participation in infrastructure" (PPI) has come to show a significant presence, particularly in developing and emerging market economies (Taguchi and Yasumura 2021). It is because they have suffered from a lack of fiscal space to deal with their infrastructure demand, and the PPI has helped fulfill the gap by mobilizing financial resources with private sectors.

Figure 3.2 shows investment by financier. As reflected in Figure 3.2 most private investment in infrastructure projects is financed by financial services institutions, but debt capital markets are also increasing (GIH 2021). In middle- and low-income countries, non-private institutions such as multilateral development banks (MDBs), export credit agencies (ECAs), governments, and others also play a significant role as financiers (GIH 2021).

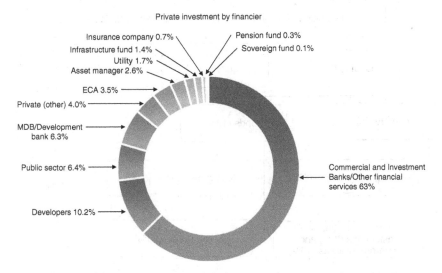

Figure 3.2 Private investment by financier. *Source:* Adapted from Global Infrastructure Hub (2021).

Project finance in the developing world is rarely 100% publicly financed and public private partnerships are far more common. The diversity of financial sources can also vary widely. To understand the massive infrastructure that is being undertaken we look in this chapter at the infrastructure endeavors in some of the poorer countries in the world, particularly in the Global South.

Sources of Financing of Private Investment in Infrastructure Projects

Figure 3.3 shows sources of financing of private investment in infrastructure projects by instrument type. Private investment is primarily financed by debt mostly loans. Funding through Green Bonds has been rising particularly in high-income countries mostly in Western Europe, North America, and Asia (GIH 2021). Green Bonds are used primarily for investment in environmental and climate friendly projects. As discussed in this chapter International Financial Institutions (IFIs) are a major source of funding for the developing countries and mobilization of capital by these institutions through loans, grants, and guarantees provides additional sources of funding for the host country government. While in high-income countries more than half of private investment in infrastructure projects is financed by the private sector alone, in middle- and low-income countries only 25% of private investment in infrastructure is financed by the private sector alone (GIH 2021).

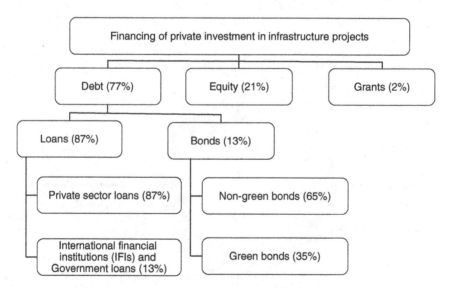

Figure 3.3 Financing of Private Investment in Infrastructure Projects. *Source:* Adapted from Global Infrastructure Hub (2021).

What Is a Developing Country?

A developing country is often defined by its economic output. There has been a lot of debate as to where to draw the line between a developed country and a developing one, which can be seen by the lack of one single meaning for the term. The IMF classifies countries with a per-person income of less than $1045 in 2020 as being low income. The IMF definition is based on per-person income, export diversification, and the degree of union with the global financial system (IMF 2021).

The United Nations has some different rules for distinguishing between developed and developing countries. For analytical purposes, the United Nations World Economic Situation and Prospects Report (WESP) (2022) classifies all countries of the world into one of three broad categories: developed economies, economies in transition, and developing economies. The composition of these groupings is intended to reflect basic economic country conditions. Several countries (in particular the economies in transition) have characteristics that could place them in more than one category; however, for purposes of analysis, the groupings have been made mutually exclusive. Within each broad category, some subgroups are defined based either on geographical location or on ad hoc criteria, such as the subgroup of "major developed economies," which is based on the membership of the Group of Seven. The six geographical regions for developing economies are as follows: Africa, East Asia, South Asia, Western Asia, and Latin America and the Caribbean.

As of 2020, the World Bank uses a different system and assigns the world's economies to four income groups – low-, lower-middle, upper-middle, and high-income countries. The classifications are updated each year on 1 July and are based on gross national income (GNI) per capita in current US$ (using the Atlas Method exchange rates) of the previous year (Hamadeh et al. 2021). The Atlas method smooths exchange rate fluctuations by using a three-year moving average, price-adjusted conversion factor.

Low Income Countries – less than $1025 GNI per capita
Lower-Middle-Income Countries – between $1026 and 3995
Upper-Middle Income Countries – between $3 996 and 12 375
High-Income Countries – those with incomes of more than $12 375

Source: World Bank GNI Classifications, World Bank Country and Lending Groups (2021).

In the UN 2021 classification system, they list 127 countries as developing economies based on the annual rate of growth of GDP. An alarming number of countries when you consider the assistance that is needed from the multilaterals and the private sector to elevate these countries eventually to transition economies.

Transition Economies

A transition economy is one that is changing from central planning to free markets. Since the collapse of communism in the late 1980s, countries of the former Soviet Union, and its satellite states, including Poland, Hungary, and Bulgaria, sought to embrace market capitalism and abandon central planning. As Table 3.1 illustrates, the transition of these nations from centrally planned to market-based economic systems was essentially complete by the early 2000s. In Europe, the transition was bolstered by a legal system which would give rights to private property, and the inclusion of the new countries in the European Union would be an irrevocable step on the way to capitalism; a democratic system would preclude the rise of a state-centered form of corporatism (Lane 2007). In Asia, a market-based system was advanced when the region experienced a massive increase in world prices for energy and mineral exports. This raised incomes in the main oil and gas exporters, Kazakhstan and Turkmenistan; brought more benefits to the most populous country, Uzbekistan; and left the poorest countries, the Kyrgyz Republic and Tajikistan, dependent on remittances from migrant workers in oil-rich Russia and Kazakhstan (Pomfret 2019). With improved infrastructure and connectivity between China and Europe (reflected in regular rail freight services since 2011 and China's announcement of its Belt and Road Initiative in 2013), a window of opportunity appears to have opened for Central Asian countries to achieve more sustainable economic futures. (Pomfret 2019).

The European Bank of Reconstruction and Development (EBRD) created a new framework for transition economies in 2016. The EBRD's new transition concept

Table 3.1 Transition economies in Europe and the former Soviet Union and Asia (pre 2019).

CEE: Albania, Bulgaria, Croatia, Czech Republic, FYR Macedonia, Hungary, Poland, Romania, Slovak Republic, Slovenia

Baltics: Estonia, Latvia, Lithuania

CIS: Armenia, Azerbaijan, Belarus, Georgia, Kazakhstan, Kyrgyz Republic, Moldova, Russia, Tajikistan, Turkmenistan, Ukraine, Uzbekistan

Asia: Cambodia, China, Laos, Vietnam

Source: International Monetary Fund (2000)

Countries Still in Transition: Albania, Armenia, Azerbaijan, Belarus, Bosnia and Herzegovina, Georgia, Kazakhstan, Kosovo, Kyrgyzstan, North Macedonia, Republic of Moldova, Montenegro, Russian Federation, Serbia, Tajikistan, Turkmenistan, Ukraine, Uzbekistan

Sources: Adapted from Myant and Drahokoupil (2010) and United Nations (2021).

argues that a well-functioning and sustainable market economy should be characterized by *six key transition qualities* – it is an economy that is:

Competitive, Well-Governed, Green, Inclusive, Resilient, and Integrated.

The Central Asian Economy

Most of the transition economies have faced severe short-term difficulties, and longer-term constraints on development. The Central Asian countries have survived for three decades since 1991 and face no pressing existential threats – in sharp contrast to the nine new independent countries formed under the Treaty of Versailles of 1919. The transition economies in Central Asia are succeeding in part because of the following (Pomfret 2021):

- Planning has been replaced by national economic systems that vary substantially in the degree of state-control over market forces.
- A new generation of leaders and people familiar with market-based economies and the world beyond the formerly centrally planned economies may be more receptive to reform. Resolution of these dilemmas will be decisive.
- Improved infrastructure in the twenty-first century has allayed the costs of landlockedness for Central Asia. Whether a country takes advantage of the window of opportunity to promote export-oriented economic growth will depend upon national reform to make economies more efficient and responsive to market incentives and to reduce the costs of doing business and of international trade.

> The lack of deep reform in the 1990s and creation of autocratic regimes with a conspicuously wealthy elite are ongoing obstacles to creation of well-functioning market-based economies. The resource boom, despite the obvious benefits of increased incomes, may have made post-boom reform more difficult as entrenched elites are content with the status quo.
>
> *(Pomfret 2019)*

The Role of the Multilaterals in Development

> Our biggest challenge in this new century is to take an idea that seems abstract – sustainable development – and turn it into a reality for all the world's people
>
> — *Kofi Annan*

The MDBs are international institutions that originated in the aftermath of World War II to rebuild war-ravaged nations and stabilize the global financial system and

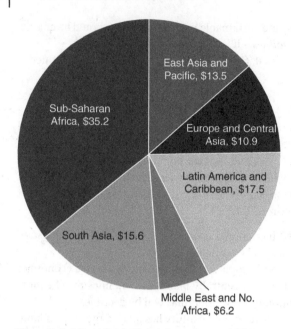

Figure 3.4 World Bank Global Commitments: $B in Loans, Grants, Equity Investments, & Guarantees. *Source:* Adapted from the *2021 World Bank Annual Report: From Crisis to Green, Resilient and Inclusive Recovery* (WB 2021a).

have been described as one of the international community's great success stories of the post-World War II era (CGD, 2016). Today, MDBs fund infrastructure, agriculture, energy, education, and environmental sustainability in developing countries typically in the form of loans and grants to promote economic and social development. Figure 3.4 highlights the World Bank commitments to the lower income regions of the world in the form of loans, grants, equity investments, and guarantees.

The United States is a member and significant donor to five major MDBs. These include the World Bank and four smaller regional development banks: the African Development Bank (AfDB); the Asian Development Bank (AsDB); the European Bank for Reconstruction and Development (EBRD); and the Inter-American Development Bank (IDB).

The socioeconomic landscape has changed dramatically and with it the demands and needs of the developing world. Developing countries now make up half of the global economy. Longer-term forecasts suggest that today's developing, and emerging countries are likely to account for nearly 60% of world GDP by 2030 (OECD 2021).

In recent years, technology and international financial aid has spread to the Global South through international and national development agencies. For example, common forms of technology include energy, transportation, and agricultural technological improvements. One of the most common is the establishment of bus rapid transit (BRT) systems particularly in Brazil beginning with the BRT system in Curitiba (Siemiatycki 2017). In 2000, Bogota opened its TransMilenio BRT system which produced a far cheaper system and created innovations that have changed the thinking about BRT over the past two decades. Now BRT systems are exploding in major cities all over the Global South including Delhi, Guangzhou, Johannesburg, Tehran, Istanbul, and Mexico City.

Measuring the Results of Development Bank Funding

> The world's multilateral development banks need to "up their ambition" in terms of the financing provided for climate-related projects.
> — *Mark Carney, United Nations Climate Envoy*

To ensure that development banks are contributing to the alleviation of poverty and improvement of quality of life several key questions must be evaluated to determine whether the MDB funded projects are meeting their projected outcomes.

- Are development initiatives making a difference and having an impact?
- How are these factors being measured and what are the results?
- How will governments know whether they have made progress and achieved these goals?
- What are the criteria for measuring progress over the years that a project is in operation?
- How will they be able to tell success from failure, or progress from setbacks?
- How will they identify obstacles and barriers?
- How are they overcoming these barriers?

The role of the MDBs is critical in continuing to expand opportunities for growth and improvement in all low income and developing cities throughout the world. Table 3.2 highlights the role and responsibilities of these banks (MDB 2021):

Project Preparation Worldwide

> Infrastructure Monitor assesses project preparation worldwide. The results confirm that project preparation capability could be substantially improved in all regions, and particularly in low-income countries. (GIH 2021)

The lack of a bankable and investment-ready pipeline of infrastructure projects is often considered one of the major bottlenecks in attracting private capital to infrastructure (GIH 2021). Enabling an investment ready pipeline has

Table 3.2 Role and contributions of the Multilateral Development Banks.

Contributions	Description
Mobilization of Private Finance	The MDB's ability to crowd-in capital from private creditors as well as other nongovernmental organizations such as the export credit agencies (ECAs)
Direct Investment	Investment in the form of loans, grants, and guarantees to the host country governments. The borrowing can originate either from financial (lending) institutions or by issuing bonds in the international capital markets. In both cases, MDBs rely on high credit ratings, allowing them to borrow at better terms than most of their borrowing members would achieve if they borrowed themselves.
Concept and Preliminary Feasibility Assessment	Assessment includes mandate, economic, technical, social, environmental relevance, and risks
Project Appraisal	Provides financial, technical, economic, social, and technical information to assist all sponsors and stakeholders. A benefit cost analysis or a return-on-investment analysis may be incorporated into the project appraisal. It is also an opportunity to reduce or eliminate risk before the project commences
Environmental Impact Assessment	The Environmental Impact Assessment (EIA) is a project review to assess the impact a project will have on the local environment including the community, its businesses, and the natural environment by evaluating the project through the EIA, a determination can be made as to whether the project is safe to move forward or if alternative option should be explored. An EIA can help prevent disasters like the BP oil spill that contaminated much of the Gulf of Mexico in 2010.
Political Risk Mitigation	One of the valuable benefits of a MDB is to reduce the likelihood that the host government will use its sovereignty to expropriate property, nationalize a project, or discriminate against a project's sponsors and investors.
Sensitivity Analysis	Assesses the contribution to the local community in terms of their return on the investment for the assets they have released for development and assurance that the project returns are equitably distributed among the participants in proportion to the risks incurred.
Ex-Ante Evaluation	Assessment of a project during the early conception and planning phases to evaluate its viability and feasibility. Detailed assessment of the application/proposal (business plan in case of private sector) that results in additional requirements, rejection, or interim approval by credit committee and/or management. Financial and legal analysis and due diligence. Assessment and articulation of expected outcomes (economic, technical, social, environmental, and fiduciary aspects).
Ex-Post Evaluation	Represents the actual results attained by the project, which is the return on investment that the project yielded

Source: Adapted from MDB Task Force on Mobilization (2021).

consistently featured as a top priority of G20 Presidencies. As described by the Global Infrastructure Hub in its *2021 Global Infrastructure Monitor Report,* project preparation facilities (PPFs), led by MDBs generally are playing an important role in supporting project preparation to develop bankable and investment-ready projects, providing both technical support and funding for this important project stage.

According to a survey conducted by Deloitte of European infrastructure investors, as shown in Box 3.1 key things that governments should do to promote private investment in the infrastructure industry, include:

Box 3.1   Unlocking Private Investment in Infrastructure

- Educate the public about private investment infrastructure
- "Unblock" planning approval processes
- Provide a pipeline in high quality infrastructure assets
- Tax stability
- Be willing to underwrite contracted risks
- Cleverly package and structure deals
- Stabilize regulatory environments

Source: Adapted from Deloitte (2016).

Country Partner Frameworks

Multilateral development projects are often funded by a combination of MDBs (leading a syndicate of commercial banks), ECSs, and equity sponsors. Most of the world's largest development banks finance projects in accordance with guidelines set forth by the bank's country sponsors and boards of directors. Investment project financing by the Bank aims to promote poverty reduction and sustainable development of member countries by providing financial and related operational support to specific projects that promote broad-based economic growth, contribute to social and environmental sustainability, enhance the effectiveness of the public or private sectors, or otherwise contribute to the overall development of member states. Investment project financing (IPF) is comprised of bank loans and bank guarantees. IPF supports projects with defined development objectives, activities, and results. IPF may also support a Multiphase Programmatic Approach (MPA) Program which supports a long-term project in smaller phases. The Bank disburses the proceeds of bank loans against specific eligible expenditures.

Global megaprojects are confronted with multiple forms of regulative frameworks, including the laws of home and host countries, legal agreements with financing firms, regulations of regional and local entities, and corporate hierarchies. On the "solution" side, project management units create rules and sanctions to buttress contractual agreements and to form the scaffolding for agreements between project and client entities.

The Bank disburses the proceeds of bank loans against specific eligible expenditures. Country's must fulfill certain requirements before they can be considered an eligible candidate. Many of the specific considerations of the Bank for lending to borrowers from the poorest member countries are set forth in Table 3.3.

Table 3.3 Country and project requirements for lending.

The World Bank Lending Requirements for Borrower's from the Poorest Member Countries
Technical analysis. The Bank assesses technical aspects of the Project, including design issues, appropriateness of design to the needs and capacity of the Borrower and any Project implementation entity, institutional arrangements, and organizational issues for the implementation of the project in the context of the long-term development objectives of the Borrower or, as appropriate, the member country
Economic analysis. The Bank undertakes an economic analysis of the project. The bank assesses the project's economic rationale, using approaches and methodologies appropriate for the project, sector, and country conditions, and assesses the appropriateness of public sector financing and the value added of bank support
Financial viability analysis. For projects supported by a bank guarantee
Financial management. The borrower or implementing entity maintains or causes to be maintained financial management arrangements that are acceptable to the bank and that, as part of the overall arrangements in place for implementing the project, provide reasonable assurance that the proceeds of the bank loan, the proceeds of the bank-guaranteed debt, or the bank-guaranteed payments are used for the purposes for which they are granted, or the payments are made. Financial management arrangements are the planning, budgeting, accounting, internal control, funds flow, financial reporting, and auditing arrangements of the borrower and entity or entities responsible for project implementation. The financial management arrangements for the project rely on the borrower's or implementing entity's existing institutions and systems, with consideration of the capacity of those institutions
Procurement. Procurement rules applicable to bank loans are set out in the procurement policy/directive, or for some borrowers the bank's administrative manual statement
Environmental and social. Environmental and social requirements applicable to IPF with concept decision, or equivalent, are set out in the environmental and social policy/directive and environmental and social standards, or OP/BP 4.03, as appropriate
Risks. The bank assesses the risks to the achievement of the project's development objectives with consideration for the risks of inaction

Source: Adapted from The World Bank (2021c).

Mobilization of Capital

> MDBs often remain the only investor with the means, risk tolerance, and ability to invest in the poorest countries during global crisis situations such as [the Covid 19 Pandemic].
>
> — *The Multilateral Development Banks and the Development Finance Institutions*

Mobilizing private capital toward addressing megaproject development efforts is crucial. Since adoption of the "From Billions to Trillions" Agenda in 2016, MDBs have focused on mobilizing private capital to meet the SDG investment needs. As described in Table 3.4 private finance for the lower income countries can be structured to include private co-financing, private direct mobilization (PDM), and private indirect mobilization. Importantly, specialized funding is also available to most countries to meet the UN's 2030 Agenda for Sustainable Goals including climate change funds and green bonds. Sponsors for specialized funding may include non-profits as well as corporate sponsors dedicated to meeting-specific goals.

One of the largest sources of capital is the assets under management (AUM) of institutional investors, such as pension funds. As an example of the size of investment pools in this class, the 20 largest pension funds had AUM in 2019 of just over $18 trillion (Williams Towers Watson 2019). Recent years have seen a broad-based increase in capital allocation by institutional investors to investment assets that meet certain environmental criteria and/or have social impact and, in some cases, track to the SDGs (MDB 2021).

The 2019 commitment data indicate that the total long-term finance mobilized by the MDBs from private investors and other institutional investors (including insurance companies, pension funds, and sovereign wealth funds) in all low- and middle-income countries of operation was $63.6B, compared with $69.4B in 2018. Of this amount, 32% was direct mobilization and 68% was indirect (MDB 2021).

> Public funding for the developing countries can come from a multitude of national and international organizations and even local municipalities in the form of concessional or non-concessional funding. Recently, national governments have been a primary source of funding with support from the development finance institutions.

World Bank Funding Structure: Benin – Country Partnering Framework

The World Bank uses country partnering as a strategy to fund projects in the poorest of countries (WB 2021b). As an example, the World Bank has held discussions with several partners that support the implementation of Benin's Agricultural Sector Development Plan. These include the African Development Bank, the United Nations Food and Agriculture Organization (FAO), the International Fund for Agriculture Development (IFAD), the German Federal Ministry of Economic Cooperation and Development (BMZ), the Netherlands Development Cooperation, the European Union, the French Development Agency (AFD), Belgium's technical cooperation Agency (Enabel), the Japan International Cooperation Agency (JICA), and the United States Agency of International Development (USAID).

The World Bank Group's strategy for Benin is governed by the Country Partnership Framework (CPF) which covers the period Fiscal Year 2019–Fiscal Year 2023. The CPF seeks to help Benin overcome impediments to achieving its goals. The proposed operations are structured around the following focus areas: (i) structural transformation for competitiveness and productivity; (ii) investing in human capital; and (iii) increasing resilience and reducing climate-related vulnerability. The program will be focused on achieving high impact outcomes while leveraging scarce resources; the IDA allocation will be selectively targeted and leveraged by drawing upon cross-group synergies, by making best use of the full array of IDA and Africa Region special initiatives, maximizing finance for development (MFD), and partnering with other international development agencies in ways that best employ the comparative advantages of each.

Sources of Public and Private Finance

Table 3.4 highlights some of the sources for private and public finance. These sources can vary from country to country and project to project, but they illustrate the various ways in which private and public finance can be mobilized, particularly when there is a gap in financing.

The World's Largest Development Banks (MDBs)

> Long-term finance is essential for financing fixed-capital investment in infrastructure and other sectors, and short-term finance is important for supporting the expansion of trade and value chains.
>
> — *The Multilateral Development Banks and the Development Finance Institutions*

Table 3.5 provides an overview of the world's largest development banks with a summary of their financial structure, areas of focus and some examples of their current projects. The financing can fluctuate based on the willingness of their sponsors to continue commitment to the bank.

Since they provide funding for the poorest countries many MDBs are located near or in the countries they service. Not all MDBs rely on countries for

Table 3.4 Sources of public and private financing for Global Development Projects.

Private finance	Public finance
Private Co-Financing Mobilization The investment made by a private entity, which is defined as a legal entity that is 1) Carrying out or established for business purposes 2) Financially and managerially autonomous from national or local government Some public entities that are organized with financial and managerial autonomy are counted as private entities. Other examples include registered commercial banks, insurance companies, sovereign wealth funds, and other institutional investors investing primarily on a commercial basis. *Private Direct Mobilization (PDM)* Financing from a private entity on commercial terms due to the active and direct involvement of an MDB leading to commitment. Evidence of active and direct involvement includes mandate letters, fees linked to financial commitment, or other validated or auditable evidence of an MDB's active and direct role leading to commitment of other private financiers. PDM does not include sponsor financing. *Private Indirect Mobilization* Financing from private entities provided in connection with a specific activity for which an MDB is providing financing, where no MDB is playing an active or direct role that leads to the commitment of the private entity's finance. PIM includes sponsor financing if the sponsor qualifies as a private entity. *Specialized Funding* Used for Climate Change and other sustainable development goals by governments and non-profit organizations.	*Multilateral Development Banks* The MDBs provide non-concessional financial assistance to middle-income countries and some creditworthy low-income countries on market-based terms. They also provide concessional assistance, including grants and loans at below market rate interest rates, to low-income countries. *Blended Concessional Financing* Explicitly uses scarce concessional financing to enable mobilization of private investment into areas it would not have otherwise gone. Since the majority (approximately two-thirds) of this volume is in low- or lower-middle-income countries, this growth is reflected in the higher number of mobilizations. *Other Development Finance Institutions* Set up to support private sector development in developing countries. They are usually majority-owned by national governments and source their capital from national or international development funds and include the export banks. *Co-Financing by Commercial and Multilaterals* Lending can be side by side or A/B loan structures that allow multilaterals to leverage liquidity while acting as lender of record (Dewar 2019). *Country Partnership Frameworks* Partnering with other IGO's to fund projects in the poorest of countries. *National Governments* Direct lending and incentives, subsidy and loan guarantee programs. Available in some countries for both domestic and foreign owned investors. *Project Bonds* The International Finance Corporation (IFC) issues bonds in a variety of markets, formats, and currencies – including global benchmarks bonds, green and social bonds, private placements, and discount notes.

Source: Adapted from MDB Task Force on Mobilization (2021).

Table 3.5 Overview of the World's Multilateral Development Banks.

Bank assets (US$)	Focus	Selected projects
China Development Bank (CDB) Assets: $2.4 trillion	CDB provides financing to Chinese firms abroad, stipulates energy-backed loans to foreign authorities and national oil companies and invests in private equity funds (CDB 2019)	*Poverty Alleviation and Development Programme* for Ping'an Township in Fengjie, Chongqing *Transportation Equipment of China Railway* aims to upgrade the equipment of China Railway and improve the transportation safety and efficiency, and includes the repair in factories, and upgrade and modification of multiple-unit trains
Shanghai Pudong Development Bank Assets: $1.228 trillion	National strategies to enhance the service capabilities of "mega finance" (Shanghai Pudong Development Bank 2021)	*Integration of the Yangtze River Delta, the coordinated development of Beijing–Tianjin–Hebei region, and the construction of the Guangdong–Hong Kong–Macao Greater Bay Area,* to strengthen the credit resources and comprehensive financial service support, fuel the country's drive for developing modern industries, and promote the optimization and upgrading of the current economic system
KfW Germany Assets: $671B	33% of loans in 2020 was used for climate and environmental protection. Financing and promotional services are aligned with the United Nations' Agenda 2030 and contribute to the achievement of the 17 Sustainable Development Goals (SDGs)	*Launch of green bonds in Latin America* KfW has launched a new fund for environmentally and climate-friendly investments in Latin America *Secure and reliable energy supply in Bangladesh.* On behalf of the German government, KfW is promoting a modernization of the power infrastructure in Bangladesh
Asian Development Bank Assets: $271B	Focus on tackling climate change, building climate and disaster resilience, and enhancing environmental sustainability, making cities more livable, promoting rural development and food security, strengthening governance and institutional capacity, fostering regional cooperation, and integration	*Cambodia: Livable Cities Investment Project.* The project will focus on enhancing urban planning, building community resilience, and providing infrastructure that will facilitate long-term sustainable and economic growth. *Updating the Revised Strategic Transport Plan for Dhaka.* The knowledge and support technical assistance (TA) will help enhance the planning capacity and coordination of the urban transport sector in Dhaka, Bangladesh

Korea Development Bank Assets: $232B	KDB is committed to promoting and disseminating the concept of sustainable development and its promotional activities in local and overseas markets to help its partners and borrowers achieve their development goals (i.e. to reduce poverty, bring prosperity, secure peace, promote democracy, shape globalization in an equitable manner and engage in environmental and climate protection	*Green Bond Projects.* The construction or expansion of renewable energy production facilities, including those that use solar or wind power and/or the manufacturing of rechargeable batteries for electric vehicles
Development Bank of Japan Assets: $164B	Create a business model that integrates investments, loans, and other business activities so as to enhance the value of managerial resources, both tangible and intangible Exercise a leadership role in building the knowledge required to create a sustainable society through ongoing surveys, research, and reporting on social issues	*Haneda Airport International Passenger Terminal.* The expansion of Haneda Airport's international passenger terminal was undertaken as a private finance initiative (PFI) project *Fukuoka Clean Energy Corporation.* This project was developed by a jointly financed special purpose company, formed by the City of Fukuoka and Kyushu Electric Power Co, to build a waste processing facility using a PFI approach *Japan Wind Development Fund.* The DBJ launched a new fund jointly established with Japan Wind Development Co to invest in wind power projects in Japan
Brazilian Development Bank Assets: $135B	Facilitating solutions that contribute with investments for the sustainable development of the Brazilian nation	*BNDES-NDB Sustainable Infrastructure Project.* The Project will contribute to the efforts of the Government of Brazil to support socioeconomic growth and development through sustainable infrastructure investments in key sectors *Curitiba's Bus Rapid Transit Rideability Improvement Project.* The loan will be used for financing the development of the East-West Corridor and the South Corridor of the Curitiba's Bus Rapid Transit (BRT) system that incorporates the Transportation Integrated Network (RIT) in Curitiba Metropolitan Region (RMC)

(Continued)

Table 3.5 (Continued)

Bank assets (US$)	Focus	Selected projects
International Finance Corporation Assets: $105B	IFC is committed to growing its climate-related investments to an annual average of 35% of its own-account long-term commitment volume between 2021 and 2025 and working with financial institutions to finance projects that will support mitigation and adaptation. Since 1956, IFC has leveraged $2.6B in capital to deliver more than $285B in financing for businesses in developing countries	*Women-inclusive Enterprises*. The proposed senior loan of up to US$100 million to Turkiye Sinai Kalkinma Bankasi A.S. (TSKB, or the Bank) aims to support Bank's lending to Women-inclusive Enterprises, which are private sector mid-cap companies in Turkey that meet an established gender equality criterion *Caribbean Green Financing Frameworks*. The project will support the Dominican Republic in engaging the financial sector, regulators and supervisors, key Ministries, and other stakeholders, as well as private sector entities in the creation and development of a national green taxonomy to enable green financing
State Development Corporation (VEB.RV) Russia Assets: $51B	In partnership with commercial banks, VEB. RV provides financing for large-scale projects to develop the country's infrastructure, industrial production, and social sphere, strengthen its technological potential and improve quality of life. In mid-2018 VEB.RF has embarked on a new business model with focus on infrastructure, high value-added industry, urban development	1) Development of the Udokan's copper deposit 2) The production of sulfuric acid at the "KuibyshevAzot" mine 3) Capacity expansion of "Shchekinoazot" to produce methanol 4) Construction of six large-tonnage vessels at the Zvezda super-shipyard near Vladivostok 5) Modernization of Novaport group's regional airports in six cities of Russia 6) Build the largest pulp and paper mill in Siberia

African Development Bank Assets: $35B	The overarching objective of the African Development Bank (AfDB) Group is to spur sustainable economic development and social progress in its regional member countries (RMCs), thus contributing to poverty reduction	*Ethiopia – Kenya Electricity Highway.* The Project objective, supporting the Eastern Africa Power Pool. EAP's mission, is to help integrate the power systems of EAPP member countries including Ethiopia, Kenya, Tanzania, Rwanda, and Uganda, to promote power trade and regional integration, contribute to social and economic development, and reduce poverty in those countries
Inter-American Development Bank Assets: $151.7B	The IADB's current strategic priorities include three development challenges: social inclusion and inequality, productivity and innovation, and economic integration; and three cross-cutting issues: gender equality and diversity, climate change and environmental sustainability, and institutional capacity and the rule of law	1) Argentina: Program to Support the National Early Childhood Plan and the Policy for Universalization of Early Childhood Education II 2) Brazil: Innovation Horizons: Digital Innovation Program for the economic and social development of the State of Paraiba 3) Chile:Digital transformation and Sustainable growth program
Islamic Development Bank Assets: $24.4B	Projects focus on sustainability, energy and innovation, water sanitation, transport, agriculture, and science and technology	Indonesia: US$261.72 million to support the strengthening of health systems and emergency preparedness Palestine: To alleviate poverty established the Economic Empowerment Fund for the Palestinian People (EEFPP) with a target financing of US$500 million US$220 million to co-finance along with the governments of Saudi Arabia and Egypt, the US$1.7B Egypt and Saudi Arabia Electricity Interconnection project to support the existing unified electricity system, which is one of the largest in the region

(Continued)

Table 3.5 (Continued)

Bank assets (US$)	Focus	Selected projects
European Bank for Reconstruction and Development Assets: $35B	The EBRD states that the purpose shall be "to foster the transition toward open market-oriented economies and to promote private and entrepreneurial initiative in the Central and Eastern European countries," and work toward three important principles: multiparty democracy, pluralism, and market economics	1) Moldova: Balti District Heating Phase 2: The Project aims to address legacy infrastructure issues and promote the system's decarbonization. Optimize the use of highly efficient combined heat and power plant 2) Belgrade Public Buildings: Support the substantial energy efficiency renovation of four public buildings located in Belgrade, Serbia. The works include the addition of thermal insulation, the repair of building envelopes, the upgrade of heating, ventilation and cooling systems, and the replacement of lighting systems
Asian Infrastructure Investment Bank Assets: $40B	Improve economic and social outcomes in Asia by scaling up financing for sustainable development and to improve the global economic governance	1) Turkey: Osmangazi Electricity Distribution Network Modernization and Expansion Project 2) Bangladesh: Sylhet–Tamabil Road 3) Indonesia: Multifunctional Satellite PPP Project 4) India: Assam Electricity Distribution System Enhancement Project
New Development Bank Assets: $18.8B	Financing infrastructure and sustainable development projects in BRICS (Brazil, Russia, India, People's Republic of China, and South Africa) and other emerging economies and developing countries	1) Brazil: The North Region Transportation Infrastructure Improvement Project – Upgrade the Carajás Railway and the Ponta da Madeira Port Terminal in the states of Pará and Maranhão. 2) South Africa: The National Non-Toll Roads Management Program

Source: Adopted from the Multilateral Development Banks Annual Reports (as available) and Sovereign Wealth Fund Institute (SWFI) (2022).

sponsorship. For example, the China Development Bank relies largely on private sources of investment. As shown in Table 3.5 the MDBs provide non-concessional financial assistance to middle-income countries and some creditworthy low-income countries on market-based terms. They also provide concessional assistance, including grants and loans at below market rate interest rates, to low-income countries.

Project loans include large infrastructure projects, such as highways, power plants, port facilities, and dams, as well as social projects, including health and education initiatives. Policy-based loans provide governments with financing in exchange for agreement by the borrower country government that it will undertake particular policy reforms, such as the privatization of state-owned industries or reform in agriculture or electricity sector policies. Policy-based loans can also provide budgetary support to developing country governments. In order for the disbursement of a policy-based loan to continue, the borrower must implement the specified economic or financial policies. Some have expressed concern over the increasing budgetary support provided to developing countries by the MDBs. Traditionally, the IMF have provided this type of support (Nelson 2020).

The World Bank is the oldest and largest of the MDBs. The World Bank Group comprises three sub-institutions that make loans and grants to developing countries: the International Bank for Reconstruction and Development (IBRD), the International Development Association (IDA), and the International Finance Corporation (IFC) established through the Bretton Woods Conference (see Table 3.5).

Financing Multilateral Development Bank's Operations

The rationale of financing of MDBs' operations is based on the combined use of its own equity, as well as resources they mobilize (borrow) from the financial markets. Many MDBs have very explicit goals of resource mobilization – for example, the EBRD, among several other institutions, has in its mandate the goal to attract additional domestic and foreign resources – as a catalytic effect of their own financing, including crowding-in private sector investments. The direct resource mobilization (borrowing) originates either from financial (lending) institutions or by issuing bonds in the international capital markets. In both cases, MDBs rely on high credit ratings, allowing them to borrow at better terms than most of their borrowing members would achieve if they borrowed themselves (Nelson 2020).

African Development Bank

The African Development Bank the AfDB was created in 1964 and was for nearly two decades an African-only institution, reflecting the desire of African governments to promote stronger unity and cooperation among the countries of their region. In 1973, the AfDB created a concessional lending window, the African Development Fund (AfDF), to which non-regional countries could become members and contribute. The United States joined the AfDF in 1976. In 1982, membership in the AfDB non-concessional lending window was officially opened to non-regional members. The AfDB makes loans to private-sector firms through its non-concessional window and does not have a separate fund specifically for financing private-sector projects with a development focus in the region.

The Asian Development Bank

The AsDB was created in 1966 to promote regional cooperation. Similar to the World Bank, and unlike the IDB, the AsDB's original mandate focused on large infrastructure projects, rather than social projects or direct poverty alleviation. The AsDB's concessional lending facility, the Asian Development Fund (AsDF), was created in 1973. In 2017, concessional lending was transferred from the AsDF to the AsDB, although the AsDF still provides grants to low-income countries. Like the AfDF, the AsDB does not have a separate fund specifically for financing private-sector projects and makes loans to private-sector firms in the region through its non-concessional window.

The European Bank of Reconstruction and Development (EBRD)

The EBRD is one of the youngest MDBs and was founded in 1991. The motivation for creating the EBRD was to ease the transition of the former communist countries of Central and Eastern Europe (CEE) and the former Soviet Union from planned economies to free-market economies. The EBRD differs from the other regional banks in two fundamental ways. First, the EBRD has an explicitly political mandate: to support democracy-building activities. Second, the EBRD does not have a concessional loan window. The EBRD's financial assistance is heavily targeted on the private sector, although the EBRD does also extend some loans to governments in CEE and the former Soviet Union.

The Changing Landscape of Development Banks

The BRICS Bank

In recent years, several countries have taken steps to launch two new MDBs. First, Brazil, Russia, India, China, and South Africa (the BRICS countries) signed an agreement in July 2014 to establish the New Development Bank (NDB), often referred to as the "BRICS Bank." The agreement outlines the bylaws of the bank and a commitment to a capital base of $100B. Headquartered in Shanghai, the NDB was formally launched in July 2015. The BRICS leaders have emphasized that the bank's mission is to mobilize resources for infrastructure and sustainable development projects in BRICS and other emerging and developing economies.

Asian Infrastructure Investment Bank

China has led the creation of the Asian Infrastructure Investment Bank (AIIB). Launched in October 2014, the AIIB focuses on the development of infrastructure and other sectors in Asia, including energy and power, transportation and tele-communications, rural infrastructure and agriculture development, water supply and sanitation, environmental protection, urban development, and logistics (Nelson 2020, p. 17). As of 19 January 2022, the total number of countries approved for membership in the AIIB is 105 that includes 46 regional members, 43 non-regional members and 16 prospective members (AIIB 2022). Members include several advanced European and Asian economies, such as France, Germany, Italy, the United Kingdom, Australia, New Zealand, and South Korea. The United States and Japan are not members. The AIIB's initial total capital was $100B.

These two institutions are the first major MDBs to be created in decades, and there is debate about how they will fit in with existing international financial institutions (Nelson 2020). Proponents of the new MDBs argue that the infrastructure and financing needs of developing countries are beyond what can be met by existing MDBs and private capital markets, and that new institutions to meet the financing needs of developing institutions should be welcomed (p. 17). Proponents also argue that the new MDBs address the long-held frustrations of many emerging markets and developing countries that the governance of existing institutions, including the World Bank and the IMF, has not been reformed sufficiently to reflect their growing importance in the global economy (pp. 17–18).

Important Questions for the Multilateral Banks

- Is the project equitable to the local community? What are the total returns on investment? Taxes? Royalties? Tariffs? Equitable Return as a Sponsor?
- Who will receive the returns and how much will be distributed? Over what period of time will the beneficiaries receive the return on investment?
- How was the discount rate determined and is the rate justifiable? A reasonable discount rate is the required return on investment (Esty 2003). Using a single discount rate with a project with falling leverage is not appropriate.

- Are the projections for the return on investment realistic? What are the uncertainties in the calculation of the investment, i.e., supply and demand for the output? Who prepared the projections? Have they been audited by an independent third party?
- Is the pricing for the project output justified? Is there an offtake purchaser for the project or product output?
- If project costs increase will the MDB's be prepared to provide additional funding? What will be the other sources of funding?

International Development Case Studies

The increasing importance of MDBs unique role in inspiring and advancing sustainable economic development through knowledge and funding is gaining further importance at a time of unprecedented rise of global challenges. To illustrate the pressure on the world's development agenda, it is necessary to look at the UN SDGs, to be reached by 2030, as agreed worldwide. The vast financing needs to achieve the SDGs by 2030 exceed by far the current level of development finance. The United Nations Conference on Trade and Development estimated that achieving the SDGs will require global investments in the range of US\$5–7 trillion per year (UNCTAD 2018). The annual investment gap in developing countries is estimated at about \$2.5 trillion, to cover mainly basic infrastructure (roads, rail, and ports; power stations; water and sanitation), food security (agriculture and rural development), climate change mitigation, and adaptation, health, and education.

World Bank Projects

> All new World Bank-financed projects are screened for climate risk. Climate change considerations are taken into account at every stage of project design and have been integrated into 100% of the World Bank's multi-year development strategies with developing country partners.
>
> — *Bernice Van Bronkhorst, Global Director,*
> *Climate Change the World Bank*

Table 3.6 highlights a few of the World Bank's projects in some of the poorest countries across Africa, Eastern Europe, and Southeast Asia that have recently completed. These projects were selected based on their exceptional results to show that megaprojects based on GDP can succeed despite political, social, and economic turmoil in the country or communities where the projects are developed.

Table 3.6 Select World Bank Projects from 2012 to 2021.

Project/Country dates /Cost	Purpose	Project finance	Results
Azerbaijan • Second Rural Investment Project • 2012–2019 • $140.1M	Improve infrastructure to promote micro-project development and women's empowerment in rural areas	World Bank through International Bank for Reconstruction & Development (IBRD) – $30M IBRD Later expanded it to $80M Govt. Contributions – $40M Beneficiaries – $10.07	Benefited 82% of rural population 33% increase in access to and use of infrastructure Travel time to schools, medical and water points and markets reduced by 53.7% Household incomes increased by 21.16% 1512 infrastructure microprojects were financed including roads, water supply systems and irrigation canals Project exceeded the target of 50% participation by women
Bosnia and Herzegovina • Infrastructure restoration and economic recovery • 2014–2019 • $100M	Restore the economy and infrastructure due to the floods in 2014	Credit granted by IDA Crisis Response Window (CRW)	Rehabilitated rural roads and non-rural roads and 35 bridges Rehabilitated schools, health facilities, public buildings, and draining ditches 26% of affected farms were benefited
Philippines • Reduce Poverty in Rural Areas • 2014–2022 • $978M	Increase Productivity and Incomes of Farmers and Fisherfolk in the Philippines	IBRD Loan – $501.25M Global Environment Faculty (GEF) grant – $7M Additional IBRD Loan in 2018 – $170M. Second additional Loan 2021 – $260M EU – $20M	Provided agricultural assets and/or services farmer and fisherfolk. Constructed/rehabilitated over 1000 kilometers of farm-to-market roads Provided potable water to some 6600 households 113% increase in real household income via technical/funding assistance Improved technical capacity for developing the agriculture sector

(Continued)

Table 3.6 (Continued)

Project/Country dates /Cost	Purpose	Project finance	Results
Ningxia, China • Halting desertification • 2012–2020 • $68.5M	Restoration of vegetation and ecological protection measures caused by the clearing of vegetation, overgrazing, and the depletion of water resources	IBRD Loan	32 351 hectares of degraded land were improved Increase of vegetation cover by 28% in the project area Seeding survival rate of re-vegetation activities exceeded 70% via training Wind erosion was reduced, due to increased vegetation cover
Benin, West Africa • Improve agriculture productivity • 2012–2019 • $1572M	Move from low productivity and largely subsistence-based activities to more productive and high value production that contributes to increased farm income and job opportunities for a rapidly growing labor force	IDA – $1272M Trust Fund – $300 000 IFC – $29M	776% increased cashew nut exports 990% increase in pineapple exports Increased yields 2.2% maize, 13.3% rice, 20% pineapple, and 37.8% cashews Constructed research infrastructures plus 200 storage facilities/warehouses and 60 drying areas plus provided scholarships for higher education to 117 students

Source: Adapted from World Bank Data (2022).

A Collective Action Perspective on the Planning of Global Megaprojects

> This year marks a once-in-a-generation opportunity for global development. The only way to seize it is through partnership. To go far, we must go together.
> — *Christine Lagarde, Former Managing Director, IMF (2015)*

Gil (2017) in his collective-action perspective on the planning of megaprojects argues that megaprojects are vast actor-networks formed to develop a new large-scale designed artefact: the infrastructure system. And that high-order decision-making within these networks is driven by the need to build interorganizational consensus at the core of the network. The need to strike a consensus on high-order development decisions results from the distribution of the direct control over the interdependent resources that are critical to develop the new infrastructure, including land, finance, planning consent, political support, and knowledge of need-in-use (pp. 259–260).

The institutional environment forces the promoter to collaborate with actors fully supportive of the goal, as well as with actors who demand a high price for their cooperation. This rules out the use of meritocracy-based authorities to resolve emerging disputes (p. 261). Megaprojects are capital intensive enterprises with emergent complexity (Miller and Hobbs 2005) and controversial and ambiguous system goals. And yet, the control of the resources necessary to achieve the goal is invariably distributed across independent actors. While these actors have heterogeneous preferences and interests, they are unlikely to commit their resources unless they directly influence the planning decisions (p. 262).

IBRD and IDA Lending by Sector

Lending in the developing world involves the commitment of the multilateral banks to a variety of sectors based on the developing country's needs. Figure 3.5 shows the lending in 2021 by the International Bank of Reconstruction and Development and the International Development Agency of the World Bank Group by sector with social protection, energy and extractions, and transportation being the largest and education and health close behind. Significantly, with the vast demand for water resources and sanitation across the developing world, the fiscal 2021 share of the total $15.6B available for the water industry was limited to 3%. The allocation of the funding available in any fiscal year is based largely on the mission of the World Bank Group which centers on two overarching goals:

1) *End extreme poverty.* . . .by reducing the share of the global population living on less than a $1.90 a day.
2) *Promote shared prosperity.* . . .by increasing the incomes of the poorest 40% of people.

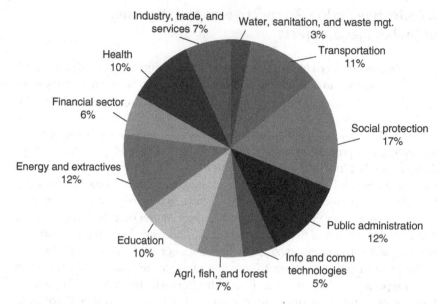

Figure 3.5 Lending by sector (fiscal 2021 share of total of $15.6B). *Source:* Adapted from World Bank Group (2021a).

Financing Climate Change Programs in Africa

Africa's High Seven Priorities to Accelerate Economic Transformation are:

1) Light up and power Africa
2) Feed Africa
3) Industrialize Africa
4) Integrate Africa
5) Improve the Quality of Life for the people of Africa
6) Internet and Interconnectivity
7) Environmental sustainability

Africa serves as an example of the challenges and achievement in mainstreaming climate change and green growth. The Climate Change Action Plan in Africa (CCAP2) is designed to incorporate the Bank's high five priorities in the Paris Agreement, the UN 2030 development Agenda, the Bank's Green Growth Framework and the lessons learned in the implementation of the first climate change action plan (CCAP1), The African Development Bank made a commitment under CCAP2 to allocate 40% of annual approvals as climate finance by 2020

Table 3.7 Africa's mobilization of climate change funds.

Africa's Climate Change Finance, 2020 Annual Report
Internal Sources of Finance
Allocate 40% of annual approvals as climate finance by 2020 ($12.3B)
Pledged ($25B) for 2020–2025
External Sources of Finance
The Clean Technology Fund financed eight projects (CTF) ($192M)
Scaling Up Renewable Energy Program (SREP) financed five project ($95M)
Forest Investment Program (FIP) four projects ($31M)
Pilot Program for Climate Resilience (PPCR) ($2.55M)
Global Environmental Facility (GEF) 41 projects/34 countries ($367M) in grant co-financing and ($177.4M) through GEF.
Green Climate Fund ($197.01M)
Africa Climate Change Fund (ACCF) ($14.5M)
Climate Development Africa Special Fund (CDSF) ($13.3M) ($42.2M) Pledged
Total Mobilization of Climate Change Funds as of 2020: 1.725B

Source: Adapted from Africa Development Bank Group (2021).

and to reach parity between adaptation and mitigation, closing the historical gap of higher mitigation. Significant progress has been made to progressively increase the Bank's climate finance performance from 9% in 2016 to 28% in 2017, 32% in 2018, and 35% in 2019, before a slight decline in 2020 (34%). As an example, Africa has mobilized climate finance from several internal and external sources as shown in Table 3.7.

The Social, Economic and Institutional Value of Megaprojects

> The principle of common but differentiated responsibilities is the bedrock of our enterprise for a sustainable world
> — *Narendra Modi, Prime Minister of India*

Speaking at the 2015 Summit on the SDGs, Prime Minister Modi reflected that developing countries need the financial and technical support of the developed world but can also be leaders in development in their own right by tailoring ways to achieve the goals to their own culture, needs, and an independent vision for their future. In other words, we cannot expect to transplant home country culture and values to the host country.

Infrastructure is understood to be a critical factor in the health and wealth of a country, enabling private businesses and individuals to produce goods and services more efficiently. With respect to overall economic output, increased infrastructure spending by the government is generally expected to result in higher economic output in the short term by stimulating demand and in the long term by increasing overall productivity. Investments in core infrastructure, defined as roads, railways, airports, and utilities, are expected to produce larger gains in economic output than investments in some broader types of infrastructure, such as hospitals, schools, and other public buildings (Stupak 2018).

Dimitriou et al. (2014) observed that "it is rational not to judge project outcomes against the first set of objectives or the first set of costs when the contexts, if not the very functions and even boundaries of such projects, may have altered dramatically." Furthermore, the meaning of sustainability in megaprojects must include concepts of economic, social, and institutional sustainability, as appropriate (Dimitriou et al. 2012).

The contributions of megaprojects, in particular infrastructure, to the social, economic, and institutional environment in which they operate are critical to understanding the full value of these massive undertakings (Flyvbjerg 2014; Flyvbjerg 2017; Greiman and Sclar 2019; May et al. 2021; Pitsis et al. 2018). These contributions include improvement in the quality of life for millions of people, mitigation of disasters, ecological protection, and reduction of poverty.

Megaprojects, especially mega-infrastructure projects, play important strategic roles in economic and social development, and their social responsibility and sustainability have been recognized by scholars in the project management field. All organizations have social responsibilities often referred to as corporate social responsibility (CSR). These responsibilities include respecting the environment, human rights, and the rights of local communities to enjoy their property without encroachment by major infrastructure. Projects have similar responsibilities, but they are often not recognized or accepted in the same way. In their empirical study, May et al. (2021) found that that megaproject social responsibility (MSR) positively affects both financial and social performance of the participating organizations.

Frameworks for Sustainable Development

The coming years will be a vital period to save the planet and to achieve sustainable, inclusive human development.

— *António Guterres, Secretary-General of the United Nations*

The United Nations defines *sustainability* as "meeting the needs of the present without compromising the ability of future generations to meet their own needs." The SDGs adopted by the United Nations in 2015 provide a comprehensive framework for sustainable development. The SDGs explicitly highlight the importance of infrastructure in achieving significant progress toward social, environmental, and economic objectives (Thacker et al. 2019). Some of these goals require significant infrastructure investment, such as the provision of universal access to drinking water, sanitation, and electricity by 2030 (GIH 2017). Infrastructure systems are expected to deliver adequate, reliable services (e.g. access to clean water, protection of the environment, and safe transportation) while promoting other socioeconomic objectives such as equity and accessibility. Though traditionally the value of mega infrastructure projects is assessed in monetary terms (Flyvbjerg 2014) more recently, methodologies such as life cycle assessment (LCA) and multi-criteria methods (MCM) have been used to account for other aspects relevant such as social and environmental performance (May et al. 2021).

For example, the Montenegro Institutional Development and Agriculture Strengthening Project (2009–2019) supported sustainable agricultural growth and rural development while advancing the country's European Union's (EU) pre-accession agenda. The project benefitted 6337 farmers, and also helped mainstream sustainable land use and natural resource management into the country's rural development policies, programs, and investments.

Sustainability Water Transfer Projects

To understand the problem of sustainability, it is helpful to look at a few megaprojects attempting to address some of these concerns. Water is an essential resource for human well-being and the functioning of ecosystems. However, increasing water scarcity is among the biggest challenges humanity is facing (Brauman et al. 2021). The gap between global water supply and demand is projected to reach 40% by 2030 if current practices continue (WRG 2021). More than 60% of the world's freshwater is shared between two or more riparian states. The sustainable and cooperative management of these transboundary water resources is crucial for access to water, sustainable development as well as regional stability and peace, as recognized by SDG 6 (Ensure availability and sustainable management of water and sanitation for all) (UN 2021).

However, many countries and basins struggle to identify the necessary funding, putting at risk the achievement of the SDGs. Financial capacity constraints faced by countries as well as limited understanding of the benefits of cooperation often hinder the mobilization of financial resources for transboundary water cooperation and basin development.

Shumilova et al. (2018) in their inventory of water transfer megaprojects (WTMP) emphasize the need to manage our hydrological systems as hybrid systems – as regional water resources for human use as well as highly valuable ecosystems, for the benefit of people and nature alike. Otherwise, we are facing an engineered water future, which may constrain alternative solutions to cope with an increasingly uneven distribution, both in space and time, of the global water resources.

The Convention on the Protection and Use of Transboundary Watercourses and International Lakes (Water Convention), serviced by the United Nations Economic Commission for Europe (UNECE 2021), helps countries in establishing or strengthening their legal and institutional framework, thereby creating an enabling environment to attract financing, and also supports countries and basins in accessing funding. The water transfer megaprojects raise some of the most important issues in building sustainability. In the water sector, megaprojects include transfer projects, large dams, navigation schemes, desalination plants, treatment plants, and ecosystem restoration projects (Sternberg 2016; Tockner et al. 2016).

Currently, there is no dedicated agency responsible for maintaining a database on water transfer projects, not even in countries where water transfer already is an important component of water supply, such as in the United States and China (Dickson and Dzombak 2017; Yu et al. 2018).

Furthermore, the world lacks internationally agreed standards to evaluate water transfer project design, performance and impacts on people and ecosystems, as have been created for large dams (*World Commission on Dams* 2000). Concurrently, the social, economic, and environmental consequences of these projects do not receive adequate attention in the decision-making process (Sternberg 2016; Zhuang 2016).

South-North Water Transfer Projects

One of the largest water transfer projects is the South North Water Transfer Project (SNWTP) in China. China's SNWTP is the world's largest interbasin transfer. Its two routes have a combined capacity of 4.5B m3 of water per year, with plans for further expansion. Water drawn from the Yangtze River and its major tributaries have been supplying the municipalities of Beijing and Tianjin with drinking water since 2013 (Eastern Route) and 2014 (Middle Route), as well as other water users in the provinces of Shandong, Jiangsu, Hebei, Henan, and Anhui (Rogers et al. 2020). Assessing the far-reaching human and physical effects of interbasin schemes such as the SNWTP therefore remains a critical task, not just during social and environmental impact assessment or cost-benefit analysis before construction, but in operation, as the impacts of such schemes continue to manifest (Rogers et al. 2020).

Wilson et al. (2017) in their review of the SNWTP found that complex regional effects are emerging as some jurisdictions take advantage of a new abundance of water, while others face the long-term consequences of displacement and constrained economic development.

The SNWTP is considered one of the grandest attempts by humans to alter their environment at regional and super-regional scales. However, despite decades of study, surprisingly large uncertainties of the project's economic, social, and environmental impacts remain. These results paint a picture of a huge project that is unlikely to be sustainable over the long term without periodic comprehensive sustainability assessments and responsive management (Wilson et al. 2017). Regardless of these uncertainties the SNWTP has been built. The SNWTP, and water sustainability in general, highlight the importance of multiscale, interdisciplinary, approaches in sustainability science. Only through continuous monitoring and periodic sustainability assessment, followed by proper and timely policy actions and collaborations (Jenkins et al. 2017) (can the project's environmental risks be minimized, its social equity be ensured, and its economic benefits be sustained) (Wilson et al. 2017).

Summary

In this new era of sustainability, it is imperative that all international development projects focus on the 2030 Agenda for sustainable development. Although it is not likely to attain all 17 goals in one project, consideration should be given to the most important goals considering the context of the project and the needs of the local community. The scale and nature of the challenges countries confront today demand new approaches to public policy and institutions. Multilateral institutions must partner with local governments and collaborate with other nongovernmental organizations, local communities, and nonprofits to ensure that all voices are at the table, but also to empower the people who will benefit from these projects upon completion. Most important mobilization of capital from the private sector will be critical to maintaining the goals and mission of improving the quality of life of the world's citizens. To attract private investment countries must provide a bankable and investment-ready pipeline of infrastructure projects.

Lessons and Best Practices

- Megaprojects in the developing world require an understanding of local customs, practices, values, and culture.
- Megaprojects must not only be fair for the investors, but equitable to the local community.
- To be successful the local community must be engaged in the project.
- The local community must also be aware that investors need to see risks reduced before they will be willing to invest particularly in former war-torn countries with continued corruption.

- Project finance in the developing world requires innovation and creativity to meet the growing demand of countries worldwide.
- Risk mitigation is essential if project owners are to attract investment.
- The role of the multilateral development banks is critical in continuing to expand opportunities for growth and improvement in all low income and developing cities throughout the world.
- One of the largest sources of capital is the assets under management of institutional investors, such as pension funds, banks, and insurance companies.
- The success of development projects must be measured in a different way than looking at profits alone – the benefits of these massive undertakings must be identified and advanced constantly throughout the life of the project and beyond.

Application of the United Nations 2030 Sustainable Goals are essential to achieve project success in the developing countries

Discussion Questions

1 How must projects in the developing world meet the growing call for climate change mitigation and sustainable development?

2 Can the sustainable development goals be prioritized? If so, how would you prioritize the goals and why?

3 How do the sources for project finance in developing countries differ from the sources of financing for the developed countries?

4 What are the requirements for obtaining a loan from the multilateral development banks and how can a borrower mitigate its risks?

5 What are the key factors in making projects bankable and delivering a pipeline of investment ready projects?

6 What are the hurdles that must be overcome by governments and the multilateral development banks to mobilize private sector funding?

7 Why is progress in alleviating the global poverty rate slowing and what can be done to ameliorate the situation?

8 Give an example of how multilateral development banks can mobilize capital through a partnering framework.

9 How does the United Nations classify developing countries and what is meant by an economy in transition?

10 What is the role of the multilateral development bank in international development and how do they add value to the social and economic development of the developing countries?

References

Africa Development Bank Group (2021). *Climate Change and Green Growth 2020 Annual Report: Mainstreaming Climate Change and Green Growth in a Time of Pandemic*. African Development Bank Climate Change and Green Growth Department.

Asian Infrastructure Investment Bank (AIIB) (2022). *Members and Prospective Members of the Bank*. https://www.aiib.org/en/about-aiib/governance/members-of-bank/index.html.

Benedek, D., Gemayel, E., Senhadji, A., and Tieman, A. (2021). *A Post-Pandemic Assessment of the Sustainable Development Goals*, IMF Staff Discussion Note. Washington, DC: International Monetary Fund Fiscal Affairs Department.

Brauman, K.A., Bremer, L.L., Hamel, P. et al. (2021). Producing valuable information from hydrologic models of nature-based solutions for water. *Integrated Environmental Assessment and Management* 18 (1): 135–147. https://doi.org/10.1002/ieam.4511.

Center for Global Development (CGD) (2016). *Multilateral Development Banking for this Century's Development Challenges: Five Recommendations to Shareholders of the Old and New Multilateral Development Banks*. Washington, DC: Center for Global Development (CGD).

China Development Bank (CDB) (2019). *2018 Annual Report*. Beijing, China: CDB.

Deloitte (2016). *A positive horizon on the road ahead?* European Infrastructure Investors Survey 2016. https://www2.deloitte.com/content/dam/Deloitte/uk/Documents/infrastructure-and-capital-projects/deloitte-uk-european-infrastructure-investors-survey-2016.pdf.

Dewar, J. (2019). *International Project Finance Law and Practice*, 3rde. Oxford, UK: Oxford University Press.

Dickson, K.E. and Dzombak, D.A. (2017). Inventory of interbasin transfers in the United States. *Journal of the American Water Resources Association* 53: 1121–1132. https://doi.org/10.1111/1752-1688.12561.

Dimitriou, H.T., Ward, E.J., and Wright, P.G. (2012). *Mega Projects Executive Summary - Lessons for Decision-Makers: An Analysis of Selected International Large-Scale Transport Infrastructure Projects*, OMEGA Project 2, OMEGA Centre, and VREF, Bartlett School of Planning, University College London, December.

Dimitriou, H.T., Low, N., Sturup, S. et al. (2014). What constitutes a "successful" mega transport project? *Planning Theory & Practice* 15 (3): 389–430.

Esty, B.C. (2003). *Modern Project Finance: A Casebook*. Hoboken, NJ: John Wiley & Sons.

Fay, M., Sungmin, H., Lee, H.I. et al. (2019). *Hitting the Trillion Mark: A Look at How Much Countries Are Spending on Infrastructure*, Policy Research Working Paper 8730. Washington, D.C: World Bank.

Flyvbjerg, B. (2014). What you should know about megaprojects and why: An overview. *Project Management Journal* 45 (2): 6–19.

Flyvbjerg, B. (2017). Introduction: The Iron Law of Megaproject Management. In: *The Oxford Handbook of Megaproject Management* (ed. B. Flyvbjerg), 1–18. Oxford: Oxford University Press.

Gil, N. (2017). A Collective-Action Perspective on the Planning of Megaprojects. In: *The Oxford Handbook of Megaproject Management* (ed. B. Flyvbjerg), 259–286. Oxford: Oxford University Press.

Global Infrastructure Hub (GIH) (2017). *Global Infrastructure Outlook (GIO): Infrastructure Investment Needs 50 Countries, 7 Sectors to 2050*. Oxford, England: Oxford Economics https://oxfordeconomics.com/recent-releases/GlobalInfrastructure-Outlook.

Global Infrastructure Hub (GIH) (2021). *Global Infrastructure Monitor (2021)*. https://www.gihub.org/infrastructure-monitor/.

Greiman, V.A. and Sclar, E. (2019). Mega-infrastructure as a dynamic ecosystem: Lessons from America's interstate system and Boston's Big Dig. *Journal of Mega Infrastructure and Sustainable Development* 1 (2): 188–200. https://doi.org/10.1080/24724718.2020.1742624.

Hamadeh, N., Van Rompaey, C., and Metreau, E. (2021). New World Bank country classifications by income level: 2021–2022. World Bank Blogs. https://blogs.worldbank.org/opendata/new-world-bank-country-classifications-income-level-2021-2022.

International Monetary Fund (IMF) (2000). *Transition Economies: An IMF Perspective on Progress and Prospects*. Washington, DC: IMF.

International Monetary Fund (IMF) (2015). *International Financial Institutions Announce $400 Billion to Achieve Sustainable Development Goals*. Washington, DC: IMF.

International Monetary Fund (IMF) (2021). *World Economic Outlook (WEO) - Q. How Does the WEO Categorize Advanced Versus Emerging Market and Developing Economies?* Washington, DC: IMF.

Jenkins, B., Gilbert, R., and Nelson, J. (2017). *The 2030 Water Resources Group: Collaboration and Country Leadership to Strengthen Water Security.* Cambridge, MA: Corporate Responsibility Initiative, Harvard Kennedy School of Government.

Lane, D. (2007). Post-Communist States and the European Union. *Journal of Communist Studies and Transition Politics* 23 (4): 461–477. https://doi.org/10.1080/13523270701674558.

May, H., Sun, D., Zeng, S. et al. (2021). The effects of megaproject social responsibility on participating organizations. *Project Management Journal* 52 (5): 418–433. https://doi.org/10.1177/87569728211015868.

MDB Task Force on Mobilization. (2021). *Mobilization of Private Finance 2019.* Multilateral Development Banks and Development Finance Institutions Joint Report. Multilateral Development Banks (MDBs) and Development Finance Institutions (DFIs).

Miller, R. and Hobbs, B. (2005). Governance regimes for large complex projects. *Project Management Journal* 36 (3): 42–50.

Myant, M., and Drahokoupil, J. (2010). *Transition Economies: Political Economy in Russia, Eastern Europe, and Central Asia.* John Wiley & Sons. ISBN: 978-0-470-91370-3.

Nelson, R.M. (2020). *Multilateral Development Banks: Overview and Issues for Congress.* Updated February 11, 2020. Washington, DC: Congressional Research Service (CRS).

Organization for Economic Cooperation and Development (OECD) (2021). *Perspectives on Global Development 2021: From Protest to Progress?* Paris: OECD Publishing https://doi.org/10.1787/405e4c32-en.

Pitsis, A., Clegg, S., Freeder, D. et al. (2018). Megaprojects redefined – complexity vs cost and social imperatives. *International Journal of Managing Projects in Business* 11 (1): 7–34.

Pomfret, R. (2019). *The Central Asian Economies in the Twenty-First Century: Paving a New Silk Road.* Princeton, NJ: Princeton University Press.

Pomfret, R. (2021). Central Asian economies: Thirty years after dissolution of the Soviet Union. *Comparative Economic Studies* 63: 537–556.

Rogers, S., Chen, D., Jiang, H. et al. (2020). An integrated assessment of China's South—North Water Transfer Project. *Geographical Research* 58 (1): 49–63.

Shanghai Pudong Development Bank (2021). *Annual Financial Report 2020.* Shanghai Pudong Development Bank, Co. RNS Number: 5485A.

Shumilova, O., Tockner, K., Thieme, M. et al. (2018). Global water transfer megaprojects: A potential solution for the water-food-energy nexus? *Frontiers in Environmental Science* 6: 150. https://doi.org/10.3389/fenvs.2018.00150.

Siemiatycki, M. (2017). Cycles in Megaproject Development. In: *The Oxford Handbook of Megaproject Management* (ed. B. Flyvbjerg), 96–117. Oxford, UK: Oxford University Press.

Sovereign Wealth Fund Institute (SWFI) (2022). *Top 51 Largest Development Bank Rankings by Total Assets.* https://www.swfinstitute.org/fund-rankings/ development-bank.

Sternberg, T. (2016). Water megaprojects in deserts and drylands. *International Journal of Water Resources Development* 32: 301–320. https://doi.org/10.108 0/07900627.2015.1012660.

Stupak, J.M. (2018). *Economic Impact of Infrastructure Investment.* Washington, DC: Congressional Research Service (CRS). 7-5700. R44896.

Taguchi, H. and Yasumura, K. (2021). Financial additionality of multilateral development banks in private participation in infrastructure projects. *Sustainability* 13: 8412. https://doi.org/10.3390/su13158412.

Thacker, S., Hall, J.W., Adshead, D. et al. (2019). Infrastructure for sustainable development. *Nature Sustainability* 2 (4): 324–331.

Tockner, K., Bernhardt, E.S., Koska, A., and Zarfl, C. (2016). A Global View on Future Major Water Engineering Projects. In: *Society-Water-Technology* (ed. R.F. Hüttl, O. Bens, C. Bismuth and S. Hoechstetter), 47–64. Heidelberg: Springer.

United Nations (2021). *Financing for Sustainable Development Report 2021.* New York: Inter-Agency Task Force on Financing for Development https:// developmentnance.un.org/fsdr2021.

United Nations (2022). *World Economic Situation and Prospects Report (WESP),* Country Classification, 144. Economic Analysis and Policy Division (EAPD) Department of Economic and Social Affairs, United Nations Secretariat (UN DESA).

United Nations Conference on Trade and Development (UNCTAD) (2018). *Financing for SDGs Breaking the Bottlenecks of Investment from Policy to Impact.* ECOSOC Chamber. Geneva: UNCTAD.

United Nations Economic Commission for Europe (UNECE) (2021). *Funding and Financing of Transboundary Water Cooperation and Basin Development.* The United Nations.

Water Resources Group (WRG) (2021). *2021 Annual Report from Dialogue to Action, the Road to 2030.* 2030 Water Resources Group. Washington, DC: World Bank.

Williams Towers Watson (2019). *Total Value of the World's Twenty Largest Pension Funds,* September. https://www.wtwco.com/en-CH/News/2019/09/ top-20-pension-funds-aum-declines-for-first-time-in-seven-years.

Wilson, M.C., Li, X.Y., Ma, Y.J. et al. (2017). A review of the economic, social, and environmental impacts of China's South–North Water Transfer Project: A sustainability perspective. *Sustainability 9* (8): 1489. Switzerland. https://doi. org/10.3390/su9081489.

World Bank (WB) (2020). *2020 Flagship Report, Poverty and Shared Prosperity: Reversals of Fortune.* Washington, DC: World Bank Group.

World Bank (WB) (2021a). *The World Bank Annual Report 2021: From Crisis to Green, Resilient, and Inclusive Recovery*. Washington, DC: World Bank.

World Bank (WB) (2021b). *Country Partnership Framework for Benin for the Period of FY19-FY23*. Open Knowledge Repository. Washington, DC: World Bank https://openknowledge.worldbank.org/handle/10986/30051?show=full.

World Bank (WB) (2021c). *Operations Manual. Development Policy Financing*. https://policies.worldbank.org/en/policies/operational-manual.

World Bank Data (2022). *How Does the World Bank Classify Countries*. Washington, DC: World Bank Group https://datahelpdesk.worldbank.org/knowledgebase/articles/378834-how-does-the-world-bank-classify-countries.

World Commission on Dams (2000). *Dams and Development. A New Framework for Decision-Making*. https://www.internationalrivers.org/sites/default/files/attached-files/world_commission_on_dams_final_report.pdf.

Yu, M., Wang, C., Liu, Y. et al. (2018). Sustainability of mega water diversion projects: Experience and lessons from China. *Science of the Total Environment* 619–620: 721–731. https://doi.org/10.1016/j.scitotenv.2017.11.006.

Zhuang, W. (2016). Eco-environmental impact of inter-basin water transfer projects: A review. *Environmental Science and Pollution Research* 23: 12867–12879. https://doi.org/10.1007/s11356-016-6854-3.

4

Leading Complex Global Projects

All things appear and disappear because of the concurrence of causes and conditions. Nothing ever exists entirely alone; everything is in relation to everything else.

— The Buddha

Introduction

Megaprojects represent a temporary endeavor that requires the development of a new legal entity. The challenges presented in any new organization but especially a megaproject require an understanding of complexity and systems thinking. Globally these large-scale projects have raised challenges regarding the possibility of financing and building these projects due to complexity. For example, in the United Kingdom, the National Audit Office infers that there is a direct cause-and-effect relationship between a projects' lack of comprehension of complexity and poor project performance (NAO 2013).

The United States passed a $1.2 trillion infrastructure bill in 2021 to create a multifaceted fund to spur the type of complicated, ambitious projects that have been stymied by decades of tentative investment and inattention from Washington. The infrastructure bill expressly includes about $16B for "major projects that are too large or complex for traditional funding programs but will deliver significant economic benefits to communities" highlighting the impact and high cost of complexity in megaprojects (U.S. Congress 2021).

In this chapter, complexity is reviewed through the analysis of several megaprojects where we learn how complexity can be used as an instrument for opportunity. The literature reflects that complexity plays a vital role in determining whether large engineering projects succeed or fail (Flyvbjerg et al. 2003;

Global Megaprojects: Lessons, Case Studies, and Expert Advice on International Megaproject Management, First Edition. Virginia A. Greiman.
© 2023 John Wiley & Sons, Inc. Published 2023 by John Wiley & Sons, Inc.

Miller and Lessard 2001). Complexity in projects is often connected with complex systems and systems engineering.

The scale, duration, cost, goals, and impact of projects internationally have grown dramatically in recent decades. These megaprojects attract widespread public interest and have an impact on the social, economic and ecological environment. However, the record of performance of these projects continues to be extremely poor (Flyvbjerg et al. 2003).

To understand the dimensions of complexity we respond in this chapter to the following questions: (i) What is the definition of complexity? (ii) What are the characteristics of complexity and its interrelationship with uncertainty, ambiguity, conflict, and risk? (iii) What are the sources of complexity? (iv) What are the impacts of complexity on the megaproject's viability, its sponsors and stakeholders, and the community at large; (v) How can we reduce complexity? and (vi) How can we organize complexity to optimize various solutions and transition from the complex to the simple?

Defining Complexity in the Context of Megaprojects

Megaprojects have been broadly described as "large-scale, complex investments that typically cost a billion dollars and up, take many years to develop and build, involve multiple public and private stakeholders, are transformational, and impact millions of people" (Flyvbjerg 2014). Importantly, as noted by scholars in the field of project management, it is, not the cost but the complexity that marks out a megaproject (Pitsis et al. 2018).

The concept of complexity from the perspective of megaprojects has been described in three characteristics: (i) interrelatedness, (ii) nonlinearity, and (iii) emergence, in terms of both structural and social elements (Damayanti et al. 2021a). The complexity arises from the interdependency of thousands of moving parts associated with funding, managing, and governing complex social and organizational relations. Involvement ranges from committed stakeholders amongst the contractors and civic authorities to those that are resistant, embedded in existing communities, social movements, and advocacy organizations (Pitsis et al. 2018).

Contributing to complexity is the existence of significant numbers of different, ambiguous, and interconnected tasks and activities to complete the project which are often hard to define due to the uncertainty that lingers over complex projects throughout their long life. It is difficult to determine the behaviors of complex systems. You can look at the system now, and make a guess where it might be next, but saying what it will be doing over a longer time period, sometimes decades, is much more difficult.

The more complex the system, the more likely emergent events will disrupt the continuity. The more complex the project the more difficult it is to predict the chain of events that any emergent event will produce. Relatively small disturbances can provoke chain reactions that can cause substantial damage and even cause the system to collapse (Miller and Hobbs 2005). For example, some megaprojects are more complex than others. Specifically, complexity in water megaprojects is embodied in structural complexity and cognitive complexity, including complex engineering schemes, limited awareness of innovative technology and objectives, strong uncertainty from the external environment, and the participation of numerous stakeholders (Mai et al. 2018). Leadership and experience will come and go throughout the project life cycle leaving a void that may not be filled for some time.

Differences in economic, social, and cultural motives can lead to chaos and conflict. Evidence shows that new developments and changes in technology increase uncertainty (Shenhar 2001). In addition, since the technology used in megaprojects is often new, developmental, or cutting-edge, its behavior and functionality are often hard to predict. In the case of an already complex product such as the F-35 Joint Strike Fighter – a very sophisticated and technical fighter aircraft – the design phase (which took almost an entire decade) witnessed countless adjustments as the underlying technologies constantly evolved.

F-35 Joint Strike Fighter. *Source:* The Washington Institute for Near East Policy.

Megaprojects are often characterized as complex without defining the underlying causes of complexity and how it impacts a project's outcome (van Marrewijk et al. 2008). Moreover, complexity remains ambiguous and ill-defined in much of the project management literature (Geraldi 2008) and there has been insufficient attention paid to early studies of systems analysis, contingency theory, and complex projects (Brady and Davies 2014).

As a comparison, we also need to look at complexity from the perspective of developing countries. Research has confirmed that it is more difficult to manage complexity in developing countries, as these nations often lack economic, political, and social stability, have unique cultural backgrounds, unstable regulatory systems and rule of law, low skilled labor and an absence of managerial skills requiring temporary leadership imported from a foreign culture (Damayanti et al. 2021b). Significantly, megaprojects are run on budgets ranging between 0.01 and 0.02% of a country's gross domestic product (GDP). Differences in economic motives, cultural perspectives, and political affiliations of stake holders are also a source of complexity.

Megaprojects have also been defined in terms of structural and dynamic complexity (Brady and Davies 2014; Maylor et al. 2008), and some scholars add to this framework one additional type, socio-political complexity (Sheard and Mostashari 2010). Project management science has borrowed greatly from systems theory to analyze and describe the functioning of projects, and it has long been accepted by the community that projects operate as complex systems (Baccarini 1996; Williams 2002). Structural complexity is characterized by the interdependence and diversity of components and results from organizational and environmental aspects (Chapman 2016). Dynamic complexity in projects highlights the question of changes and evolutions over time and focuses on the dynamic relations between the internal components of the project and between the project components and the environmental components (Geraldi 2008). Socio-political complexity has been defined as multiple stakeholders, views, human interactions, and dynamic governance.

Characterizing complexity is not easy, nor can it be easily contained or mitigated. In Table 4.1, five types of complexity commonly found in megaprojects are described: (i) structural, (ii) dynamic, (iii) technological, (iv) governance, and (v) socio-political. Each of these types of complexity have been analyzed in the literature with sometimes different meanings and characteristics.

Complexity in Relation to Uncertainty, Ambiguity, Conflict, and Risk

As shown in Figure 4.1, uncertainty, ambiguity, and conflict are all elements of complex systems, and must be strategically managed to improve project outcomes. The more complex a project is the greater the uncertainty. Complexity also impacts the conflicts in a project by the very fact that as the number of project sponsors, and stakeholders and institutions increase so do competing interests and cultural divides which in turn can lead to confusion, ambiguity and often chaos when there is no clear direction or process to

Table 4.1 Complexity characteristics, concerns, and mitigation.

Type of complexity	Characteristics	Megaproject concerns	Mitigation/ Management
1) Structural Chapman (2016) Davies and Mackenzie (2014) Damayanti et al. (2021a) Kasser and Zhao (2014) Miller and Hobbs (2005) Pitsis et al. (2018)	Size, diversity of components, interrelatedness, emergence, nonlinearity, institutional framework	Internal/external coordination, long- versus short-term orientation, tolerance for ambiguity v. predictability, conflicts, chaos and disruption	Linking program and strategic management, decomposing project into levels of systems integration, with clearly defined interfaces, project and operational integration, development of relationship competencies, developing cross functional teams
2) Dynamic Brady and Davies (2014) Geraldi (2008) Loch et al. (2008) Maylor et al. (2008)	Unpredictable situations, uncertainty, emergent events associated with interaction among project components	Decision-making authority, dealing with uncertainty in an environment of rapid change, flexibility of governance structures	Facilitating flexibility to enable complexity to be managed, creating a learning environment, valuing and promoting collaboration, opening communication channels between project levels, reducing distance between organizational levels, focus on agility
3) Technological Mai et al. (2018) Maylor et al. (2013) Rebentisch et al. (2017) Shenhar (2001)	Engineering and design complexity, systems integration, integration of people	Lack of knowledge, industry maturity, resources and capability, convergence, time to innovate, safety and health, interconnected networks	Integrated teams and disciplines, generation of options for engineering and design, shared objectives, development of people competencies, interface planning and risk mitigation, incentivizing innovation, quality assurance

(Continued)

Table 4.1 (Continued)

Type of complexity	Characteristics	Megaproject concerns	Mitigation/ Management
4) Governance Locatelli et al. (2014) Miller and Lessard (2001) Pitsis et al. (2014, 2018)	Multiple sponsors and governance structures, changes in leadership, decision making complexity, multiple regulators, laws, regulations, process and standards	Change in strategy and priorities, capacity of self-regulation, politicization of project, lack of coordination among regulatory bodies, immature legal systems, changing contractual requirements	Governance transformation from a program to a systems perspective, consistent and continuous definition of requirements and enforcement of contracts, integrated regulatory approval process across agencies, explicit and accessible procedure for dispute resolution, escalation criteria, ex post and ex ante evaluations
5) Socio-political Lee et al. (2017) Damayanti et al. (2021a) Sheard and Mostashari (2010)	Multiple stakeholders, countries and cultures, human interaction and dynamic governance	Political sensitivities, competing interests, resistance to project, cultural divides, individualism v. collectivism, conflicts, diversity, ethics	Building intraorganizational and interorganizational relations and trust, multiple and diverse communication channels, stakeholder engagement and participation of community members and Indigenous population

follow. Globalization by its very nature tends to increase the likelihood that projects that already have characteristics of complexity will only increase because of the many unknowns that exist when many cultures come together. For instance, the project finance may involve multiple lenders and sponsors from many different countries that have different understandings of the contractual and legal environment, supply chain structures add complexity when suppliers from multiple countries are involved, or the engineers and contractors may represent a diversity of different experiences, standards, and practices. Complexity is the number of interdependent relationships that occur within a system. An increase in complexity increases the amount of unknowns

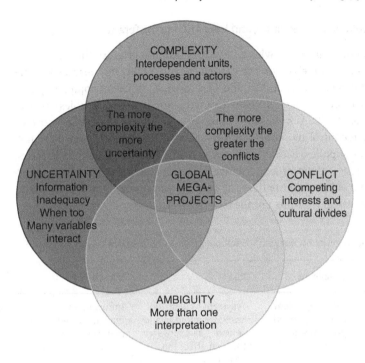

Figure 4.1 Characteristics of complexity.

that exist within a system. Put differently, uncertainty refers to the components, relationships, and interactions we do not fully comprehend or of which we may not even be aware. Complexity and uncertainty are thus strongly related (Salet et al. 2013).

Characteristics of Complex Projects

Megaprojects are challenging to manage and difficult to predict or foresee, possible outcomes due to the competing elements of:

Complexity – Interdependent units, [systems], processes, and actors

Uncertainty – Information inadequacy when too many variables interact – unknowable, unmeasurable, and uncontrollable

Ambiguity – More than one interpretation – increases uncertainty and creates conflicts

Conflict – Competing interests and cultural divides – increases uncertainty and impacts trust

How are Complexity, Uncertainty, and Risk Alike or Different?

Though risk is often grouped with complexity and uncertainty they have very distinct meanings. With risk, you can predict the probability of a future outcome as a risk is knowable, measurable, and controllable. Whereas complexity and uncertainty create environments where the interaction of many variables can lead to unpredictable results. Yes, we cannot stop or prevent a hurricane from occurring, but we can control the impact of hurricanes through emergency planning, food, and water supplies, accelerated escape plans, safe shelter, storm barriers, cyber security precautions, and alternative communication sources. With uncertainty we are not able to identify the risks because they are unknowable so, therefore, we cannot mitigate or prevent the risk (Table 4.2).

Table 4.2 Comparison of elements of complexity, risk, and uncertainty.

Attributes	Complexity	Risk	Uncertainty
Definition	Interdependent units, [systems], processes, and actors	A system to identify, assess, respond to and control risk exposures that can affect the creation or preservation of business value	Information inadequacy when too many variables interact
Characteristics	Interaction of many variables that can lead to unpredictable results	Knowable, measurable, and controllable	Unknowable, unmeasurable, and uncontrollable
Strategies	Up front planning and exploration of options	Designed to identify specific threats, assess unique vulnerabilities, and implement risk reduction efforts	Megaprojects continue to evolve despite uncertainty. As uncertainty is understood risks are identified and transferred to the risk register
Focus	On decision making at the strategic and operational level	The mitigation of exposures	Discovery driven planning; probing for knowledge
Approach	Reduction of complexity and optimization of choices	Follows risk management plan or framework. Reports on potential exposures	Uncertainty reduction approach; reduce by seeking information. Reports on the future

Miller and Lessard (2007) define risk as the possibility that events will turn out differently than anticipated. While risk can be described in statistical terms, uncertainty represents situations which are not fully understood in terms of causal forces and potential outcomes. So, odds are known for risks but not for uncertainty. Megaprojects involve both high risk and uncertainty; yet some refer to all such cases as risk (Kardes et al. 2013).

Uncertainty

Uncertainty like complexity creates management challenges for megaprojects that go beyond the traditional project management concerns of cost, scope, and schedule though they are likely to have an impact on all three. Uncertainty may mean that there is insufficient information to even decide which alternative may be viable and whether proceeding with a particular alternative makes sense without the evidence needed to evaluate such a major decision.

Though uncertainty is a distinct concept from complexity, one can feed off the other as uncertainty can create complexity and complex projects often have a lot that is unknown at least initially. This relationship between complexity and uncertainty was demonstrated in the Crossrail Project in London:

> It is important to note that at an early stage, any cost estimate can only be an approximation based on the information that is known at the time. Given the complexity of Crossrail there were aspects in which the areas of uncertainty exceeded the elements that could be determined and measured. The areas of uncertainty were evaluated in a risk model to produce a probability-based forecast of overall cost. (Buck 2017)

Lenfle and Loch (2017) found that the first of the three major causes of megaproject failure is: "Underestimation of or refusal to acknowledge uncertainty." Projects suffer from unforeseen uncertainty that cannot be identified during planning. Because they have information gaps, they cannot create the necessary contingencies. Uncertainty can arise from large events or from the interaction among stakeholders through complexity. Uncertainty requires more flexible and emergent approaches to managing megaprojects.

Experience has found that uncertainty that is unforeseen can be diagnosed through discovery-driven planning or by systemically probing for knowledge gaps (Lenfle and Loch 2017; McGrath and McMillan 2000).

Uncertainty refers to the possibility of emergent indeterminate events. Under conditions of low uncertainty, approximate forecasts still can be made. High uncertainty entails conditions of indeterminacy where future events are neither identifiable nor amenable to calculation.

> The strategies for managing risk and uncertainty are quite different. Traditionally, the project management community has focused on what economists call risk and has not tended to deal explicitly with uncertain, emergent, and unforeseen events (Miller and Hobbs 2005).

How should one cope with complexity and uncertainty in mega infrastructure projects? Several solutions have been offered by scholars in the project management field: (i) The development of an institutional environment with more formal process and procedure to address the challenges brought on by complexity; (ii) the shaping of a learning environment in order to deal with uncertainty and emergent properties; and (iii) balancing the generation and the reduction of a variety of policy options in order to select a limited number of feasible options and to bridge the strategic exploration and the operational processes of decision making (Salet 2021).

Thus, an increase in complexity often means that it is more difficult to comprehend the effect of influencing one element; hence there is increased uncertainty. Put differently, uncertainty refers to the components, relationships, and interactions we do not fully comprehend or of which we may not even be aware. Complexity and uncertainty are thus strongly related (Salet 2021).

The Sources and Impact of Megaproject Complexity

As shown in Figure 4.2 megaprojects have characteristics such as uncertainty, interdependence on process and people, unpredictability, volatility, and emergent events that lead to effects on the project that may not be controllable. Projects have become more complex because of the increasing factors that are considered sources of complexity. Complexity comes from:

- The need for technological innovation,
- Large numbers of project participants from multiple employers, representing distinct cultures and companies, and
- a volatile environment subject to rapid change

All three of these factors affect project outcome. This complex environment influences all aspects of a megaproject's long-life cycle from project selection to project planning, coordination, and control; it can also affect the selection of an appropriate project governance structure and the strategic alignment of the megaproject's goals with the vision and goals of the sponsoring organizations. A different strategy is required for complex projects. Complexity also impacts decision making, the ability to build resilience, trust, and to stabilize a project with much uncertainty, ambiguity, and conflict.

Characteristics

Effects on project

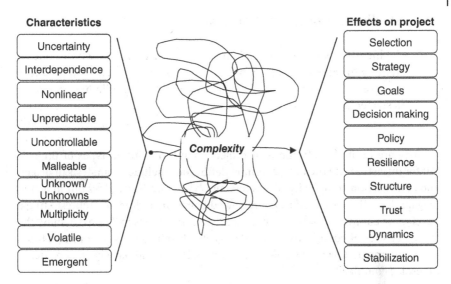

- Uncertainty
- Interdependence
- Nonlinear
- Unpredictable
- Uncontrollable
- Malleable
- Unknown/ Unknowns
- Multiplicity
- Volatile
- Emergent

Complexity

- Selection
- Strategy
- Goals
- Decision making
- Policy
- Resilience
- Structure
- Trust
- Dynamics
- Stabilization

Figure 4.2 Complexity characteristics and effects.

The origins of complexity theory applied to project management leads to early research by Baccarini (1996), Gidado (1993), and Morris (2002). The importance of complexity to the project management process is widely acknowledged and can have influence on many aspects of project selection, planning and development, and control. Importantly, complexity can have an impact on governance, and governance can make a project more complex. As recognized by Miller and Hobbs (2005) in their study of large capital projects, there is a sharp contrast between the binary, hierarchical, and static nature of corporate principal-agent governance relations, and the time dependent co-determination found in the network relations typical of the governance of megaprojects. The governance regimes must adapt to the specific project and context, deal with emergent complexity, and change as the project development process unfolds (p. 42, 48).

As described by San Cristóbal et al. (2019) complexity can affect the selection of an appropriate project organization form and experience requirements of management personnel; and it can affect different project outcomes (time, cost, quality, safety, and risk). As projects have become more complex, there has been an increasing concern about the concept of project complexity and the application of traditional tools and techniques developed for simple projects that have been found to be inappropriate for complex projects. Identifying and characterizing distinct aspects of project complexity to understand more efficiently the stakes of project management complexity can be of great support in assisting the global project management community (San Cristóbal et al. 2019).

Case Studies of Complex Projects

> Engineering is the art of modeling materials we do not wholly under-
> stand, into shapes we cannot precisely analyze, so as to withstand forces
> we cannot properly assess, in such a way that the public has no reason
> to suspect.
>
> — *Dr. E.H. Brown (1967)*

There are multiple definitions of project complexity, for all types of projects but for infrastructure the most common definitions include an analysis of design and construction complexity. To understand how complexity plays out in actual megaprojects five case studies are reviewed below, including one representing a developing country project in Indonesia. The case studies in Table 4.3 show the characteristics of complexity in these projects, the challenges the projects faced and solutions that were developed or in hindsight should have been considered to resolve complexity.

Hong Kong–Zhuhai–Macao Bridge. *Source:* The Transport and Housing Bureau.

Table 4.3 Case studies: complex project characteristics, challenges, and solutions.

Project cost and timeline	Description	Characteristics	Challenges	Solutions
Hong Kong Zhuai-Macao Bridge China US$17B 2009–2018	The Hong Kong–Zhuhai-Macau Bridge is a 55 km bridge–tunnel system consisting of a series of three cable-stayed bridges, an undersea tunnel, and four artificial islands. It is both the longest sea crossing and the longest open-sea fixed link in the world	The complexity of the mega-project is mainly derived from the political attribute of "One country, Two systems" and the project attribute of its scale as a "mega project" (Zhu et al. 2018). Therefore, the investment and financing modes of the Hong Kong-Zhuhai-Macao Bridge needed negotiations between the three governments and even coordination from the central government (Mai et al. 2018).	Complex geological and topographical conditions considering prevailing winds and tidal forces. Unique political, social, and natural environment Delays, ballooning construction costs and other engineering challenges	A multilevel decision-making management structure of "three-tier and two-level coordination mechanism" was established according to the expertise of decision-making affairs, power allocation needed, and attribution of solutions in the decision-making process. The principal-agent relationship in the establishment and decision-making process of the Hong Kong–Zhuhai–Macao Bridge must have a dynamic flexibility. basic legal decisions, investment and financing decisions, engineering schemes, public affairs management matters, and matters of project companies. Joint action is required to resolve disputes (Zhu et al. 2018; Mai et al. 2018).

(*Continued*)

Table 4.3 (Continued)

Project cost and timeline	Description	Characteristics	Challenges	Solutions
F-35 Joint Strike Fighter 2001–Present	Designed to meet the bulk of the needs of the US military throughout the first half of the twenty-first century. Designed and built by prime-contractor Lockheed Martin and partners Northrop Grumman and BAE Systems, the Pratt & Whitney and the GE Rolls-Royce, assisted by innumerable subcontractors. Provides electronic warfare and intelligence, surveillance, and reconnaissance capabilities.	Global supply network includes over 1000 companies. Funding is provided by the United States the United Kingdom and seven other countries. Project involves complex design and changes in technology (Gertler 2020)	The complexity added cost. Rising costs-imposed delays. Delays gave developers more time to add yet more complexity to the design. Those additions added more cost (Gertler 2020) The design phase (which took almost an entire decade) witnessed countless adjustments as the underlying technologies constantly evolved. Cost for each Jet rose from $50M in 2001 to more than $113M in 2010 (Gertler 2020)	A RAND Corporation study found that the fundamental concept behind the F-35 program – that of making one basic airframe serve multiple services' requirements – may have been flawed. Congress may wish to consider how the advantages and/or disadvantages of joint programs may have changed as a consequence of evolutions in warfighting technology, doctrine, and tactics (Lorell et al. 2013)

| LGV Mediterranee €3.8 billion. 2001 | A 250 km-long (160-mile) French high-speed rail line connecting the regions of Provence-Alpes-Côte d'Azur and Occitanie to the LGV Rhône-Alpes, and from there to Lyon and the north of France. Construction costs rose to €3.8 billion; the line entered service in June 2001. The commencement of service on this line has led to a reversal of the respective airplane and train markets: by making Marseille reachable in three hours from Paris – the train now handles two-thirds of all journeys on that route | Social and political complexity | Strong opposition could lead to risk in future projects. To improve the handling of the social and political risk in subsequent MTPs in France, the state voted for a law on 2 February 1995, committing itself to a greater level of environmental protection

This introduced public debate concerning the building of major infrastructure projects from the outset of the decision-making process (Dimitriou et al. 2014) | The performance of the LGV Mediterranee – in terms of handling environmental, social, and political risks – could have been improved by undertaking earlier consultation on the route (Dimitriou et al. 2014) |

(Continued)

Table 4.3 (Continued)

Project cost and timeline	Description	Characteristics	Challenges	Solutions
Ranstad Rail 1995–2008 The Hague	The building of a light rail network in the Rotterdam–The Hague metropolitan area in the west of the Netherlands to connect the cities of Rotterdam, The Hague and Zoetermeer	Multiple stakeholders at the local, regional and national level	Key issue was the cost of building a new rail infrastructure without available financing from the public or private sector. The large number of authorities and interest groups meant that decision-making was protracted, similarly with the limited compatibility of the infrastructure to be assimilated under the RandstadRail banner	Informal collaboration between local authorities. The involvement of the national rail operator provided an opportunity to convert existing underused heavy rail tracks rather than build new infrastructure (van Der Bijl et al. 2018)
Kecamatan Development Project (KDP) 2006 Indonesia	The KDP's objectives are to raise rural incomes, strengthen Kecamatan and village government and community institutions, and to build public infrastructure through labor intensive methods (Adler eta al. 2009)	The project was ridden with corruption, social unrest, and poverty-stricken rural communities in need of infrastructure and institutions	The key challenge is the changing of the mindset of both government and World Bank officials from a focus on the necessities of agency supply to one in which projects respond to community demand (Adler et al. 2009)	Delivery of development resources to rural communities "using local representative community forums . . . wherein villagers, not government officials or external experts, determine the form and location of small-scale development projects via a competitive bidding process"

Strategic Management of Complexity

In this section, diverse ways of managing complexity are summarized from theory and practice. Though strategically managing complexity is not limited to these frameworks and methodologies, it is a good starting point to understanding that complexity, like risk management, quality management and other project disciplines, needs to be understood and managed in a way that optimizes the benefits of complexity for the project owners, the project sponsors and those that will ultimately benefit from the project outcomes.

1. Managing Complexity through Systems Integration

> Surely, we are being presented with one of the greatest triumphs of science and engineering, destined deeply to influence the future of the world.
> — *Niels Bohr*

As described by Niels Bohr, recipient of the Noble Prize in Physics in 1922, the Manhattan Project was a research and engineering project to harness the energy of the atom. Named for its first sites in New York City, it had profound and far-reaching consequences. It accelerated changes in science and set off a continuum of reactions outside the laboratory that have been felt in international power politics, agriculture, protest movements, medicine, the presidency, photography, ecology, warfare, economics, popular culture, research ethics, attitude toward science, government, and the future. It impacted aspects of our world and lives both foreseeable and unanticipated. The Manhattan project provides an excellent example of managing complexity through trial and error and exploring parallel options until sufficient information becomes available (Lenfle 2011). The Manhattan project had no parallels in history.

In large engineering projects, systems integration first emerged in the 1950s and the 1960s to deal with the design and integration of weapons systems projects such as Atlas, Titan, and Polaris Ballistic Missiles (Hughes 1998). In the Polaris case the problem was to coordinate and integrate multiple branches of government and the dozens of firms involved (Sapolsky 2003). Systems engineering (SE) has been defined as "the science of designing complex systems in their totality to ensure that the component subsystems making up the system are designed, fitted together, checked and operated in the most efficient way" (Jenkins 1996). Since Jenkins first set forth his definition of SE, systems engineers seem to have been busy creating more complex models and processes. This observation can be mapped into the holistic approach to problem solving where the undesirable situation is the failure of systems engineering to manage the complexity of the systems development environment (Kasser and Zhao 2014). Nevertheless, systems engineering has been a major tool used for solving complex projects for decades and is essential to build airports, tunnels, roadways, bridges, defense systems, and much more. Systems engineering is the art and science of developing an operable system capable of meeting requirements within often opposed constraints to produce a coherent whole that is not dominated by the perspective of a single discipline (Rebentisch et al. 2017).

Dynamic and Structural Complexity

When problems fundamentally dynamic are treated statically, delays, and cost overruns are common. Experience suggests that the interrelationships between the project's components are more complex than is suggested by traditional approaches. These, traditional approaches, using a static approach, provide project managers with unrealistic estimations that ignore the nonlinear relationships of a project and, thus, are inadequate to the challenge of today's dynamic and complex projects.

Complex projects require decision making in an evolving and unstable environment and demand an exceptional level of management and the application of the traditional tools and techniques developed for ordinary projects have been found to be inappropriate for complex projects. Developing a strategy for a complex megaproject requires a strategy unto itself. Due to the inherent complexity of multi-stakeholder projects, projects must consider multiple stakeholders' interest in their goal setting. Projects cannot directly adopt only one uniform and explicit goal or method communicated by a top management representative of a single parent organization. In fact, the project must carefully position itself to its environment, and the goals and management methods of the project must be carefully matched with the situation at hand and the context. Such approaches are contained in the concept of project strategy (Artto et al. 2008).

Davies and Mackenzie (2014) suggested that organizations cope with complexity by decomposing a project into distinct levels of systems integration with clearly defined interfaces between levels and component systems. Complexity should be managed at a system of systems level (which they call "meta systems integration" to convey the idea of standing above and looking across the overall program and system of systems). Managing structural and dynamic complexity on London's Heathrow Terminal 5 and London's 2012 Olympics provides two different approaches to managing complexity, yet both also emphasize the approaches to governance (Brady and Davies 2014).

Dynamic Complexity

In a study of complexity in the 2012 Summer London Olympics, "dynamic complexity" was described as a "tight loose approach" which enabled the Olympic Delivery Authority (ODA) to achieve a highly consistent approach across the whole program while facilitating flexibility to enable dynamic complexity to be managed (Brady and Davies 2014; Davies and Mackenzie 2014). The concept of dynamic complexity addresses the unpredictable situations and emergent events that occur over time, which are associated with interactions among components of a system and between the system and its environment. Dynamic complexity is therefore associated with different types of uncertainty influencing the progress of a project (Loch et al. 2008). Dealing with uncertainty in an era of rapid change and unpredictability requires a different approach than traditional project management can offer, especially when organizations are trying to accomplish something that has not been done before and might not be done again (Brady and Davies 2014).

Structural Complexity

In the Heathrow Terminal 5 Project, structural complexity was defined as arrangement of components and subsystems into an overall system architecture. The breakdown of the overall program into individual projects helped deal with the structural complexity. Although some projects corresponded to individual venues, others such as logistics and security provided program-wide services. Managing the numerous stakeholders involved in the program called for a strong governance structure involving multiple layers of assurance and reporting (Brady and Davies 2014).

2. Parallel Approaches to Project Complexity

Addressing complex and uncertain projects particularly megaprojects require exploration and learning throughout the long life of the project. No one approach will offer an ideal solution. Recent work demonstrates that, when confronted with unforeseeable uncertainties, managers can adopt either a learning, trial-and-error-based strategy, or a parallel approach. As an example, studying the case of the Manhattan Project, demonstrates that managers must not necessarily choose between solutions, but can also combine them or add new ones during the project life. (Lenfle 2011). Megaprojects often raise situations of uncertainty for which the project cannot plan or prepare. One way of confronting unforeseeable uncertainties consists of trying different approaches in parallel to find out which one works best (Lenfle 2011).

3. Uncovering the Unknown Unknowns

Unknown unknowns is a common term in strategic planning and project management. A major challenge in project management is dealing with the uncertainties within and surrounding a project that give rise to outcomes that are unknown.

The project management literature provides little insight and frameworks to deal with these unexpected events that can cause serious damage or catastrophic loss.

Ramasesh and Browning (2014) in their analysis of unknowns-unknowns referred to in the literature as (unk-unks), created a conceptual framework for tackling knowable unknown unknowns in project management. They broke unknown unknowns into two categories:

1) Unknowable unk unks: These are unexpected surprises that cannot be anticipated. No amount of action by a project manager will be able to convert unknowable unk unks to known unknowns. For example, COVID-19, the tsunami in the Indian Ocean in 2004, that disrupted many projects or Hurricane Katrina that caused over 1800 fatalities and $125B in damage in late August 2005 could not have been predicted. Not even the world's greatest underwriters at Lloyds of London and Munich Re saw them coming at least not to the extent of the impact.

2) Knowable unk unks: These unk unks could be foreseen by the project manager but for some reason (e.g. barriers to cognition) are not (yet) seen. Many studies of project failures suggest that a large amount of unk unks could have been anticipated given due diligence by the project manager.

Ramasesh and Browning (2014) fill a gap in the project management literature by presenting a framework for recognizing and reducing specific areas in a project likely to contain these unknown unknowns. This framework conceptualizes six main factors – relating to both project design and behavioral issues that can increase the likelihood of unk unks in a project. These are: (i) complexity, (ii) complicatedness, (iii) dynamism, (iv) equivocality, (v) mindlessness, and (vi) project pathologies. From this analysis they develop eight propositions, each of which covers five important subsystems (product, process, organization, tools, and goals) through which unknown unknowns can emerge in a project. They propose five concepts for increasing awareness of these situations: (i) develop system thinking; (ii) build experiential expertise; (iii) become a high-reliability organization (HRO) and embrace preoccupation with failure and learn from surprising outcomes. Because megaprojects are surrounded with complexity developing tools to manage the complexity and uncertainty is a strategic necessity.

Qualitative Assessment of Vulnerabilities to Unknown Unknowns

Loch et al. (2008) developed an assessment for dealing with uncertainty. Loch looked at Escend Technologies, a Silicon Valley start-up company, for the qualitative assessment of a project's vulnerability to unknown unknowns and outlined the following steps to uncover unknown unknowns that could become knowable.

Uncovering the Unknown Unknowns (Loch et al. 2008)
1) First, identify the problem structure: goals, external influences, activities, and causality of activities and effects.
2) Second, break the problem into pieces – e.g., product modules, process activities, and stakeholders.
3) Third, examine (through what could be a highly iterative and gradual process) the complexity and uncertainty of each piece to identify the major risks (known unknowns) that need managing and the knowledge gaps that point to areas of potential unknown–unknowns.
4) Fourth, consider managing the selected pieces of the project in parallel by using different project management methods – e.g., an "option play" that treats different modes of project activity as options contingent on the outcomes (much like options in the financial securities market).

4. Using the Front End of Projects to Reduce Complexity

Recognition in the Early Conception Phase that Complex Projects will cost more and take longer than originally budgeted due to all the missing parts being essential to ensure success during implementation. Essentially, due to complexity, decisions are often made before the project is ready and during a time when there is generally still much uncertainty. The literature is clear that megaprojects are often over budget and behind schedule because the budget was produced early in the up-front planning phases when there was still much uncertainty in the project and the data was not available to assess a realistic outcome. Since megaprojects have long durations, 10–20 years is not unusual, by the time the project is shovel-ready inflation and unknowns may have caused the project to double in price.

Scholars have determined that the front-end of the project has the highest potential for reducing uncertainty and complexity by shrinking the number of alternatives, but often the focus is on deriving detailed information (Samset and Volden 2016). A common notion for reducing uncertainty at the front-end is defining clear goals and objectives for projects (Samset and Volden 2016). However, Giezen et al. (2015) mention that, to some extent, ambiguity in goals and objectives opens space for applying different views to decisions according to future changes.

Following a systematic literature review (SLR), Babaei et al. (2021) found five essential characteristics emerged from analyzing front-end definitions: (i) Exploratory nature, (ii) Generating managerial information, (iii) Shaping a feasible concept, (iv) Terminating with a decision, and (v) Uncertainty. Through their research, they then investigated the remedies for managing the issues at the front-end of infrastructure projects and derived six remedies for managing the front-end issues, namely: (i) Using more qualitative data and less detailed quantitative analysis for decision-making, (ii) Involving external stakeholders' views in decision-making, (iii) Generating reliable estimations and controlling the quality of the results, (iv) Applying lessons learned, (v) Increasing the skills and competencies of the front-end actors, and (vi) Promoting transparency and accountability and defining clear roles and responsibilities in project governance.

5. Organizational and Social Complexity

Complexity requires more elaborate forms of organization and formal processes and governance structures. Megaprojects should focus not just on intraorganizational structure but more importantly on interorganizational relationships and trust building among the multiple organizations involved in planning, designing, engineering, and constructing the project. Megaprojects are known for their vast size, political challenges, high costs, long duration, technical and procedural complexity, and millions of diverse and changing stakeholders. Coordination and understanding between the actors of these mega projects is essential to their success. However, this is particularly difficult to achieve in this context, as these stakeholders are extremely diverse and generally have few opportunities to work together. To achieve this cohesion, a major, yet often invisible, element is trust. The delivery of a megaproject requires bringing together multiple interdependent organizations that must collaborate and cooperate to achieve successful outcomes (Winch 2014). This distinction is important when seeking to understand the uncertainty associated with complex projects (Chapman 2016).

Organizational complexity arises from the participation of key project delivery partners coming from different national frameworks who must find a way to resolve their differences so they can work effectively together to resolve multiple challenging technical, contractual, and political issues (Levitt and Scott 2017). Empirical research has shown that firms can build trust and develop a shared project identity by defining early on the conceptual design of the project collectively using teams made up of representatives from the client and multiple delivery organizations (p. 113).

6. Focus on Decision Making at the Strategic and Operational Level

Because of the very long-time span of the process of a mega infrastructure project, changes in social and political conditions are inevitable (Altshuler and Luberoff 2003). One of the most frequent findings in empirical studies on decision making in mega infrastructure projects is that decision-making processes are organized in a manner that is far too reduced to enable adequate decision making on complex issues (Priemus et al. 2008). Highly complex mega infrastructure projects are frequently treated as simple processes of decision making, risking not only the occurrence of errors but also the neglect of the strategic potential of alternative options and the potential offered by recombination and enrichment with other trajectories of policymaking (Priemus et al. 2008).

Acceptance of complexity and uncertainty would require preserving flexibility in the decision-making and implementation process – in other words, the possibility to take decisions and implement actions later when more is known about the relevant project conditions. In the case of LGV Méditerranée, this would have meant postponing final decisions regarding the route as far as possible so that better alternatives could emerge from debate.

Within such a long timeframe it is not unusual to have recurring rounds of decision making, each involving different actors. During building the Big Dig from conception to substantial completion there were eight Governors which meant that the administration and the federal and state government infrastructure was constantly changing (Greiman 2013). If there is a lot of consensuses across government as to the purpose and goals of the megaproject change will be expected, if there is resistance and opposition change may not be possible.

Managing Complexity in High-Reliability Organizations

A HRO is an organization that has succeeded in avoiding catastrophes despite a high-level of risk and complexity. Specific examples that have been studied, most famously by researchers Karl Weick and Kathleen Sutcliffe, include nuclear power plants, air traffic control systems, and naval aircraft carriers. Recently healthcare organizations have moved to adopt the HRO mindset as well. In each case, even a minor error could have catastrophic consequences. According to Weick and Sutcliffe (2015) the so-called HROs demonstrate particular characteristics in the way they operate: anticipating problems (being aware of what is happening in the work system; being alert to ways in which an incident could occur; looking beyond simplistic explanations for incidents); and containing problems (being prepared to deal with contingencies; using relevant expertise regardless of where it is situated within the organizational hierarchy).

In their research on "high-reliability organizations" Weick and Sutcliffe (2015) analyzed how highly regulated and standardized organizations such as nuclear plants, aircraft carriers, and firefighting units achieve resilience in a complex environment. They found that technical checklists were not the key to success but instead a list of cultural features they define as "mindful organizing." They identified five principles of mindful organizing that include preoccupation with failure, a reluctance to simplify, a sensitivity to operations, a commitment to resilience and a deference to expertise as a shared set of values that foster resilience through constant communication and recalibration in the face of unknowable risks (Weick and Sutcliffe 2015). Contrasting the mechanisms used to manage high reliability projects with the complexity inherent in megaprojects may illuminate solutions that otherwise may not be evident.

Source: Adapted from Weick and Sutcliffe (2015).

7. Community Consultation and Participation

Involvement of powerful Indigenous and interest groups and individuals is not only essential in complex projects with multiple options, but it is sometimes required particularly in highly regulated industries such as nuclear power and other energy projects that have environmental and health impacts and complex procedures and technology. Continuous scrutiny and oversight of the project to manage emergent events is critical in complex projects.

Experience has found that unforeseeable uncertainty can be diagnosed through discovery-driven planning or by systemically probing for knowledge gaps (Lenfle and Loch 2017; Loch et al. 2006; McGrath and McMillan 2000). Bottom-up approaches to decision making can also help in managing complex projects in volatile environments.

For example, The Kecamatan Development Project (KDP) in Indonesia highlighted in Table 4.3, focused on a model of participatory development projects designed on social rather than economic theory (Adler et al. 2009). As the world's fourth largest country, Indonesia was recovering from widespread corruption and social unrest in the mid-1990s, the Asian financial crisis, and the 2004 Tsunami. The goal was to deliver development resources to rural communities "using local representative community forums . . . wherein villagers, not government officials or external experts, determine the form and location of small-scale development projects via a competitive bidding process" (p. 11). The major mission of KDP was to provide "a more efficient and effective mechanism of getting valued development resources to a designated target group (in this case, the rural poor)" (p. 12). Using the local villagers most familiar with the environmental needs of the local population, and the political and economic landscape, was a strategy which made sense from economic and cultural perspectives (p. 12). A major study suggests that such processes have been effective in this regard with respect to enhancing the capacity of KDP participants, specifically participants in other development projects, to constructively manage everyday disputes (Barron et al. 2007). The key challenge is the changing of the mindset of both government and World Bank officials from a focus on the necessities of agency supply to one in which projects respond to community demand (Guggenheim 2006). Projects that have explicit and accessible procedures for managing disputes arising from the development process are less likely to cause conflict (Barron et al. 2007, pp. 21–22). Developing a framework for dispute resolution from the outset of the project is critical to community involvement and acceptance (Greiman 2011).

8. Managing Change, Conflicts, and Resistance

One of the key factors in complex megaprojects is recognizing that change is inevitable and that when and how that change will occur is often unpredictable. For instance, what was originally thought of as a transportation project may become an urban-development or a landscape-preservation project too, as was the case with the HSL South (Salat et al. 2013). Or an interstate highway project may develop into a transportation and environmental improvement project within an inner city as reflected in the Big Dig (Greiman and Sclar 2019).

A megaproject can take many twists and turns throughout its long life before settling on an option. Researchers have expressed concern that conflicting preferences in project selection may be avoided and screened off from the "relevant options" of decision making, rather than using conflicts to improve deliberation about alternatives (Salet et al. 2013). It is not easy to deal with conflicts; they may be irreconcilable, but they also can lead to better decisions in challenging and turbulent situations.

In the face of complexity, strategic decision making should focus on the organizational mission and alignment of the project with that mission with the goal of developing multiple options that can be explored by those who will benefit most from a rationale decision-making process.

Summary

In this chapter, we looked at complexity in megaprojects through its characteristics and impacts and learned that megaprojects require specialist skills to understand and manage complexity. We contrasted complexity with uncertainty, risk, ambiguity and conflict and recognized that they represent different disciplines that require different strategic approaches. The various methods and systems for managing complexity were also reviewed including system engineering, upfront engagement of stakeholders, strategic decision making, community consultation and participation and change management to provide insight into the dynamic nature of megaprojects. In our next chapter, we will explore governance to understand better how a megaproject organizes for complexity and delivers in an increasingly unpredictable and unstable environment.

Lessons and Best Practices

1) Understanding that managing complexity is an active and ongoing activity that requires exploration, openness, and continuous discovery.
2) Development of systems to manage structural, dynamic, technical, and social complexity.
3) Realizing the difference between complexity, uncertainty, ambiguity, risk, and chaos and developing a plan and strategy to manage each.
4) Being prepared to act when opportunity strikes based on good decision-making practice and awareness.
5) Exercising sound judgment when presented with different options through exploration of all reasonable alternatives and analyzing the pros and cons of each option.
6) Avoid making decisions too early or too late. If you decide too early, you may not have explored all your alternatives. If you decide too late, you may miss opportunities.
7) Avoid irreversible commitments until enough information is presented to make rationale commitments.
8) Plan for the reality that a megaproject is a journey that evolves and changes over time.
9) Realizing that complexity is managed at the strategic level of the organization, while risk is managed at the tactical level.
10) Infusing the project organization with the properties of cohesion and resilience requires an approach that goes beyond the risk management approach currently used in practice.

Discussion Questions

1 Why is it important that complexity, uncertainty, and risk be managed with different strategies?

2 What are the key considerations in the reduction of complexity?

3 Why is getting to the simple so difficult and why do Weick and Sutcliffe recognize that it is not always the best solution?

4 Why is preoccupation with failure an important attribute for High Reliability Organizations?

5 Provide an example of a project from real life that chose the wrong strategy culminating in project failure. Explain what went wrong in the project selection process.

6 Why is strategic alignment important in project selection?

7 Explain the statement by (Miller and Hobbs 2005), "The longer the development time, the higher is the likelihood that projects will be affected by turbulence."

8 Why is community involvement so important at the early stages of the project?

9 How can participatory governance be encouraged?

10 Why do scholars suggest that megaprojects should be anchored into its institutional environment?

References

Adler, D., Sage, C., and Woolcock, M. (2009). *Interim Institutions and the Development Process: Opening Spaces for Reform in Cambodia and Indonesia*, 4. Brooks World Poverty Institute, U. of Manchester.

Altshuler, A. and Luberoff, D. (2003). *Mega-Projects: The Changing Politics of Urban Public Investment*. Washington, DC and Cambridge, MA: Brookings Institution Press and Lincoln Institute of Land Policy.

Artto, K., Kujala, J., Kietrich, P., and Martinsuo, M. (2008). What is project strategy? *International Journal of Project Management* 26 (1): 4–12. https://doi.org/10.1016/j.ijproman.2007.07.006.

Babaei, A., Locatelli, G., and Sainati, T. (2021). What is wrong with the front-end of infrastructure megaprojects and how to fix it: A systematic literature review. *Project Leadership and Society* 2: 10032. https://doi.org/10.1016/j.plas.2021.100032.

Baccarini, D. (1996). The concept of project complexity—A review. *International Journal of Project Management* 14 (4): 201–204.

Barron, P., Diprose, R., and Woolcock, M. (2007). *Local Conflict and Development Projects in Indonesia: Part of the Problem or Part of a Solution?* World Bank Policy Research Working Paper 4212. Washington, DC: World Bank.

Brady, T. and Davies, A. (2014). Managing structural and dynamic complexity: A tale of two projects. *Project Management Journal* 45 (4): 21–38.

Brown, E.H. (1967). *Structural Analysis*, vol. I, 398. New York, NY: John Wiley & Sons.

Buck, M. (2017). *Crossrail Project: Finance, Funding and Value Capture for London's Elizabeth Line.* Institution of Civil Engineers.

Chapman, R.J. (2016). A framework for examining the dimensions and characteristics of complexity inherent within rail megaprojects. *International Journal of Project Management* 34 (6): 937–956. https://doi.org/10.1016/j.ijproman.2016.05.001.

Damayanti, R.W., Hartono, B., and Wijaya, A.R. (2021a). Complexity, leadership, and megaproject performance: A configuration analysis. *Journal of Industrial Engineering and Management* 14 (3): 570–603.

Damayanti, R.W., Hartono, B., and Wijaya, A.R. (2021b). Clarifying megaproject complexity in developing countries: A literature review and conceptual study. *International Journal of Engineering Business Management*, 13, January. https://doi.org/10.1177/18479790211027414.

Davies, A. and Mackenzie, I. (2014). Project complexity and systems integration: Constructing the London 2012 Olympics and Paralympics Games. *International Journal of Project Management* 32 (5): 773–790.

van Der Bijl, R., Van Oort, N., and Bukman, B. (2018). *Light Rail Transit Systems: 61 Lessons in Sustainable Urban Development.* Amsterdam, The Netherlands: Elsevier, B.V.

Dimitriou, H.T., Low, N., Sturup, S. et al. (2014). What constitutes a "successful" mega transport project? *Planning Theory and Practice* 15 (3): 389–430. https://doi.org/10.1080/14649357.2014.935084.

Flyvbjerg, B. (2014). What you should know about megaprojects and why: An overview. *Project Management Journal* 45 (2): 6–19.

Flyvbjerg, B., Bruzelius, N., and Rothengatter, W. (2003). *Megaprojects and Risk: An Anatomy of Ambition.* Cambridge, England: Cambridge University Press.

Geraldi, J. (2008). The balance between order and chaos in multi-project firms: A conceptual model. *International Journal of Project Management* 26: 348–356.

Gertler, J. (2020). *F-35 Joint Strike Fighter (JSF) Program.* Washington, DC: Congressional Research Service (CRS).

Gidado, K. (1993). *Numerical Index of Complexity in Building Construction to Its Effect on Production Time.* UK: University of Brighton.

Giezen, M., Salet, W., and Bertolini, L. (2015). Adding value to the decision-making process of megaprojects: Fostering strategic ambiguity, redundancy, and resilience. *Transport Policy*, Elsevier 44 (C): 169–178.

Greiman, V.A. (2011). The public/private conundrum in international investment disputes: Advancing investor community partnerships. 32 *Whittier L. Rev.* 395. https://heinonline.org/HOL/LandingPage?handle=hein.journals/ whitlr32&div=17&id=&page=.

Greiman, V.A. (2013). *Megaproject Management: Lessons on Risk and Project Management from the Big Dig.* Hoboken, NJ: Wiley.

Greiman, V.A. and Sclar, E.D. (2019). Mega infrastructure as a dynamic ecosystem: Lessons from America's interstate system and Boston's big dig. *Journal of Mega Infrastructure and Sustainable Development* 1 (2): 188–200. https://doi.org/10.108 0/24724718.2020.1742624.

Guggenheim, S. (2006). Crisis and Contradictions: Understanding the Origins of a Community Development Project in Indonesia. In: *The Search for Empowerment: Social Capital as Idea and Practice at the World Bank* (ed. A. Bebbington, M. Woolcock, S. Guggenheim and E. Olson), 127. Boulder, CO: Kumarian Press.

Hughes, T.P. (1998). *Rescuing Prometheus.* New York: Pantheon Books.

Jenkins, G.M. (1996). *Systems Engineering: A Unifying Approach In Industry And Society.* London, UK: C.A. Watts.

Kardes, I., Ozturk, A., Cavusgil, S.T., and Cauvusgil, E. (2013). Managing global megaprojects: Complexity and risk management. *International Business Review* 22: 905–917.

Kasser, J. and Zhao, Y. (2014). Managing complexity: The nine-system model. *INCOSE International Symposium* 24: https://doi.org/10.1002/j.2334-5837.2014. tb03174.x.

Lee, C., Won, J.W., Jang, W. et al. (2017). Social conflict management framework for project viability: Case studies from Korean megaprojects. *International Journal of Project Management* 35 (8): 1683–1696.

Lenfle, S. (2011). The strategy of parallel approaches in projects with unforeseeable uncertainty: The Manhattan case in retrospect. *International Journal of Project Management* 29 (4): 359–373. https://doi.org/10.1016/j.ijproman.2011.02.001.

Lenfle, S. and Loch, C. (2017). Has Megaproject Management Lost Its Way? Lessons from History. In: *The Oxford Handbook of Megaproject Management* (ed. B. Flyvbjerg), 21–38. Oxford, England: Oxford University Press.

Levitt, R.E. and Scott, W.R. (2017). Institutional Challenges and Solutions for Global Megaprojects. In: *The Oxford Handbook of Megaproject Management* (ed. B. Flyvbjerg), 96–117. Oxford, England: Oxford University Press.

Locatelli, G., Mancini, M., and Romano, E. (2014). Systems engineering to improve the governance in complex project environments. *International Journal of Project Management* 32 (8): 1395–1410. https://doi.org/10.1016/j.ijproman.2013.10.007.

Loch, C.H., DeMeyer, A., and Pich, M.T. (2006). *Managing the Unknown: A New Approach to Managing High Uncertainty and Risk in Projects.* Hoboken, NJ: Wiley.

Loch, C.H., Solt, M.E., and Bailey, E.M. (2008). Diagnosing unforeseeable uncertainty in a new Venture. *Journal of Product Innovation Management* 25 (1): 28–46.

Lorell, M.A., Kennedy, M., Leonard, R.S. et al. (2013). *Do Joint Fighter Programs Save Money?* Santa Monica, CA: RAND Project Air Force http://www.rand.org/pubs/monographs/MG1225.html.

Mai, Q., An, S., Lin, H., and Gao, X.L. (2018). Complexity and adaptive organization of mega project: The case of Hong Kong-Zhuhai-Macau bridge. *Journal of Management Science* 31 (3): 86–99. https://doi.org/10.3969/j.issn.672-0334.2018.03.008.

Maylor, H., Vidgen, R., and Carver, S. (2008). Managerial complexity in project-based operations: A grounded model and its implications for practice. *Project Management Journal* 39 (1): S15–S26.

Maylor, H.R., Turner, N.W., and Murray-Webster, R. (2013). How hard can it be? Actively managing complexity in technology projects. *Research-Technology Management* 56: 45–51.

McGrath, R.G. and MacMillan, I.C. (2000). *The Entrepreneurial Mindset: Strategies for Continuously Creating Opportunity in an Age of Uncertainty.* Cambridge, MA: The Fellows of Harvard College.

Miller, R. and Hobbs, B. (2005). Governance regimes for large complex projects. *Project Management Journal* 36 (3): 42–50. https://doi.org/10.1177/875697280503600305.

Miller, R., and Lessard, D. (2001) Understanding and managing risks in large engineering projects. *International Journal of Project Management*, 19 (8) 437–443. https://doi.org/10.1016/S0263-7863(01)00045-X.

Miller, R., and Lessard, D.R. (2007) *Evolving Strategy: Risk Management and the Shaping of Large Engineering Projects.* MIT Sloan Research Paper No. 4639-07. https://ssrn.com/abstract=962460 or http://dx.doi.org/10.2139/ssrn.962460.

Morris, P.W.G. (2002). Science, objective knowledge and the theory of project management. *Civil Engineering* 150 (2): 82–90.

National Audit Office (NAO) (2013). *Planning for Economic Infrastructure*, Report by the Comptroller and Auditor General HC, vol. 595. UK: HM Government.

Pitsis, T.S., Sankaran, S., Gudergan, S., and Clegg, S.R. (2014). Governing projects under complexity: Theory and practice in project management. *International Journal of Project Management* 32 (8): 1285–1290. https://doi.org/10.1016/j.ijproman.2014.09.001.

Pitsis, A., Clegg, S., Freeder, D. et al. (2018). Megaprojects redefined – Complexity vs cost and social imperatives. *International Journal of Managing Projects in Business* 11 (1): 7–34. https://doi.org/10.1108/IJMPB-07-2017-0080.

Priemus, H., Flyvbjerg, B., and van Wee, B. (ed.) (2008). *Decision-Making on Mega-Projects: Cost-Benefit Analysis, Planning and Innovation.* Cheltenham, England: Edward Elgar Publishing.

Ramasesh, R.V. and Browning, T.R. (2014). A conceptual framework for tackling knowable unknown unknowns in project management. *Journal of Operations Management* 32 (4): 190–204.

Rebentisch, E.S., Nelson, R.M., Townsend, S.A. et al. (ed.) (2017). *Integrating Program Management and Systems Engineering: Methods, Tools, and Organizational Systems for Improving Performance*. Hoboken, NJ: Wiley.

Salet, W. (2021). Public norms in practices of transitional planning: The case of energy transition in the Netherlands. *Sustainability* 13: 4454. https://doi.org/10.3390/su13084454.

Salet, W., Bertolini, L., and Giezen, M. (2013). Complexity and uncertainty: Problem or asset in decision making of mega infrastructure projects? *International Journal of Urban and Regional Research* 37 (6): 1984–2000. https://doi.org/10.1111/j.1468-2427.2012.01133.x.

Samset, K. and Volden, G.H. (2016). Front-end definition of projects: Ten paradoxes and some reflections regarding project management and project governance. *International Journal of Project Management* 34 (2): 297–313. https://doi.org/10.1016/j.ijproman.2015.01.014.

San Cristóbal, J.R., Diaz, E., Carral, L., Fraguela, J.A., and Iglesias, G. (2019) Complexity and project management: Challenges, opportunities, and future research. *Complexity 2019*. Wiley and Hindawi online. https://doi.org/10.1155/2019/6979721.

Sapolsky, H. (2003). Inventing Systems integration. In: *The Business of Systems Integration* (ed. A. Prencipe, A. Davies and M. Hobday). Oxford: Oxford University Press.

Sheard, S.A. and Mostashari, A. (2010). A complexity typology for systems engineering. *INCOSE International Symposium* 20 (1): https://doi.org/10.1002/j.2334-5837.2010.tb01115.x.

Shenhar, A.J. (2001). One size does not fit all projects: exploring classical contingency domains. *Management Science* 47 (3): 394–414.

United States Congress (2021). *H.R.3684 - Infrastructure Investment and Jobs Act.117th Congress (2021–2022)*. Washington, DC: House of Representatives.

Van Marrewijk, A., Clegg, S.R., Pitsis, T., and Veenswijk, M. (2008). Managing public – Private megaprojects: Paradoxes, complexity and project design. *International Journal of Project Management* 26 (6): 591–600.

Weick and Sutcliffe (2015). *Complexity Theory: Managing the Unexpected*. Hoboken, NJ: John Wiley & Sons.

Williams, T. (2002). *Modelling Complex Projects*. Hoboken, NJ: John Wiley & Sons.

Winch, G.M. (2014). Three domains of project organizing. *International Journal of Project Management* 32 (5): 721–731.

Zhu, Y., Shi, Q., Li, Q., and Yin, Z. (2018). Decision-making governance for the Hong Kong-Zhuhai-Macao Bridge in China. *Frontiers of Engineering Management* 5 (1): 30–39. https://doi.org/10.15302/J-FEM-2018087.

5

Global Megaproject Governance

Everyone with all those good intentions came to help Indonesia rebuild from the tsunami, but the coordination problem was very big, because they came with their own way of doing business; they came with the inflexibility of their own governance.

— Sir Mulyani Indrawati

Introduction to Global Governance of Megaprojects

As expressed by Sir Mulyani Indrawati, governance can best be designed by the people who will benefit most from the project's development. However, often in the building of the great projects of the world, we are not prepared to govern due to the urgency and uncertainty that surrounds these massive endeavors. For policymakers and scholars, global governance is one of the defining characteristics of the current international moment (Barnett and Duvall 2005).

Megaprojects represent a diversity of views and visions that result from the many organizations and stakeholders that come together to find solutions for some of the world's most pressing challenges. A review of the definition of governance in megaprojects reveals that there is no universal agreement neither on how a particular megaproject should be governed, nor on the meaning of governance in these large-scale operations with multiple owners, sponsors, developers, and other stakeholders. Governance can mean many things in the project management context. To understand better how megaproject governance can be structured and developed for success, in this chapter we explore the following questions:

- What is the meaning and purpose of governance?
- What are the multiple governance structures being used in global megaprojects?
- Which structures have been most effective?

Global Megaprojects: Lessons, Case Studies, and Expert Advice on International Megaproject Management, First Edition. Virginia A. Greiman.
© 2023 John Wiley & Sons, Inc. Published 2023 by John Wiley & Sons, Inc.

- What are the essential decision-making processes and methods?
- Who should decide and how do you hold the decision makers accountable?
- Why is the active participation of all stakeholders vital to project success?
- How do you maintain a transparent and ethical environment in a megaproject?

Global Megaproject Governance

> . . . We will fight for the ideals and Sacred Things of the City both alone and with many . . . We will strive increasingly to quicken the public sense of public duty; that thus . . . we will transmit this city not only not less, but greater, better, and more beautiful than it was transmitted to us.
>
> — *Oath of office sworn by young men of classical Athens upon induction into the military academy*

The term governance originated from the Greek word kyberman, meaning "to steer or guide." From its Greek origins, it moved to Latin, where it was known as gubernare and then migrated to France as governer. There are many definitions of governance in the corporate governance literature, and one of the most frequently used is from the Organization for Economic Cooperation and Development (OECD). Corporate governance, as defined by the G20/OECD (2015), involves: a set of relationships between a company's management, its board, its shareholders, and other stakeholders. Corporate governance also provides the structure through which the objectives of the company are set, and the means of attaining those objectives and monitoring performance are determined. The purpose of corporate governance is to help build an environment of trust, transparency, and accountability necessary for fostering long-term investment, financial stability and business integrity, thereby supporting stronger growth, and more inclusive societies.

As distinct from corporate governance, which involves the relationship of the shareholders, the board, and the CEO, project governance structures can look very different, and no two structures are ever alike. Project governance, as it applies to portfolios, programs, projects, and project management, "coexists within the corporate governance framework" (Müller 2009; Müller et al. 2015). Table 5.1 provides a high-level comparison of corporate v. project governance.

The Association of Project Management (APM) (2019) provides a broad definition of *program governance* as:

> The framework of authority and accountability that defines and controls the outputs, outcomes and benefits from projects, programs and portfolios. The mechanism whereby the investing organisation exerts financial and technical control over the deployment of the work and the realization of value. (APM 2019)

Table 5.1 Corporate governance v. project governance.

Corporate governance	Project governance
Centralized and top-down mandates	Horizontal or Bottom-up Compacts
Detailed upfront plan	Continuous planning based on learning
Requirements agreed to in advance	Requirements iteratively prioritized
Centralized control throughout the life cycle	Decentralized decision-making at each phase
Board of Directors as oversight authority	Board of Directors, Commissions, Task Force, Advisory Board as oversight
Board members are usually inside directors or outside directors	Board members are usually gray (non-executive) directors coming from the sponsors of the project
Success is measured by profits to shareholders	Success is measured by benefits realized by stakeholders
Integrity is demonstrated through Corporate Social Responsibility (CSR)	Integrity is demonstrated through Megaproject Social Responsibility (MSR)
Value is measured by the business case	Value is measured by contributions to the wider eco system

The 2017 World Development Report: *Governance and the Law* defines better governance as "the process through which state and non-state groups interact to design and implement policies, working within a set of formal and informal rules that are shaped by power" (WDR 2017).

The Report defines three essential elements of effective policy: *commitment, coordination,* and *cooperation.* As three core functions to produce better governance outcomes, institutions need to:

- *Bolster commitment:* to policies in the face of changing circumstances.
- *Enhance coordination:* to change expectations and elicit socially desirable actions by all.
- *Encourage cooperation*: effective policies help promote cooperation by limiting opportunistic behavior such as tax evasion-often through credible mechanisms of rewards or penalties.

Megaproject Governance from the Literature

Table 5.2 highlights some of the recent literature on project governance as well as the literature that has withstood the test of time and is still valid despite the many changes that have occurred in megaproject governance. Change has resulted from the expansion of megaprojects globally and the need to ensure that all projects are

Table 5.2 Literature review on the characteristics of governance.

Characteristic	Source
"We still know little about how governance arrangements actually work in practice, and which approaches to assurance are most appropriate in which contexts."	Winch and Leiringer (2016, p. 7)
Flexibility is a key characteristic of successful governance at various levels with different forms.	Müller et al. (2015)
The literature considers that "project governance" is one of the key aspects in the delivery of megaprojects Joslin and Müller (2016). This shows that project success is significantly correlated with project governance.	Müller (2009), Joslin and Müller (2016), and Locatelli et al. (2017)
. . . "the confluence point where the competing interests of the temporary project organisation and the more permanent parent organizations must be resolved."	McGrath and Whitty (2015, p. 755)
. . . it represents an overarching business function that acts across different stakeholder levels. Of particular interest is the interaction between narratives of governance as mobilized by practicing managers in owner organizations and the projects in which they invest.	Biesenthal and Wilden (2014)
Project governance is a concept that helps to align different objectives and is thus an important factor to delivering megaprojects successfully.	Biesenthal (2016)
Those who view project governance as internal to a specific project believe the governance structure of a project performs a role that is equivalent to the role of the top management team in firms – a role of oversight and coordination.	Zhai et al. (2017)
Support of the public and the commitment of individuals working for the project was strengthened by actively and broadly promoting the societal importance of the focal project.	Zhai et al. (2017)
Governance is considered the art of addressing complexity, the goal of which is to resolve complexity in order to ensure the ordering and collective action of organizations with low cost and high efficiency.	Brahm and Tarziján (2016)
The allocation of control rights in projects has been overlooked for a long time but has recently attracted wide attention and resulted in notable achievements in public-private partnership (PPP) projects.	Ma et al. (2020)
If researchers in project governance give a greater emphasis on the relationship between project governance and external stakeholders, it could help project investors and project teams achieve better support from the external stakeholders and improve the short term and long-term prosperity of the project and the organization.	Derakhshan et al. (2019)

Table 5.2 (Continued)

Characteristic	Source
Complexity is a distinguishing feature of water megaprojects. It exists in the top-level governance structure and is further aggravated in the relationships between government and project owner. As its central goal, the methods of project governance can allow practitioners to resolve the myriad complexities of such complex and costly undertakings (Ma et al. 2020).	Ma et al. (2020)
Previous research on project governance mainly focuses on control and trust as the main mechanism of governance; in fact, the relationship between project subjects is a breakthrough.	Xie et al. (2019)
The significance of good governance to alleviate pressure points in the project is widely recognized. This area is a foundational aspect of megaprojects and is a cohesive element of many of the disparate aspects that make up the megaproject's successes or failures.	Sanderson (2012)
Apart from the fiscal and financial areas of impact, once special purpose entities are in place, they play a central role in the governance of PPMs.	Medda et al. (2013)

operating in a sustainable and responsible way so that the goals of the project are met including global climate change mandates, environmental improvements and anti-corruption, and inclusive citizen-centered governance. While a few researchers contend that project governance is nested within project management as a subfield (Ahola et al. 2014), the majority of scholars hold that project management is a component of project governance at the project level (Turner 2020). To grasp the multi-level and multidimensional concept of governance, some research focuses on the connections between project governance and institutional settings (Müller et al. 2015). In recent decades, the theme of project governance has been gaining influence from multiple disciplines covering organizational studies (Clegg et al. 2021), public administration (Warsen et al. 2018), urban development (Valverde and Moore 2019), and environmental management (Sparrevik et al. 2018).

Purpose of Governance

> The speed of decision making is the essence of good governance.
> — *Piyush Goyal, Minister of Railways and Coal, Government of India*

Since 2001, several high-level corporate collapses have brought to light shareholder demands that organizations strengthen their governance systems and pay closer attention to concerns of the stakeholders. In response to Enron, Tyco, Wirecard, WorldCom, and Europe's Enron (Parmalat), there have been more rigid oversight and regulatory requirements and a broadening corporate governance agenda to include stakeholders' concerns as well as shareholder accountability. Governance structures have been used on projects for the following primary purposes:

- To ensure the organization receives a maximum return on investment.
- To direct and control its operations and strategic activities.
- To respond to the legitimate rights, expectations, and desires of its shareholders and stakeholders.
- To formalize organizational learning.
- To empower people to make decisions or to facilitate decisions.
- To monitor the delivery of benefits through progress reports and audits and reviews at various phases in the project's life cycle.
- To evaluate performance before permitting the project to progress.
- To assure delivery of value to local communities and society in general.

Megaprojects as Temporary Institutional Structures

Megaprojects face choices over what type of governance structure will best serve the goals and demands of the project. Megaprojects are usually governed by temporary institutional governance structures that borrow resources and technical capacity from their parent organizations through a secondment process. In the case of a public project, the government agency with authority for infrastructure development typically appoints members to a governing board, while in private infrastructure projects, the governing board is often represented by the project's sponsors. In private infrastructure projects, consortiums are formed such as that of Bechtel, Arup, Systra, and Halcrow, established to run the $10.4B Channel Tunnel Rail Link project linking the United Kingdom to Europe's high-speed rail network (Davies et al. 2009).

A consortium is essentially a contract, between the government owner and the private-sector management consultant, that allocates risk, accountability, and decision making among the participants. In Crossrail, Europe's largest megaproject, a key part of the governance structure was the signing of a Project Sponsor Agreement (between the two main funders), the Department for Transport (DfT) and Transport for London (TfL) followed by a Project Development Agreement (between the Sponsoring Board and the Delivery Body) and a Stakeholder Agreement. This triangulation was built on the previous governance model used by the Channel Tunnel Rail Link project (HS1). These agreements brought the key parties together for the building of the project.

Boston's Big Dig program was unique in that the management of construction was carried out by a temporary joint venture, Bechtel/Parsons Brinckerhoff

(B/PB), represented by the project's designer, Parsons Brinckerhoff, and the project's contractor, Bechtel Corporation, while the major supporting functions reported directly to the public owner. The joint venture was overseen by a board of control made up of representatives of each joint-venture partner, while the project was overseen by a government appointed Board of Directors (Greiman 2013).

Governance as a Dynamic Regime

In megaprojects, the governance structure is complex and dynamic. Project governance is conceptualized in the literature as an oversight and control function. The structures are stable, but the activities being overseen are dynamic and changing. Governance scholars contend that governance regimes themselves must be dynamic – that they can change to adapt to the emerging context (Miller and Hobbs 2005, p. 48). Significantly, public infrastructure projects do not always meet the expectations of its stakeholders. Most are delivered too late and above budget, and do not meet agreed quality standards. Is this due to poor or inadequate governance structures? The subject of governance is rarely discussed in the public discourse, yet it is essential to understand how governance may contribute to the success or failure of projects and how more effective governance frameworks can better meet stakeholder expectations and an improved return on investment to society.

An important part of understanding governance is recognizing that projects must continually change to meet the demands of its stakeholders and the needs of its customers. For instance, an organization may start out with a strong centralized governance framework but may become more decentralized as the organization evolves, requiring decision making to be delegated to a lower level of the organization, where the technical knowledge and expertise can be applied first-hand. In accordance with PMI taxonomy, megaprojects are in reality "programs" consisting of hundreds of related projects and other work that require alignment with the strategic goals of the parent organization or the project sponsors (Greiman 2013).

The Megaproject Conundrum: Too Much or Not Enough Governance

The debate over how much governance is enough continues to dominate the literature even if not directly. For example, flexibility is a key characteristic of successful governance at various levels with different forms (Müller et al. 2015). The push for more flexibility in governance is evident in the success of the Agile methodology for project management which incorporates adaptive development approaches and has recently been incorporated into the new Project Management Institutes Body of Knowledge (PMI 2021). Adaptive approaches are useful when requirements are subject to a high level of uncertainty and volatility and are likely to change throughout the project which is often the case in megaprojects when multiple stakeholders can influence the scope and approach to the project.

Establishing project governance is not a simple task. Significant investment on this task needs to be made when embarking on a new project. It is often challenging to quantify what the benefits are when it comes to investing in the creation of the project governance framework. While the traditional governance tends to put in place a rather formal governance structure, agile governance promotes in principle a more self-steering control by the project team members.

Developing a Megaproject Governance Framework

Since the foundation of every organization is its governance structure the question of how to best structure a megaproject with multiple employers, numerous stakeholders, thousands of designers, engineers and contractors, multiple government agencies and oversight authorities and millions of citizens impacted by the project development is a major challenge facing the project sponsors.

Miller and Lessard (2007) emphasize the significance of project shaping in their study of large engineering projects. Projects they argue are shaped not selected. They have highlighted the need for design criteria when developing a governance regime for a megaproject that would permit transformation of the governance structure as the project unfolds over a long period of time. They argue that since the governance structure will undoubtedly change, there is a need for flexibility in project structures rather than a single project governance structure.

The conventional approaches to project governance need to be critically reviewed in light of recent development and ongoing scientific debates about sustainability, social protection, climate change and innovation. Governance structures must transition from the traditional top-down approach to project governance toward a more agile, adaptive, and sustainable approach to governance at the strategic, operational, and tactical levels of the project organization.

Figure 5.1 highlights the five steps involved in megaproject development beginning with the project conception, to the selection process, to the feasibility analysis to stakeholder engagement and finally the government approval process. If any of these steps are not appropriately implemented, then the entire governance framework will fail. Once the megaproject is approved the implementation of the megaproject governance begins. In the next section of the chapter, we will describe the five implementation steps for establishing a governance framework as described in Figure 5.1: (i) Deciding on the governance structures; (ii) Developing the megaproject social responsibility (MSR)

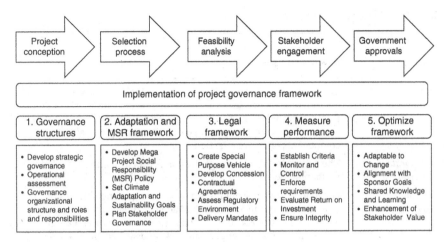

Figure 5.1 Governance framework development.

and adaptation governance framework; (iii) Implementing the legal framework; (iv) Measuring the performance of the governance framework; and (v) optimizing the framework.

Step 1: Selecting the Governance Structures

The first step in developing a governance framework for any organization is deciding the structures that will best enable the organization to implement its strategy and accomplish its goals and objectives. There is no one model that fits the needs of every organization, and multinational enterprises tend to rely on various governance structures to meet the needs of its complex organization. For organizations with a broad portfolio of projects and programs, different governance structures may be used at different times in the life cycle. The need for alternative governance structures arises from the difficulties of hierarchical coordination and the competing interests and values that must be considered. These structures will need to be managed both individually and as part of a larger integration. In Chapter 6, we will explore in more detail the integration of megaproject organizational structures and the integration of systems that support these structures.

Table 5.3 provides an overview of the various governance structures designed to comply with the changing expectations of the international community and social and environmental norms. These governance frameworks are not intended to be used as a singular framework and can be combined or adapted to a particular megaproject. For example, local governance frameworks may exist at the strategic level of the organization, while a program organization may exist at the operational level of the organization, and projects at the tactical level of the project organization.

Table 5.3 Governance structures.

Governance frameworks	Governance goals
Enterprise governance	Enterprise governance is the structure and relationships that control, direct, or regulate the performance of an enterprise and its projects, portfolios, infrastructure, and processes (Wilson 2009). In essence, it sets the strategy for governance throughout all components of an organization.
Corporate governance	The system by which an organization is overseen and controlled by its shareholders.
Portfolio governance	Portfolio governance guides portfolio management activities in order to optimize investments and achieve organizational strategic goals (PMI 2016). Portfolio-level decisions can result in programs being added and accelerated, cut back, and slowed, deferred, or even canceled.
Program governance	The structure by which related projects are integrated, coordinated, and managed among all stakeholders in alignment with the strategic goals of the sponsoring organization(s). Program governance determines whether benefits achievement is occurring within the stated parameters so component changes can be proposed and executed when necessary (PMI 2016).
Project governance	The framework, functions, and processes that guide project management activities in order to create a unique product, service, or result and meet organizational strategic and operational goals (PMI 2016).
Partnership governance	The process of multisectoral engagement often structured as public–private partnerships involving collaboration between a government agency and a private-sector company.
Joint Venture (JV)	The cooperation of two or more entities for a finite period in which each agrees to share profit, loss, and control in a specific enterprise. The parties to the JV can be private or public entities. These arrangements are generally governed by contract.
Participatory governance	A structure in which stakeholders or external groups or committees are involved in the project's decision-making or oversight
Local governance	Local governance is a broad concept and is defined as the formulation and execution of collective action at the local level. It is a form of participatory governance. It encompasses the direct and indirect roles of formal institutions of local government and government hierarchies, as well as the roles of informal norms, networks, community organizations, and neighborhood associations in pursuing collective action by defining the framework for citizen–citizen and citizen–state interactions, collective decision making, and delivery of local public services (Shah and Shah 2006).
Architecture governance	The principles, standards, guidelines, contractual obligations, and regulatory framework within which goals are met at an enterprise wide-level.

Defining the Roles and Responsibilities of the Project's Governance Structures

> No man is good enough to govern another man without that other's consent.
> — *Abraham Lincoln*

Strategy Development

Figure 5.2 shows the traditional hierarchy of conventional portfolio governance. It begins with a strategy that is developed by the project sponsors. This high-level strategy sets the vision and goals for the project and helps attract investors and government or private sector sponsors by convincing the potential sponsors of the feasibility and merits of the project but most importantly the potential return on investment. If the projected return is not high enough it will be difficult to attract sponsors until they are convinced the project can meet its goals.

Enterprise Portfolio Management Office

It is important to note that often a vision is tested before going to the actual stage of portfolio management. Portfolio management is the next level on the ladder where the concept is formalized and the support for the project is developed. This will include a cost benefit analysis, a feasibility study, an environmental impact statement and often additional documents to prove the value of the project. The project selection process at the portfolio level may take months or on complex megaprojects often years to convince the broad stakeholder groups that it is the right project.

Program Management Office

Once the right project has been selected with the right justification the difficult task of developing a program begins. At this stage, you are involved with the resolution of political issues such as permitting rights, land access, and resettlement of the people who may be in the path of the project. Programs can be organized

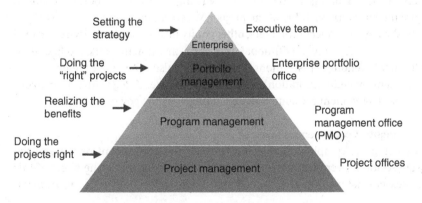

Figure 5.2 Project organization governance frameworks.

structurally through a Program Management Office (PMO) which provides support and oversight for the projects under its jurisdiction. The role of the PMO depends on the responsibility it assumes. This responsibility may include development of standards, contracts, processes, policies and other guidance for all projects within the program so that the projects are managed in a uniform manner. The program may be overseen by a program board, a task force or an independent, interdisciplinary oversight board and the program manager typically reports up to the sponsors CEO or chairman of the board.

Project Offices

On a typical megaproject you may have hundreds of contractors each managing a separate contract under the control of a management consultant, construction manager at risk or an executive delivery team. It is through these project teams that the project is delivered, and the goals and objectives of the projects are achieved. These teams generally have oversight from the operational or strategic level of the organization either through a program manager or a director of operations that reports directly to the executive team or the board of directors of the sponsoring organization.

The Major Actors in the Project Organization

The Project Company and the Project Promoters   Privately financed infrastructure projects are usually carried out by a joint venture of companies including construction and engineering companies and suppliers of heavy equipment interested in becoming the main contractors or suppliers of the project. The companies that participate in such a joint venture are referred to as the "promoters" of the project. Those companies will be intensively involved in the development of the project during its initial phase and their ability to cooperate with each other and to engage other reliable partners will be essential for timely and successful completion of the work. Furthermore, the participation of a company with experience in operating the type of facility being built is an important factor to ensure the long-term viability of the project. Where an independent legal entity is established by the project promoters, other equity investors not otherwise engaged in the project (usually institutional investors, investment banks, bilateral or multilateral lending institutions, sometimes also the government or a government-owned corporation) may also participate. A legislative framework and project agreement are sometimes encouraged by the government.

Control Rights: Government as Investor

Recognizing the complex relationship between the government owner and the private sector delivery team is a challenge in all megaprojects. The effect of this complex coexistence on the control rights allocation is even more unusual

(Ma et al. 2020). The government as project investor has strong control over the procurement, implementation, construction, management, and operation of projects, but as the actual operator and manager, the project owner does not in practice hold substantial control rights. The project owner believes it should hold more control rights to manage projects in order to guarantee competitiveness in the marketplace, while the government considers itself to have absolute control over the whole project to prevent moral hazard behaviors by the project owner. Therefore, there are divergences between the two top-level participants, creating problems falling within the scope of project governance theory.

The Lenders

Owing to the magnitude of the investment required for a privately financed infrastructure project, loans are often organized in the form of "syndicated" loans with one or more banks taking the lead role in negotiating the finance documents on behalf of the other participating financial institutions, mainly commercial banks. Commercial banks that specialize in lending for certain industries are typically not ready to assume risks with which they are not familiar. For example, long-term lenders may not be interested in providing short-term loans to finance infrastructure construction. Therefore, in large-scale projects, different lenders are often involved at different phases of the project. For example, in international projects, a multilateral bank may be the organization involved in mobilizing capital as discussed in Chapter 2.

Step 2: Adaptation Sustainable Governance and Megaproject Social Responsibility (MSR) Framework

As described by Ma et al. (2021) megaprojects, especially mega-infrastructure projects, play very important strategic roles in economic and social development, and their social responsibility and sustainability have recently attracted widespread attention (Flyvbjerg 2014). Zeng et al. (2015) describe MSR as "the policies and practices of stakeholders through the whole project lifecycle that reflect responsibilities for the well-being of the wide society." To address these problems, a series of important social and environmental concerns in megaprojects has been proposed, such as anti-corruption, ecological protection, disaster mitigation, immigrant settlement, occupational health and safety, pollution control, and poverty eradication (Zeng et al. 2017).

While corporate governance is the system of financial control through the board of directors, the term governance has also been used relationally. Relational governance aims at influencing networks to create innovation, reciprocity, trust, and self-organization for organizations that require collective action, such as megaprojects (Gil 2017). Project governance that includes multiple firms and other agencies is more complex than the corporate governance of a single firm or organization. First, there is no necessary alignment between the many corporate governance doctrines that might be involved. Second, the project duration might require an overall code of governance separate from those of the firms involved. Third, there may well be stakeholders to govern who are not themselves directly involved as project partners in the governance of the project.

MSR Governance comes in many forms as shown in Table 5.4 and is a constantly evolving paradigm that can change as the project progresses particularly because the project's stakeholders will continue to evolve both internally and externally to the project creating new opportunities for the project to grow and develop through self-organizing and innovation.

Adaptation Governance

As recognized by the United Nations Conference on Trade and Development (2021), with the growing intensity of major extreme events, adaptation is considered a priority for megaprojects globally, but particularly in the Global South where the impact may be the greatest. Institutional reforms that are required must build toward moving away from the principles of a regulatory, market-enabling state, and toward a developmental green state which would be in control of its own long-term priorities in climate adaptation and economic trajectories (p. 156).

Table 5.4 Emerging models of sustainable governance.

Type	Description
Adaptation governance	Sensitivity to climate change is a challenge to humanity that requires an integrated, anticipatory, and strategic approach to governance (UNCTAD 2021)
Relational governance	Relational governance aims at influencing networks to create innovation, reciprocity, trust, and self-organization for organizations that require collective action, such as megaprojects (Gil 2017; Sanderson 2012, p. 441)
Governance by commitment	When governance is created by the stakeholders themselves to live up to an established mission and value system (Clegg et al. 2021)
Governance by participation	Participatory governance has been defined as a process through which citizens and state officials interact to express their interests, exercise their rights and obligations, work out their differences, and collaborate to undertake development activities and offer services (Basurto et al. 2013)
Polycentric governance	Polycentric Governance is an intuitive approach to structure large arenas of consensus-oriented collective action (Gil et al. 2017) including ecological, social and environmental endeavors

Trade has an important role to play in shaping sustainable development paths. Facilitating climate adaptation in developing countries through trade agreements will require green technology transfers without restrictive patents, appropriate SDT in environmental goods and services so that providers of these goods and services in the developing world can have a level playing field (p. 156).

The lack of a truly global response to the pandemic "makes more urgent the need for a common action of climate" (Wolf 2021). Some seventy-five years ago, the Marshall Plan helped deliver shared prosperity among the war-torn economies. Today, climate change is a challenge to humanity that requires a similar integrated, anticipatory, and strategic approach. A global, green-oriented structural fund would support realignment of developing countries and deliver funding for both adaptation and mitigation initiatives as an urgent priority (p. 156).

Relational Governance
In recent years, megaprojects have recognized that a one size fits all governance structure does not work in in large-scale projects because the context of projects is so different across geography, culture, and industry.

While corporate governance is the system of financial control through the board of directors, the term governance has also been used relationally. Relational governance aims at influencing networks to create innovation, reciprocity, trust, and self-organization for organizations that require collective action, such as megaprojects (Gil 2017). Project governance that includes multiple firms and other agencies is more complex than the corporate governance of a single firm or organization. First, there is no necessary alignment between the many corporate governance doctrines that might be involved. Second, the project duration might require an overall code of governance separate from those of the firms involved. Third, there may well be stakeholders to govern who are not themselves directly involved as project partners in the governance of the project.

The significance of good governance to alleviate pressure points in projects is widely recognized. This area is a foundational aspect of megaprojects and is a cohesive element of many of the disparate aspects that make up the megaproject's successes or failures. "We must give proper attention to the ways in which project governing happens in a situated, relational sense, rather than just focusing solely on governance as a set of pre-designed structures" (Sanderson 2012, p. 441).

Governance Through Commitment
Governance by Commitment is best illustrated by the United Nations 2030 Agenda for Sustainable Development. Though not mandated by law the sustainable development goals have been adopted by large inter-governmental organizations including the World Bank, the European Bank of Reconstruction and Development, the Inter-American Development Bank among many other

financial institutions. In turn, these commitments have been memorialized in agreements with national and local governments as well as public private arrangements. Principles have served international development for many years as a governance tool and are evident also in John Ruggie's Guiding Principles on Business and Human Rights adopted by the United Nations (UN 2011). The three pillars of the Principles are to protect, respect, and remedy described as follows:

- Make a public commitment to respect human rights.
- Identify, prevent, mitigate, and account for, damage or damage caused to human rights.
- Dispose of procedures for remedying the negative consequences on human rights they cause or contribute to causing.

Source: The United Nations Guiding Principles on Business and Human Rights (2011).

Governance by Participation

In recent years, "participatory governance" has emerged as an important concept within the governance domain. It is a policy that encourages participation of local citizens to implement locally based, pro people development initiatives. International aid agencies have been pursuing this agenda with an aim to provide greater legitimacy to development projects for economic growth in developing countries (Waheduzzaman and As-Saber 2015). Participatory governance is also referred to as citizens governance because of the large number of local citizens who are involved in the governance either represented by interest groups or local stakeholder organizations.

Polycentric Governance

Polycentric governance is a crucial concept in the work of Elinor and Vincent Ostrom (Ostrom 2012). Ostrom's work was based on the principle that shared resources are well managed by those communities that benefit the most from them and that regulation should be addressed at the local level, through the farmers, communities, local authorities, and NGOs. Polycentrism is a form of collective action that can be used to implement major programs including environmental, ecological, social systems and other societal endeavors. It is a complex form of governance with a distributed form of leadership and citizenship that protects the integrity of the system and multiple centers of decision-making, where each center has a certain degree of autonomy (McGinnis et al. 2020). The idea of polycentric governance rests on eight pillars: (i) existence of multiple decision centers; (ii) autonomy of decision-making authorities; (iii) different decision centers have/share overlapping jurisdictions; (iv) decision centers are engaged into processes of mutual adjustment; (v) there are emergent patterns of behavior, an emergent order, that are shared across decision centers; (vi) low entry and exit costs; (vii) existence of an overarching system of rules, values, norms; (viii) existence of means for effective coordination at all levels (whether at the level of a decision center or the system as a whole) (McGinnis et al. 2020). Polycentrism is often used as the front end of projects to reduce complexity and create a nested structure of groups with delineated authority (Gil et al. 2017). In other words, support from those with the greatest interest in the project.

Step 3: The Legal Framework for Governance

A megaproject with its multiple employers, sponsors, government agencies, contractors, and other participants requires the careful structuring of legal entities to protect it from the enormous, risk, uncertainty, and complexity that it faces throughout its long life. These entities not only safeguard from liability but also provide structure to foster good governance and clear roles, responsibilities, and accountability to the general public often the beneficiaries of the project's outcomes. Although we will look into more detail at the contractual obligations of these entities in later chapters a brief overview is provided here.

Special Purpose Vehicle (SPV) as a Governance Tool

Since megaprojects are undertaking a new endeavor that has never been done before and they involve multiple actors that often are working together for the first time, it requires the establishment of a new entity commonly referred to as a special purpose entity (SPE). A SPE, also called a special purpose vehicle (SPV), is a subsidiary created by a parent company to isolate its financial risks. Established as a standalone temporary organization, megaprojects can be led by a client team, prime contractor, or some form of temporary alliance, joint venture, or coalition of multiple parties (owners, sponsors, clients, contractors, suppliers, and other stakeholders) that work jointly on a shared activity for a limited period of time in an uncertain environment (Jones and Lichtenstein 2009; Merrow 2011).

Each megaproject is usually decomposed into many smaller interrelated projects and organized as a program. A large organization – the client, prime contractor, and/or delivery partner – is established to coordinate and integrate the efforts of numerous subgroups and suppliers involved in project activities (Davies and Mackenzie 2014; Davies et al. 2009; Merrow 2011). This organization manages the overall program and the interfaces between projects; deals with external suppliers through separate contracts; and is accountable for meeting time, cost, and quality performance goals.

Apart from the fiscal and financial areas of impact, once SPEs are in place, they play a central role in the governance of megaprojects. Because of the multiple stakeholders there is a need to forge a shared modus operandi for the new SPE; once these new relationships have been successfully established, they reap rewards in terms of a shared knowledge and a more efficient design and construction phase than if the project was constructed separately by the individual companies.

Concession agreements are usually established alongside of the SPE to set forth the rights and obligations of the Grantor and the Grantee. The concession agreement, sometimes called a delivery agreement is the main contract that binds the major parties into a commitment that may last as long as 20–30 years in large development projects such as oil and gas projects, dams and tunnels, green energy, and transportation, or water transport projects.

Engineering, Procurement and Construction Contracts (EPC Contracts)

Though there are many types of construction contracts used in project financings, the majority of project sponsors and all project finance lenders strongly prefer turnkey contracts. Turnkey contracts are based on the idea that when construction is complete, all you need to do is turn the key. From the perspective of the project company, project sponsor, and project lender, the most desirable type of turnkey construction contract in project financings are Engineering, Procurement, and Construction Contracts, or EPC Contracts. With EPC Contracts the contractor is obligated to design, engineer, build and deliver the project for a fixed price by a specified date, according to the construction documents, plans, specifications, shop drawings, and other support documents.

EPC contracts and turnkey contracts are similar, but they are not interchangeable. In addition to all the construction risks and responsibilities born by the contractor in turnkey contracts, with EPC Contracts all the project design, engineering and procurement risks and responsibilities are transferred to the contractor for essentially all purposes.

Legal Framework for Contractual Governance Transparency

The establishment of an appropriate and effective legal framework is a prerequisite to creating an environment that fosters private investment in public infrastructure. It is a critical part of governance and often determines whether or not a project can attract enough investment to be considered viable. For countries where such a legal framework already exists, it is important to ensure that the law is sufficiently flexible and responsive to keep pace with the developments in various infrastructure sectors.

A transparent legal framework is characterized by clear and readily accessible rules and by efficient procedures for their application. Transparent laws reduce uncertainty and complexity, creates predictability, allowing potential investors to estimate the costs and risks of their investment and thus to offer their most advantageous terms. Transparent laws and administrative procedures may also foster openness through provisions requiring the publication of administrative decisions, including, when appropriate, an obligation to state the grounds on which they are based and to disclose other information of public relevance. They also help to guard against arbitrary or improper actions or decisions by the contracting authority or its officials and thus help to promote confidence in a country's infrastructure development program. Transparency of laws and administrative procedures is of particular importance where foreign investment is sought since foreign companies may be unfamiliar with the country's practices for the award of infrastructure projects.

Fairness

The legal framework is both the means by which governments regulate and ensure the provision of public services to their citizens and the means by which public service providers and their customers may protect their rights. A fair legal framework considers the various (and sometimes possibly conflicting) interests of the government, the public service providers, and their customers and seeks to achieve an equitable balance between them. The private sector's business considerations, the users' right to adequate services, both in terms of quality and price, the government's responsibility for ensuring the continuous provision of essential services, and its role in promoting national infrastructure development are but a few of the interests that deserve appropriate recognition by the law.

Long-term Sustainability

An important objective of domestic legislation on infrastructure development is to ensure the long-term provision of public services, with increasing attention being paid to environmental sustainability. Inadequate arrangements for the operation and maintenance of public infrastructure severely limit efficiency in all sectors of infrastructure and result directly in reduced service quality and increased costs for users. From a legislative perspective, it is important to ensure that the host country has the institutional capacity to undertake the various tasks entrusted to public authorities involved in infrastructure projects throughout their phases of implementation. Another measure to enhance the long-term sustainability of a national infrastructure policy is to achieve a correct balance between competitive and monopolistic provision of public services.

Public Outreach

Nature makes human development possible but our relentless demand for the earth's resources is accelerating extinction rates and devastating the world's ecosystems.

— Joyce Masuya, Acting Executive Director,
United Nations Environment Programme (UNEP)

Under some regulatory schemes such as the building of nuclear reactors, public outreach is required before approval can be given by the public authorities. This is to ensure that before substantial funds to the tune of billions of dollars are spent, all public concerns have been aired and responded to in a manner that at least satisfies stakeholders that they have been heard. Sometimes public hearings provide the opportunity for local citizens and stakeholders to present their concerns and air their grievances and oppositions. It also provides an opportunity to address differences and find ways to convince those that oppose the project of its opportunities and benefits for improvement of life in the local community which in turn can reduce resistance and sometimes encourage participation.

For instance, a critical aspect of global development is ensuring that there is engagement with Indigenous communities who own, manage, use or occupy 25% or more of the global land area as recommended by the Intergovernmental Science-Policy Platform on Biodiversity and Ecosystem Services (IPBES 2019). To align locally led plans with national policy frameworks requires a multilevel governance architecture "to avoid policy gaps between local action plans and national policy frameworks and to encourage cross-scale learning between relevant departments or institutions in local and regional governments" (Corfee-Morlot et al. 2009). Examples of multilevel governance linking in climate policy can be found in many cities and countries around the globe including Japan, Finland, New Zealand, Portugal, Sweden, Cape Town, Oregon in the United States and Barcelona (pp. 88–99).

Approval of Public Authorities

Depending on the administrative structure of the host country, privately financed infrastructure projects may require the involvement of several public authorities, at various levels of government. For instance, the competence to lay down regulations and rules for the activity concerned may rest in whole or in part with a public authority at a level different from the one that is responsible for providing the relevant service. It may also be that both the regulatory general legislative and institutional framework and the operational functions are combined in one entity, but that the authority to award government contracts is centralized in a different public authority. For projects involving foreign investment, it may also happen that certain specific competences fall within the mandate of an agency responsible for approving foreign investment proposals. Recent international experience has demonstrated the usefulness of entrusting a central unit within the host country's administration with the overall responsibility for formulating policy and providing practical guidance on privately financed infrastructure projects. Such a central unit may also be responsible for coordinating the input of the main public authorities that interface with the project company. It is recognized, however, that such an arrangement may not be possible in some countries, owing to their particular administrative structure.

Step 4: Measuring Performance

The performance criteria in a long-term project will only have meaning if they are clear and measurable and reflect the project's goals, mission, and the key performance requirements of the contract. To ensure that the public interest is protected, the performance of all project participants must be monitored and controlled, and all contract requirements must be enforced. Contract provisions must provide for incentives and disincentives, and they must be applied in accordance with the contract. When schedules slip and costs rise, it indicates that either the governance is weak, or nonexistent or has impaired accountability. To ensure that a governance framework is functioning at the highest level, constant evaluation is needed to make sure that safeguards are in place to prevent projects from spiraling out of control. For projects to succeed, they must be adaptable to change, and the governance frameworks in which they operate must be adaptable to the changing needs of the organization.

Mechanisms for Measuring Governance Performance

Good governance is driven by benchmarks and processes that are measurable and can be evaluated. Three mechanisms for measuring governance performance are summarized here; however, in reality, projects generally measure performance on all aspects of the project on a daily basis through such tools as milestone management, root-cause analysis, and earned value. If projects are substantially behind schedule, or costs are escalating out of control, it is most likely the result of a weakness or failing in the governance framework that needs to be evaluated through the following analytical tools:

1) *Auditing*, as an analytical tool, is a good device for measuring governance performance and infrastructure quality. Governance on Boston's Big Dig was measured through more than a dozen external government agencies that conducted regular audits, as well as independent auditors hired by the program owner and the Commonwealth of Massachusetts. Changes were implemented based on these recommendations, including the restructuring of the project to better address governance.

2) *External regulation* is another tool that drives good governance. As an example, the Sarbanes-Oxley Act was implemented in 2002 to set new or enhanced standards for all U.S. public company boards, management, and public accounting firms (Sarbanes 2002). The bill was enacted as a reaction to a number of major corporate and accounting scandals including those affecting Enron, Tyco Peregrine Systems, and WorldCom.

3) *Staged development* is a third way to measure governance performance through focused reviews at key decision points in the project life cycle. This process was introduced in the United Kingdom in 2001 by the Office of Government Commerce (OGC). It has been used successfully on large projects in the United States, Australia, and New Zealand.

Step 5: Optimization Framework for Governance

An important part of understanding governance is recognizing that projects must continually change to meet the demands of its stakeholders and the needs of its customers. To ensure that a governance framework is functioning at the highest level, constant evaluation is needed to make sure that safeguards are in place to prevent projects from spiraling out of control. The structure necessary at the planning stage of a project may look different from the structure at the implementation stage. For example, at the beginning of the project it is critical to have expertise in strategy and infrastructure policy planning and financing, while at the implementation stage critical governance skills include technical expertise and experience in how the plans and policy will be carried out. Some important questions that arise in assessing the need for governance optimization include the following:

1) Is the governance structure transparent, and does it instill an ethical culture?
2) Can performance be measured in a timely way, and is it effectively managed?
3) Is project oversight too narrowly focused?
4) Have expected project behaviors been effectively communicated?
5) Is the requisite expertise available through the project employees, or is additional independent oversight required?
6) Are inconsistent decisions emanating from the project's governance structures?
7) Are project incentives delivering the expected results?
8) Are responsible participants being held accountable?

Project Governance Between Developed and Developing Countries: Considerations for Improving Governance

Though there are no global guidelines or standards on how megaproject governance should be structured, the practices do vary widely, and consideration must be given to both national practice and international norms. As discussed in various sections of this book we must look to these international norms and practices for guidance in governing projects in the international arena. Whereas most experts recognize the substantial differences in the construction sector between developed and developing countries, very little is known about how and to what extent construction project governance actually differs between the two contexts.

Scholars and practitioners recognize some significant differences in project governance between developed and developing countries. However, the tendency has been to read these differences as "problems to be fixed," and consequently to identify "solutions" to the project challenges in developing countries. This tendency has been exacerbated by a lack of empirical data on the differences in project governance between the two contexts. In response to these challenges, Lizarralde et al. (2013) conducted a comprehensive study using three research methods. The results showed significant differences in the exercise of power and authority, the uses and causes of informality, and the role of stakeholders in projects. Some specific findings of this study include:

- The client organization's top management plays a more influential role (formally and informally) in projects in developing countries.
- Leadership styles tend to differ between the two contexts: hierarchy, adaptation, and trust are more important in developing countries.
- Procurement units, municipalities, and users play a greater role in projects in developed compared to developing countries. Less authoritarian client organizations can be expected in developed countries.
- Similar levels of informal communication are used in both contexts. Informal communication is perceived positively in both contexts, yet for different reasons. However, formal and informal relations between participants differ.

Culture and Governance

Many authors from different disciplines such as systems theory, systems thinking, management theory, sociology, and project management have highlighted the criticality of culture (Henrie and Sousa-Poza 2005; UNCTAD 2021). Trying to obtain accurate data on a complex social system requires understanding the perceived meaning of certain actions. Culture has been defined diversely by different theorists and researchers. Hofstede developed a cultural dimension theory to deal with cross-cultural communications. He defines culture as "the collective programming of the mind distinguishing the members of one group or category of people from others" (Hofstede 2001).

Different perspectives on organizational culture are well documented in the literature, and a few studies have focused on how project culture and project design support successful cooperation between partners in megaprojects (van Marrewijk et al. 2008). Other scholars have studied the cultural micro-practices of governance as it actually takes place (van Marrewijk and Smit 2016). Edgar Schein has laid a theoretical and empirical foundation for the thought that, "on the one hand, organizational founders and leaders have a profound influence in creating and shaping an organizational culture; whereas, on the other hand, once a culture has settled in, it becomes increasingly difficult for new generations of leaders to change it" (Ybema et al. 2011).

Corruption and Governance: Developing Anti-Corruption Policy

Corruption is the enemy of development, and of good governance. It must be got rid of. Both the government and the people at large must come together to achieve this national objective.

— *Pratibha Patil*

Corruption in Megaproject Governance

Though 140 countries have signed the UN Convention against Corruption, it remains one of the key issues for public policies. It is one of the major impediments to the development of emerging countries and to further improve the quality of life in developed countries (Loosemore and Lim 2015). The eradication of corruption is one of the key challenges that the world faces. Vee and Skitmore (2003) show that ethical behaviors in the construction industry are promoted by ethical guidelines and policies of private organizations and professional bodies together with the leadership of public sector procurement agencies. Corruption does not simply lead to extra cost and delay, but also increases the transaction costs showing the need to set a certain procurement and controlling system (Locatelli et al. 2017).

Corruption in Construction Projects in Developing Countries

We define corruption as the abuse of entrusted power for private gain.

— *Transparency International*

The World Bank Group considers corruption a major challenge to its twin goals of ending extreme poverty by 2030 and boosting shared prosperity for the poorest 40% of people in developing countries. According to the indexing of Transparency International, construction is one of the most corrupt industries among the various economic sectors. Recent research has reported the following as leading causes and effects and factors in eradication of this corruption in global projects.

Cause of Corruption
Infrastructure projects (IP) are most vulnerable to corruption owing to a great amount of capital involved which triggers a surge in corruption risks in construction project management (Kingsford and Chan 2018).

Impact of Corruption
Creation of a monopoly increased operational and procurement costs, lowered quality construction, and a decrease in direct foreign investments were found to be the most important ill effects of the corruption (Khadim et al. 2021).

Factors in Eradication of Corruption
Professional standards, transparency, the fairness of punishment, procedural compliance, and contractual compliance (Khadim et al. 2021).

Extensive research by Khadim et al. (2021) examined the negative and positive impacts of corruption on infrastructure projects in the developing country of Pakistan. The results supported the statement as the majority of construction practitioners strongly agree that corruption is widespread in infrastructure projects. It was established that corruption cannot be attributed to a single contractual party, and everyone is responsible for anomalies. Moreover, it was also revealed that different parties tend to blame each other for wrongdoings and irregularities.

Case Studies in Governance: What Causes Governance Failure?

Governance failure is as well-known as governance success, and without an understanding of the root causes of these failures we can expect that we will continue to see many more in the future. A failure of governance in a megaproject can have ramifications that can last for years and cause considerable damage including termination of the project.

Megaproject governance fails in different ways and for different reasons including an absence of knowledge, a lack of transparency, unexpected events surfacing in unexpected ways, the failure of leadership, strategy, policy, capacity, and, quite commonly, simply a failure of communication. As we have discussed in this chapter, awareness, early problem resolution, and developing an ethical and transparent culture should support detecting governance failure in an early stage.

In this section, we look at two examples of governance gone wrong, one in the developing country of Bangladesh and the other on a NASA funded project supported by the international community. We then look at two examples of governance gone right on a micro project in Azerbaijan and an Agricultural Productivity and Diversification Project in Benin. In reviewing these cases four important questions arise: (i) Could these governance failures have been prevented? (ii) If so, what should be done in the future to prevent these failures? (iii) In the case of project success, what are the factors that contributed to good governance? (iv) How can we improve governance structures to ensure our projects will have positive outcomes?

Governance Gone Wrong – Bangladesh

In recent years, "participatory governance" has emerged as an important concept within the governance domain. It is a policy that allows participation of local citizens to implement locally based pro people development initiatives. International aid agencies have been pursuing this agenda with an aim to provide greater legitimacy to development projects for economic growth in developing countries. In response, the government of Bangladesh has been trying to implement participatory governance policies for aid-assisted development projects for the last three decades. However, empirical studies reveal that the level of participation of local citizens in development projects has hardly been improved despite such attempts. Relying on six aid-assisted project-based case studies, Wahed Waheduzzaman and Sharif As-Saber (2015) explored the reasons for such a failure and found out that the dysfunctional political system and corruption in Bangladesh have compromised the role of the state in ensuring any meaningful participation of ordinary citizens in local-level development activities.

The government has restructured regional government bodies and prepared policies following the conditionality guidelines of the various aid agencies to foster participation of local citizen groups in local development programs (Waheduzzaman and As-Saber 2015). Despite such initiatives, meaningful citizen participation in regional development programs has hardly been achieved. The strong control of government bureaucrats over regional government affairs is the main reason that leads to this failure (Basurto et al. 2013; Khan 2000).

The study revealed that citizens' direct participation needs to be encouraged and ensured to embolden participatory governance and democracy in regional Bangladesh. In doing so, monitoring the formation of local management committees needs to be strengthened so that the project management committees are properly formed, representing the targeted users in aid-assisted development programs. The study also suggests that the formation of a local management committee is not the end, it is only a means to foster participatory governance in local level development projects.

A new form of monitoring and evaluation system could be feasibly developed through a consensus among all IAAs and the government of Bangladesh. This may nevertheless require rigorous and stringent conditionality to be followed by local governments as well as the project management committees to ensure effective and unbiased participation of ordinary citizens in the project governance process.

Governance Gone Wrong – James Webb Space Telescope

NASA's James Webb Space Telescope, the agency's successor to the famous Hubble telescope, was launched on 25 December 2021, from Kourou, French Guiana on a mission to study the earliest stars and peer back farther into the universe's past than ever before. This mission is happening more than 30 years after the James Webb Space Telescope (JWST), was first envisioned and sketched. The telescope is 14 years behind schedule and 20 times over budget. "We've worked as hard as we could to catch all of our mistakes and test and rehearse," said John Mather, the Nobel Prize-winning astrophysicist who has been the chief scientist of the NASA-led project for 25 years. Now, he said, "we're going to put our zillion-dollar telescope on top of a stack of explosive material" and turn things over to fate (JWST 2022).

An excellent example of governance gone wrong in this case is when an independent review team actually blamed the governance structure for this project failure. The review team said the budget was flawed because of a failure to conduct a bottom-up estimate and account for known threats. The review team asserted that leadership at Goddard Space Flight Center (GSFC) and NASA Headquarters failed to independently analyze the project's performance and recognize the flawed baseline; thus, as costs on the project increased, so did the funding. The project was ultimately confirmed with the flawed budget, and the failure to meet the launch schedule was interpreted by NASA leadership as a cost-control issue on the part of the contractor rather than a fundamentally broken estimate. The lack of an operational and effective cost and programmatic analysis capability at HQ was a contributing factor. Among the findings of the Independent Review Panel in assessing the problem in this case were the following (JWST 2010): Lack of clear lines of authority and accountability contributed to a lack of executive leadership in resolving the broken JWST life cycle cost baseline. The flawed budget should have been discovered as part of the center's execution responsibility, but the interpretation of the agency governance policy on the center's role in this regard is ambiguous and not uniformly interpreted within the agency. Ongoing, regular independent assessment and oversight processes at the agency were missing. To fix the problems in the JWST Project, NASA actually changed the governance structure by moving the JWST management and accountability from the Astrophysics Division to a new organizational entity at HQ that would have responsibility only for the management and execution of JWST. Various managers were reassigned to other jobs, and new managers took over the leadership roles.

Source: Adapted from James Webb Space Telescope (2010).

The James-Webb Space Telescope. *Source:* Adobe stock.

Governance Gone Right – Azerbaijan Rural Investment Project

Azerbaijan, one of the worlds' poorest countries, successfully developed a $140M rural investment project from 2012 to 2019 to improve infrastructure to promote micro-project development and women's empowerment in rural areas. The project was funded through the International Bank for Reconstruction and Development (IBRD) along with contribution from the national government and the beneficiaries of the project. The project successfully accomplished the following goals:

Benefited 82% of rural population
33% increase in access to and use of infrastructure
Travel time to schools, medical and water points, and markets reduced by 53.7%
Household incomes increased by 21.16%
1512 infrastructure microprojects were financed including roads, water supply systems and irrigation canals
Project exceeded the target of 50% participation by women

Source: Adapted from The World Bank (2019).

Benin Agricultural Project. *Source:* iAgua.

Governance Gone Right – Benin Agricultural Productivity and Diversification Project

Item	Activity
Background	The project became effective 12 March 2012, and closed on 30 September 2021. It was made up of two phases: Initial project from March 2012 to March 2017; and an additional financing covering April 2017 to September 2021. The total investment from IDA in this project was US$91 million (US$46 million for the initial financing and US$45 million for the Additional Financing). There were no IFC contributions to the project. At closing, the project mobilized some private sector financing within the targeted value chains through the matching grant scheme and the guarantee funds in the amount of about US$5 million.
Private sector support	The private sector was attracted to and incentivized by this Project by: (i) sensitizing and informing private agribusinesses and entrepreneurs about the investment opportunities available along the various segments of the targeted value chains; (ii) facilitating access to financial services along targeted value chains, through the use of matching grants, credit facilitation and guarantee funds; (iii) developing partnerships between the project and selected microfinance institutions and commercial banks that had significant presence in the areas targeted by the project. These financial institutions benefited from capacity strengthening aimed at improving their knowledge of the targeted value chains, to facilitate decision making in financing the business plans.
Project targets and benefits	This project has achieved and even surpassed the targets of its outcome indicators and had a significant impact on the livelihoods of beneficiaries. This project has done well in terms of reduction of poverty and adhering to climate change policy and other sustainability goals as shown by the various monitoring reports and project impact assessment conducted at completion.
Poverty reduction and shared prosperity	The project targeted the poorest areas, as well as the areas affected by the 2010 flood (55 out of 77 communes of Benin), the food crops (corn and rice, soybean, vegetables), cash crops (cashew, pineapple, and cotton) and livestock (poultry and small ruminants), aquaculture produced by a significant number of the poor, smallholders, and affected farmers in these areas of Benin.

Item	Activity
	The project interventions impacted the beneficiaries' livelihoods positively and significantly. The project contributed to improving employment, incomes and the living conditions of the beneficiaries as indicated in the various monitoring reports, the project completion report, as well as the 2021 impact evaluation reports. The project has created 1079 permanent jobs and about 3145 temporary jobs. Average incomes of project beneficiaries improved by 253 586 FCFA, 620 875 FCFA, 1 500 000 FCFA.
Gender: women beneficiaries	The project reached 40.25% of women-beneficiaries, against the end-target of 40%. About 47% of the matching grants beneficiaries were women and invested in processing activities in which women are heavily involved (cashew, apple juice, pineapple juice, and fish processing). The matching grant schemes included training tailored to women beneficiaries, focused on helping subprojects grow into successful SMEs, led either by women entrepreneurs or to create more jobs for youth and women. Women were also the key beneficiaries of the nutrition-related activities (promotion of nutrient-rich food crops, such as soybean, iron-fortified cowpea, sweet potatoes, vegetable gardening, nutrition education and awareness) supported by the project. Through these activities alone, the project trained 7 708 women out of 13 869 small and marginal farmers, developed eleven (11) gardening farms, and distributed 70 000 seedlings of moringa, papaya, and banana, as well as vegetable seeds (tomato, pepper, okra, watermelon, cabbage, and leafy vegetables) and funded subprojects on gardening farms, leading to a better integration of nutrition-rich food in their beneficiary diet. By promoting women participation in the project activities and in decision-making, the project contributed significantly to foster women participation in the labor market in the communities involved, as well as to improve food security, safety, and nutrition among vulnerable groups, including women and children.
Climate smart agriculture	In addition to capacity building to raise awareness on climate change and its impact on agriculture and appropriate measures to increase resilience and reduce climate-related vulnerabilities, the project financed the adoption of improved technologies for the development of food security (including nutrient-rich food and vegetables) and export-oriented value chains (aquaculture, maize, rice, cashew, and pineapple), and supported the adoption of readily available improved technologies and climate smart farm management practices to immediately boost farm productivity and incomes.

Summary

As evidenced in this chapter, technologically complex projects require multiple governance structures that are brought together through a centralized governing board or structure such as a program management office or an enterprise framework of governance. The important first step is the identification of benchmarks that must be brought to bear when developing a governance framework. Traditional hierarchical structures will fail in projects that require innovative solutions with multiple stakeholder interests, as risk must be balanced with the need for technological advancement.

The alignment of corporate, project, and political structures is essential to achieve success. Alignment becomes all the more important on megaprojects where there may be an absence of tools or structures that fit within every scenario, thus requiring the need to look to higher authorities for strategic guidance. Given the increasing internal pressures of project alignment and return on investment, as well as the external pressures of government compliance, stakeholder demands, and the continued pursuit of enhanced shareholder value, governments and corporate entities are increasingly searching for the ideal project governance framework for their organizations. Emerging models for governance were reviewed in this chapter including adaptation governance, relational governance, governance by commitment, governance by participation and polycentric governance. Today, it is essential to incorporate into every megaproject governance framework sustainable structures to advance MSR as well as the ecological concerns related to the environment through adaptation and participative governance models.

Lessons and Best Practices

1) As you conduct your project activities, ask yourself whether the direction of your results (products, recommendations, etc.) are consistent with the enterprise's governance structure (Wilson 2009).
2) Be willing to challenge governance if you decide it is not valid or is detrimental to achieving desired results.
3) Though high-level governance practices, such as policies or standards, are created with the intent of being applicable across the organization, generally they cannot account for every possible situation.
4) Strike a balance between the governance practices that work for most parties and situations with those appropriate for your program.
5) If governance practices need to be changed, consider how to augment, adjust, or refine existing practices while satisfying local objectives, rather than recommending dramatic changes that may be almost impossible to achieve.

6) When recommending changes, ensure that the intent of the governance concepts is honored and that the proposed revisions are applicable to and suitable for programs and enterprises beyond your local environment or situation.

7) A megaproject must maintain a clear statement of strategy and vision of governance that is planned during the conception stage.

8) Comprehensive information on the project's or the sponsor's corporate responsibility policies, including the policy objectives for each CSR area with quantified progress toward their achievement, is critical for effective program management. Create a megaproject social responsibility (MSR) policy if one does not already exist.

9) Scrutinize the performance of management in meeting agreed goals and objectives.

10) Ensure the integrity of the financial information and that financial controls and risk management systems are robust and defensible.

11) Governance and strategy should be discussed with key stakeholders to determine the appropriate roles and responsibilities and decision-making authority of all project participants.

12) Structure accountability on an organizational level with a single point of contact to provide leadership and drive the project forward.

13) Ensure that the owner or project sponsor retains decision-making responsibility throughout the life of the project, since the sponsor owns both the project budget and the business case for the project. This is particularly important where projects are funded entirely by the government owner.

14) Understand the important distinction between project governance and organizational governance, and clearly separate the decision making of each structure.

15) If the project governance structure is not given clear direction, the authority of both the project and the organizational governance structure is compromised, resulting in decisions that may not be enforceable.

16) Engage project stakeholders at a level that is commensurate with their importance to the organization and in a manner that fosters trust. Provide organization structure to manage project stakeholders in a consistent forum.

17) Establish an independent Board of directors consisting of sponsor representatives and outside experts to oversee the project that is not compromised by other goals of the larger organization and has sufficient time to devote to the multiple and complex issues faced by the project.

Discussion Questions

1 What does governance structure mean on a public project?

2 Why is it important?

3 How can you measure performance of a governance framework?

4 What should be the role of a governance board?

5 What is the critical role of the project sponsor/owner (and the relative roles of sponsor, portfolio manager, and program manager in governing and managing projects)?

6 What are the critical success factors that should be measured by a project board?

7 Who owns the benefits on a project, and who has strategic responsibility for realizing these benefits?

8 What is the role of the strategic plan on a project, and who should have responsibility for the strategic plan?

9 Define the effective role of a governance structure, including systems and tools.

10 What are some frameworks that would assist in building more sustainable projects and improving quality of life for those vulnerable communities around the world?

References

Ahola, T., Ruuska, I., Artto, K., and Kujala, J. (2014). What is project governance and what are its origins? *International Journal of Project Management* 32: 1321–1332.

Association of Project Management (2019). *APM Body of Knowledge*, 7the. Buckinghamshire, England: Association of Project Management.

Basurto, X., Gelcich, S., and Ostrom, E. (2013). The social–ecological system framework as a knowledge classificatory system for benthic small-scale fisheries. *Global Environmental Change* 23: 1366–1380.

Barnett, M. and Duvall, R. (2005). Power in international politics. *International Organization* 59 (1): 39–75.

Biesenthal, C. (2016). *Governance in Megaprojects: A Pragmatic Perspective.* OPUS. Open Publications of UTS Scholars. https://opus.lib.uts.edu.au/handle/10453/74225.

Biesenthal, C. and Wilden, R. (2014). Multi-level project governance: Trends and opportunities. *International Journal of Project Management* 32: 1291–1308.

Brahm, F. and Tarziján, J. (2016). Toward an integrated theory of the firm: The interplay between internal organization and vertical integration. *Strategic Management Journal* 37: 2481–2502.

Clegg, S.R., Skyttermoen, T., and Vaagaasar, A.L. (2021). *Project Management: A Value Creation Approach*, 1ste. London: Sage Publications.

Corfee-Morlot, J., Kamal-Chaoui, L., Donovan, M.G. et al. (2009). *Cities, Climate Change and Multilevel Governance*, OECD Environmental Working Papers No. 14. Paris: Organization for Economic Cooperation and Development (OECD).

Davies, A. and Mackenzie, I. (2014). Project complexity and systems integration: Constructing the London 2012 Olympics and Paralympics Games. *International Journal of Project Management* 32 (5): 773–790. ISSN 0263-7863, https://doi.org/10.1016/j.ijproman.2013.10.004.

Davies, A., Gann, D., and Douglas, T. (2009). Innovation in megaprojects: Systems integration at Heathrow Terminal 5. *California Management Review* 51 (2): 101–125.

Derakhshan, R., Turner, J.R., and Mancini, M. (2019). Project governance and stakeholders: A literature review. *International Journal of Project Management* 37 (1): 98–116.

Flyvbjerg, B. (2014). What you should know about megaprojects and why: An overview. *Project Management Journal* 45 (2): 6–19. https://doi.org/10.1002/pmj.21409.

Gil, N. (2017). A Collective-Action Perspective on the Planning of Megaprojects. In: *The Oxford Handbook of Megaproject Management* (ed. B. Flyvbjerg), 259–286. Oxford, UK: Oxford University Press.

Gil, N., Lundrigan, C., Pinto, J.K., and Puranam, P. (2017). *Megaproject Organization and Performance The Myth and Political Reality*. Newtown Square, PA: The Project Management Institute (PMI).

Greiman, V.A. (2013). *Megaproject Management: Lessons on Risk and Project Management from the Big Dig.* Hoboken, NJ: John Wiley & Sons.

Henrie, M. and Sousa-Poza, A. (2005). Project management: A cultural literary review. *Project Management Journal* 36 (2): 5–14. https://doi.org/10.1177/875697280503600202.

Hofstede, G. (2001). *Culture's Consequences: Comparing Values, Behaviors, Institutions, and Organizations Across Nations*, 2nde. Thousand Oaks, CA: Sage.

International Bank for Reconstruction and Development (IBRD) (2019). *Azerbaijan Rural Investment Project (AZRIP): Updated Project Information Document.* Washington, DC: The World Bank.

IPBES (2019). Executive Summary, XVIII. In: *Global Assessment Report on Biodiversity and Ecosystem Services of the Intergovernmental Science-Policy Platform on Biodiversity and Ecosystem Services* (ed. E.S. Brondizio, J. Settele, S. Díaz and H.T. Ngo). Bonn, Germany: IPBES Secretariat, 1144 pages. https://doi.org/10.5281/zenodo.3831673.

James Webb Space Telescope (JWST) (2010). Independent Comprehensive Review Panel (ICRP) Final Report, 29 October 2010. The Aerospace Corporation. https://www.nasa.gov/pdf/499224main_JWST-ICRP_Report-FINAL.pdf

James Webb Space Telescope (JWST) (2022). JWST Mission: Live updates. (March 16) National Aeronautics and Space Administration (NASA). https://www.space.com/news/live/james-webb-space-telescope-updates.

Jones, C. and Lichtenstein, B. (2009). Temporary Inter-organizational Projects: How Temporal and Social Embeddedness Enhance Coordination and Manage Uncertainty. In: *The Oxford Handbook of Inter-Organizational Relations* (ed. S. Cropper, C. Huxham, M. Ebers and P.S. Ring), 231–255. Oxford, UK: The Oxford University Press https://doi.org/10.1093/oxfordhb/9780199282944.003.0009.

Joslin, R. and Müller, R. (2016). The relationship between project governance and project success. *International Journal of Project Management* 34 (4): 613–626. https://doi.org/10.1016/j.ijproman.2016.01.008.

Khadim, N., Jaffar, S.T.A., Musarat, M.A., and Ilya, U. (2021). Effects of corruption on infrastructure projects in developing countries. *International Journal on Emerging Technologies* 12 (1): 284–295.

Khan, Z.R. (2000). Strengthening Local Government in Bangladesh: Recent Experience and Future Agenda. In: *Strengthening Local Government: Recent Experience and Future Agenda* (ed. R. Sobhan), 20–25. Dhaka, Bangladesh: Centre for Policy Dialogue.

Kingsford, O.E. and Chan, A.P. (2018). Barriers affecting effective application of anticorruption measures in infrastructure projects: Disparities between developed and developing countries. *Journal of Management in Engineering* 35 (1): 04018056.

Lizarralde, G., Tomiyoshi, S., Bourgault, M. et al. (2013). Understanding differences in construction project governance between developed and developing countries. *Construction Management and Economics* 31 (7): 711–730. https://doi.org/10.1080/01446193.2013.825044.

Locatelli, G., Mariani, G., Sainati, T., and Greco, M. (2017). Corruption in public projects and megaprojects: There is an Elephant in the room! *International Journal of Project Management* 35 (3): 252–268.

Loosemore, M. and Lim, B. (2015). Inter-organizational unfairness in the construction industry. *Construction Management and Economics* 33 (4): 310–326.

Ma, T., Ding, J., Wang, Z., and Skibniewski, M.J. (2020). Governing government-project owner relationships in water megaprojects: A concession game analysis on allocation of control rights. *Water Resource Management* 34: 4003–4018. https://doi.org/10.1007/s11269-020-02627-z.

Ma, H., Sun, D., Zeng, S.X. et al. (2021). The effects of megaproject social responsibility on participating organizations. *Project Management Journal* 52 (1): https://doi.org/10.1177/87569728211015868.

McGinnis, M.D., Baldwin, E.B., and Thiel, A. (2020). When is polycentric governance sustainable? Using institutional theory to identify endogenous drivers of dysfunctional dynamics. [Paper Presentation] Ostrom Workshop Colloquium Series, 14 September 2020. Indiana University, Bloomington.

McGrath, S.K. and Whitty, S.J. (2015). Redefining governance: From confusion to certainty and clarity. *International Journal of Managing Projects in Business* 8 (4): 755–787.

Medda, F.R., Carbonaro, G., and Davis, S.L. (2013). Public private partnerships in transportation: Some insights from the European experience. *IATSS Research* 36 (2): 83–87.

Merrow, E.W. (2011). *Industrial Megaprojects: Concepts, Strategies, and Practices for Success.* Hoboken, NJ: John Wiley & Sons.

Miller, R. and Hobbs, B. (2005). Governance regimes for large complex projects. *Project Management Journal* 36 (3): 42–50. https://doi.org/10.1177/875697280503600305.

Miller, R. and Lessard, D. (2007). *Evolving Strategy: Risk Management and the Shaping of Large Engineering Projects.* Massachusetts Institute of Technology (MIT), Sloan School of Management, Working Papers. https://doi.org/10.2139/ssrn.962460.

Müller, R. (2009). *Project Governance: Fundamentals of Project Management.* Farnham, UK: Ashgate Publishing Group.

Müller, R., Pemsel, S., and Shao, J. (2015). Organizational enablers for project governance and governmentality in project-based organizations. *International Journal of Project Management* 33 (4): 839–851. https://doi.org/10.1016/j.ijproman.2014.07.008.

Organization for Economic Cooperation and Development (OECD) (2015). *G20/OECD Principles of Corporate Governance.* Paris: OECD Publishing http://dx.doi.org/10.1787/9789264236882-en.

Ostrom, E. (2012). Nested externalities and polycentric institutions: Must we wait for global solutions to climate change before taking actions at other scales? *Economic Theory* 49 (2): 353–369.

Project Management Institute (2016). *Governance of Portfolios, Programs, and Projects: A Practice Guide.* Newtown Square, PA: PMI.

Project Management Institute (2021). *A Guide to the Project Management Body of Knowledge (PMBOK)*, 7the. Newtown Square, PA: PMI.

Sanderson, J. (2012). Risk, uncertainty and governance in megaprojects: A critical discussion of alternative explanations. *International Journal of Project Management* 30 (4): 432–443.

Sarbanes-Oxley Act of 2002 (2002) H.R.3763 – 107th Congress (2001–2002).

Shah, A. and Shah, M. (2006). Local governance in developing countries: Public sector governance and accountability series. *International Journal of Business and Social Science* 2 (16): 76–89.

Sparrevik, M., Magerholm, A., Wangen, H.F., and DeBoer, L. (2018). Green public procurement – A case study of an innovative building project in Norway. *Journal of Cleaner Production* 188: 879–887. https://doi.org/10.1016/j.jclepro.2018.04.048.

Turner, R. (2020). How does governance influence decision making on projects and in project-based organizations? *Project Management Journal* 51 (6): 670–684. ISSN 8756-9728.

United Nations (2011). *Guiding Principles on Business and Human Rights: Implementing the United Nations "Protect, Respect and Remedy" Framework*. and Geneva: United Nations Human Rights, Office of the High Commissioner, United Nations.

United Nations Conference on Trade and Development (UNCTAD) (2021) Trade and Development Report 2021 from Recovery to Resilience: The Development Dimension Report. The Secretariat of the United Nations Conference on Trade and Development.

Valverde, M. and Moore, A. (2019). The performance of transparency in public-private infrastructure project governance: The politics of documentary practices. *Urban Studies* 56: 689–704.

van Marrewijk, A. H., Clegg, S., Pitsis, T., and Veenswijk, M. (2008). Managing Public-Private Megaprojects: Paradoxes. *Complexity and Project Design, International Journal of Project Management* 26 (6): 591–600.

van Marrewijk, A. and Smits, K. (2016). Cultural practices of governance in the Panama Canal Expansion Megaproject. *International Journal of Project Management* 34 (3): 533–544.

Vee, C. and Skitmore, C. (2003). Professional ethics in the construction industry. *Engineering, Construction and Architectural Management* 10 (2): 117–127.

Waheduzzaman, W. and Sharif As-Saber, S. (2015). Politics and policy in achieving participatory governance in a developing country context. *Politics and Policy* 43 (4): 474–501. https://doi.org/10.1111/polp.12121.

Warsen, R., Nederhand, J., Klijn, E.H. et al. (2018). What makes public-private partnerships work? Survey research into the outcomes and the quality of cooperation in PPPs. *Public Management Review* 20: 1165–1185. https://doi.org/10.1080/14719037.2018.1428415.

Wilson, W.L. (2009). *Conceptual Model for Enterprise Governance*, Ground System Architectures Workshop. Lockheed Martin.

Winch, G.M. and Leiringer, R. (2016). Owner project capabilities for infrastructure development: A review and development of the 'strong owner' concept. *International Journal of Project Management* 34 (2): 271–281.

Wolf, M. (2021). The G20 has failed to meet its challenges. *Financial Times* 13 (July 14): 17.

World Bank (2021). Benin Agricultural Productivity and Diversification Project Poverty and Impact Evaluation Report. International Bank for Reconstruction and Development (IBRD) The World Bank, Washington, DC.

World Development Report (WDR) (2017). *Governance and the Law*. Washington, DC: The World Bank.

Xie, L., Han, T., and Skitmore, M. (2019). Governance of relationship risks in megaprojects: A social network analysis. Hindawi. *Advances in Civil Engineering* 2019: https://doi.org/10.1155/2019/1426139.

Ybema, S.B., Yanow, D., and Sabelis, I.H.J. (ed.) (2011). *Organizational Culture – Vol. 1*, The International Library of Critical Writings on Business and Management, No. 15. Edward Elgar.

Zeng, S.X., Ma, H.Y., Lin, H. et al. (2015). Social responsibility of major infrastructure projects in China. *International Journal of Project Management* 33 (3): 537–548.

Zeng, S., Ma, H., Lin, H. et al. (2017). The societal governance of megaproject social responsibility. *International Journal of Project Management* 35 (7): 1365–1377.

Zhai, Z., Ahola, T., and Le, Y. (2017). Governmental governance of megaprojects: The case of EXPO 2010 Shanghai. *Project Management Journal* 11 (1): 37–50. https://doi.org/10.1177/875697281704800103.

6

Integrated Project Organizations and Public Private Partnerships

I believe nuclear energy in Jordan will be done in such a way where it is a public-private partnership so everyone can see exactly what's going on.
— Abdallah II of Jordan

Introduction

In this chapter, we explore the dynamics of global megaprojects including project integration, systems integration, and the integration of organizations through the development of public private partnerships.

Integration is a reality of all large complex systems whether they be united in a mutual endeavor such as preservation of ecosystems or major partnerships or alliances between nations, governments or military organizations to accomplish a mutual goal. Megaprojects by their very nature are complex organizations in terms of the sheer number and variety of interdependent components including systems, people, and processes that must come together to work as a unified whole to meet a common objective. This chapter is about the manner of integrating these components through various structures including integrating the project organizations (IPOs), public–private partnership (PPP) integration, system of systems integration of megaproject disciplines such as systems engineering and program management, stakeholder integration, and the development of a team of teams.

The exploration of the features of successful integration, the impact of effective integration on program performance, and developing integration of competencies in people are a primary focus of this chapter. To fully understand integration in megaprojects it is essential to review the public private partnership models which rely on good integration as a key to their success to learn what makes some PPPs successful while others face many challenges that cannot be overcome and ultimately may result in a

Global Megaprojects: Lessons, Case Studies, and Expert Advice on International Megaproject Management, First Edition. Virginia A. Greiman.
© 2023 John Wiley & Sons, Inc. Published 2023 by John Wiley & Sons, Inc.

failed or terminated project. Both the question whether PPPs are a success and the question what makes PPPs successful cannot be easily answered as the criteria for judging success can vary widely. Scholars have identified a long list of critical success factors for PPP performance that include risk allocation and sharing, strong private consortium, political support, community/public support, and transparent procurement (Osei-Kyei and Chan 2015). The research also reflects two significant characteristics of successful PPP performance. One stream of research emphasizes the importance of contractual conditions: a well-written contract, the possibility to impose sanctions, clear performance indicators, and the allocation of risks (Koppenjan 2005; Pollitt 2002). Another stream emphasizes the importance of relational characteristics such as trust, informal communication, and openness (Huxham and Vangen 2005; Poppo and Zenger 2002; Warsen et al. 2018, 2019).

Based on a comparative analysis of nine case studies concerning public private partnerships in the Netherlands, Koppenjan (2005) identified three patterns of partnership formation. The first is the successful formation of partnerships resulting in enriched projects. The second pattern is that of early interaction resulting in ambitious proposals for which there is no support. The third pattern shows ineffective market consultations followed by unilateral public planning, leading to stagnating contract negotiations. These patterns are caused by a number of generic factors. For example, an important explanation for stagnation is the lack of interaction. As a result, public and private parties will (i) fail to reach a common understanding, (ii) will be unable to contribute to the enrichment of the project content, and (iii) will fail to develop mutual trust.

Despite the dynamics and uncertainties in PPP projects, a good relationship and a shared understanding on agreements regarding the realization (and sustainability) of the project might help dealing with the issues partners face in PPPs. However, many PPPs do not succeed as planned. For example, a Special Report by the European Union Court of Auditors revealed the following:

> PPPs have the potential to achieve faster policy implementation and ensure good maintenance standards, however, the audited projects were not always effectively managed and did not provide adequate value for money. Potential benefits of PPPs were often not achieved, as they suffered delays, cost increases and were under-used, and resulted in €1.5B in ineffective spending. This was also due to the lack of adequate analyses, strategic approaches towards the use of PPPs and institutional and legal frameworks. With only a few Member States having consolidated experience and expertise in implementing successful PPP projects, there is a high risk that PPPs will not contribute to the expected extent to the aim to implement a greater part of EU funds through blended projects including PPPs.
>
> (EU, Court of Auditors, 2018)

Part I: Project Organization Integration: A New Mindset for Systems Engineering and Program Management

Almost every large system is initiated with an intention of building something, which has new value to its stakeholders. Yet, turning this simple statement into reality is often problematic since the concept of value is not always explicitly defined or structurally incorporated in the formal frameworks of program planning and systems engineering (Patanakul and Shenhar 2009).

In 2011, the International Council on Systems Engineering (INCOSE) and the Project Management Institute (PMI) allied to change the mindsets of program managers and systems engineers. Both organizations believed that the two disciplines had established silos between them that blocked collaboration. The joint whitepaper, *Toward a New Mindset: Bridging the Gap Between Program Management and Systems Engineering* (Langley et al. 2011), identified the following challenge (Rebentisch et al. 2017):

> While program management has overall program accountability and systems engineering has accountability for the technical and systems elements of the program, some systems engineers and program managers have developed the mindset that their work activities are separate from each other rather than part of the organic whole (p. 24).

Regardless of who was in authority, whose inputs were more respected and accepted, or who better understood the path forward, the silos focused each discipline on advancing its own approach toward delivering solutions to meet customer needs. The whitepaper goes on to say:

> Historically, program managers and systems engineers have viewed the stakeholder problem entirely from within their own disciplinary perspectives [. . .]. As a result, the two groups have applied distinctly different approaches to the key work – managing the planning and implementation, defining the components and their interactions, building the components, and integrating the components.
>
> *(Langley et al. p. 25)*

Integration of Multiple Stakeholders' Interests: Singapore Case Study

A significant case where integration of the program management team and the systems engineering team resulted in positive outcomes involved the issue of water supply in Singapore. The Singapore water supply is managed in totality. Collecting rainwater, purchasing water, purifying water utilizing reverse osmosis, and desalination were all considered. Approaches included even incentivizing consumers to change their habits by making drains and canals recreational areas to encourage the public not to dispose of waste in their drains. By managing sewage and drainage together with water, environmental as well. By carefully adjusting organizational boundaries, Singapore has managed to reduce silo thinking and parochial interests. The relationships between the industry innovators, government, suppliers and users, and technology innovators create opportunities for Singapore's water management. This demonstrates how multiple stakeholder interests can be combined and integrated to create a viable water management solution.

Continuous improvements through the use of technology and elimination of waste, such as reducing water that is not accounted for in the system, help to assure the sustainability of an adequate supply of water for a growing Singapore population. The importance of relationships between the stakeholders is also recognized. Industry innovators, political leadership, suppliers, and consumers are all involved; the program has been able to incentivize this diverse group to work together for a common goal, i.e., assuring the sustainability of an adequate supply of water for Singapore into the future (Chia 2008).

Through the years, Singapore has embarked on an integrated, effective and cost-efficient way to meet the nation's water needs with investments in research and technology to treat, recycle, and supply water. Singapore believes that everyone in the country has a stake in water – as a necessary resource, an economic asset, and an environmental treasure.

Singapore Water and Waste Works. *Source:* letkat/Adobe stock.

> Today, Singapore is internationally recognized as a model city for integrated water management and an emerging Global Hydrohub – a leading centre for business opportunities and expertise in water technologies. Singapore's holistic approach to water management can be distilled into three key strategies:
> - Collect every drop of water
> - Reuse water endlessly
> - Desalinate seawater
>
> *Source:* Adapted from Singapore's National Water Agency.

Utilizing systems engineering and taking into consideration the systemic structures and culture helped Singapore achieve its first milestone of supplying its own water resources. Singapore has been able to overcome the shortfall that would have come about with the expiry in 2011 of the first water agreement with Malaysia (Chia 2008). By 2060 when Singapore's second water agreement expires with Malaysia, Singapore plans to expand the current water capacity to meet 55% of Singapore's future water demand (Singapore PUB (2022).

Part II: The Structure of Organizations as Systems of Systems

All organizations are developed through an eco-system of related and unrelated and formal and informal projects, programs, networks, alliances, partnerships, and functional operations. This eco-system defines the organization's structure. Because organizations are evolving systems, the structures within the organization continue to grow and change as the strategic goals of the organization change. Significant time has been spent on the development of new organizational paradigms that are "characterized by . . . decentralized decision-making, greater capacity for tolerance of ambiguity, permeable internal and external boundaries, empowerment of employees, capacity renewal, self-organizing units, and self-integrating coordination mechanisms" (Campagnolo and Camuffo 2010).

To understand organizational and program integration better from a systems engineering perspective, early in 2011 a gathering of researchers from what would eventually come to be called the Consortium for Engineering Program Excellence (CEPE) at the Massachusetts Institute of Technology (MIT) and industrial partners met to explore the application of Lean principles to program management. The participants in this effort quickly expanded to include PMI and INCOSE with the goal to evaluate the level of integration and collaboration between program managers and chief systems engineers. Partnering with MIT's Consortium for Engineering Program Excellence, the organizations conducted a series of studies to explore among other factors:

- How integrated were the practices, tools, and approaches used by chief systems engineers and program managers in a project organization?
- Did critical links exist where they were needed?
- Were common practices, such as risk management, managed in intersecting or parallel paths?
- Were practices, tools, and approaches evaluated and benchmarked to identify opportunities for improvement?

Box 6.1 Highlights some of the insights gathered from the research.

The difficulty of governing megaprojects with multiple interconnected systems leads to greater uncertainty and turbulence within the wider organization. The complexity of a project can be defined as a system in terms of the number and variety of components and interdependencies among them (Shenhar and Dvir 2007). Oehmen et al. (2015) in their study of program complexity from a system engineering perspective, define complex programs as "characterized by feedback loops and unforeseen emergent behavior that can spiral out of control, but are fundamentally still tractable by structure (if costly and time consuming) analysis" (p. 7).

Box 6.1   Insights on Organizational Attributes for Integration

Adjusting organizational boundaries is critical for avoiding silo thinking and parochial interests.

The components of the organization may not necessarily be interdependent or have related objectives, but the organization must ensure that the strategies of its component parts – stakeholders, programs, teams, operations – align with the organization's overall strategy.

The integration of systems thinking and program management requires taking into consideration the systemic structures, the cultural factors, and the relationship of the organization with its external stakeholders.

Use relationships, not rules, to encourage behavior.

Best practices focus on how teams perform, and not just on the individuals within the teams.

Team members must be willing and incentivized to actually share knowledge to integrate the skills and experiences from different disciplines.

Source: Rebentisch et al. (2017)/John Wiley & Sons.

Characteristics of Successful Program Integration: The Systems Perspective

Systems integration techniques originally were developed to manage and integrate complex weapons and space exploration projects (Hughes 2004; Morris 2000) but are now used in all types of projects including infrastructure, transport, agriculture, sports, and energy. Systems integration may be undertaken by the project organization or a prime contractor or can be implemented on a public project by the public sponsor(s) or an external company specializing in systems management.

Systems Integration on the Battlefield

Retired Navy Captain, Eugene Razzetti (2020) describes the relationship between systems integration and program management this way:

> System Integration is an indispensable subset of Program Management. Programs become products, and products find their way to the Battlespace. They must perform as required (and sustainably) in the Battlespace; U.S. lives and missions may depend on it. If that product, be it a weapons system, platform, or a piece of communication equipment, fails – someone may not make it home alive.

Moreover, as described by Razzetti (2020), if a weapon system or platform has reached the Battlespace, it had better reflect the absolute best doctrines and efforts of proactive DoD program management. "Fight with what you have," the unfortunate cliché in the briefing room remains the unhappy battle cry at the front. Program managers and contractors must recognize and support what warfighters already know all too well, that:

- Weapon systems or platforms have vital end uses in support of mission commitments.
- Successful end use requires urgency of arrival in-theater, and sustainability of operations in-theater.
- U.S. lives and U.S. missions depend on those weapons systems and platforms.

Systems Integration in Interorganizational Projects

Systems integration refers to the work undertaken across organizational boundaries in interorganizational projects to integrate the systems that these projects deliver. As the world's first dedicated systems integrator firm, for example, Ramo-Wooldridge Corporation, in California, USA, worked across organizational boundaries on the

Atlas missile defense project (following World War II) with the responsibility for coordinating "the work of hundreds of contractors and development of thousands of sub-systems" (Mahnken 2008, p. 38). Yet separating responsibilities for systems integration from those of project management has not been successful on later projects (Hughes 1998). Systems integration, and the associated practices of interface management, are a natural locus for project management (Morris 2000).

Despite numerous failures, the literature also describes successful programs resulting from good systems governance and integration such as the Øresund Bridge linking Denmark and Sweden (INCOSE 2015, pp. 40–42); the NASA MARS Pathfinder program (Nicholas and Steyn 2008) and the I-15 Reconstruction Program in Salt Lake City (FHWA 2011).

Miller and Lessard (2001) in their study of 60 large civil engineering programs (LEPs) describe the organizational structure of these programs, the shaping of a program, the program's institutional framework, and the capacity of self-regulation and governance. The up-front effort involved in shaping the program so it can survive turbulence improves the chances of success, but this requires leadership and systems thinking (Miller and Lessard 2001). Müller (2009) discusses governance at the program and organizational level linking program and strategic management. As noted by Locatelli et al. (2014), for programs delivered in complex environments, the governance needs to be transformed from a corporate/program perspective to a "systems perspective." As the authors describe, system governance increases the likelihood of program success.

An excellent example of systems governance is the Super Hornet Project developed by the U.S. Military that successfully exceeded cost, schedule, and technical requirements based on the integration of systems engineers and program managers (see Box 6.2).

F/A-18E/F Super Hornet Strike Aircraft.

Box 6.2   The F/A18E/F Super Hornet Case Study

The F/A18E/F Super Hornet case study illustrates in some detail a program where integration was a core philosophy, driven by an understanding that the cost of not being integrated was unacceptably high. There was awareness that the traditional ways that teams had worked together in the past would not have met a critical schedule driven need to provide key strike capabilities to the Navy fleet. What emerged was a program that subordinated the individual, functional, and organizational identities to the needs of the overall program.

The mantra of the program leaders was "the airplane is the boss" and was used to ensure that individuals and teams worked to make the right choices and act as needed to ensure overall success. The numerous organizations involved in the program were deliberate about their intent to have a more integrated, high functioning relationship between program management and systems engineering. A number of practices and approaches was employed to encourage increased integration. Senior leadership played a defining role in establishing the vision for greater integration and was willing to expend time, resources, and leadership capital over an extended period of time to ensure that the entire program participated in enacting that vision.

The following were among the approaches used to increase integration:

1) Develop a high-level plan to transition to integrated product teams
2) Matrix the functional organization across the integrated project teams
3) Co-locate program managers and system engineers in the same building
4) Create a culture of rapid decision making
5) Master shared information
6) Early identification and mitigation of risk
7) Promotion of collaborative work

Source: Adapted from Rebentisch et al. (2017).

Up-Front Planning Stage

The most important stages for developing a "systems perspective" are during the up-front planning or earliest stages of the program planning (IEEE 2005; Miller and Hobbs 2005). For example, in the planning of the California High Speed Rail, the government has addressed the importance of a statutory oversight authority and a legislatively mandated risk management plan to integrate diverse organizations into one integrated team.

Closely related to systems governance is "systems thinking" – a method developed to understand how systems influence one another within the whole. Systems thinking is what distinguishes systems engineering from other types of engineering, and is the underlying skill required to perform systems engineering (Beasley and Partridge 2011).

Case Study: The International Space Station: A Model in Systems Integration

The International Space Station (ISS) is perhaps the most famous of all systems engineering programs known for its convergence of science, technology, and human innovation. It demonstrates new technologies and makes research breakthroughs not possible on Earth. The space station has been continuously occupied since November 2000 and during that time more than 251 people from 19 countries have visited the space station with 155 from the United States and 52 from Russia (NASA 2021). This program crosses cultural barriers and raises all kinds of issues relevant to the integration of systems engineers and program managers including organizational integration, knowledge sharing, leadership, and team trust. Over the years NASA has had to develop expertise in systems engineering while keeping up with all of the technology issues and challenges faced by this daunting laboratory in space.

International Space Station. *Source:* dimazel/Adobe Stock.

In relation to the major cost and schedule issues, the systems engineering challenge on the ISS was equally monumental. NASA had to quickly learn how to adapt its system engineering approaches to include an awareness of those of the international partnership (Stockman et al. 2010). Essentially, NASA had to learn how to operate as a "managing partner" to accommodate its international partners (IPs) including different perspectives on approaches, designs, and operational risk and safety. A major integration effort was involved in developing the partnership agreements, allocating costs and usage rights, and determining operational control. Under the new ISS partnership, NASA was concerned about maintaining schedule and cost on the ISS program because failures would not be tolerated by the U.S. Congress. Initial program strategy was for no IP to be on the critical path, which would allow NASA more control to reduce risk. As it turned out, however, the Russians ended up providing the first two major modules that were at the front of the critical path. NASA was the first IP among equals, with each board chaired by the NASA representative. In cases where consensus could not be reached, the NASA representative technically had the right to decide for the board; however, this right was rarely used in practice.

NASA had to solve many major systems engineering challenges. It had to figure out how to coordinate and integrate all of the IPs and their highly integrated modules. While it was not easy, NASA eventually worked out a process that addressed the concerns of multiple countries with differing cultural and engineering approaches to major program development and execution. NASA's lessons learned report issued the following recommendations for NASA systems engineers (Stockman et al. 2010):

- "Systems engineering involves communications, critical to international partnerships, so before worrying about technical interfaces, make sure the integrated product teams and communication bandwidth between partners are optimal. This fundamentally includes face-to-face meetings, so regular international travel is a large and essential part of the systems engineering cost" (p. 87).
- "In an International Space Station like project where so many different countries and companies contribute hardware and software, the interfaces must be extremely simple" (p. 87).
- "Maintaining a high level of competent and experienced personnel over a two-decade long program requires strategic level planning and execution of workforce planning . . ." (p. 87).
- All of the above required not only integrated teams, but an integrated organizational structure as shown in Figure 6.1. Boeing plays a critical role on this team as the lead systems engineer for the program. Currently, the overall program team is managed through an ISS Control Board Structure (Figure 6.1). The ISS team uses top-level control boards and panels to manage the ISS hardware and software configuration along with any operational products. At the very top of the process is the Space Station Control Board (SSCB) that manages the multilateral control of the configuration. A NASA Space Station Program Control Board exercises control

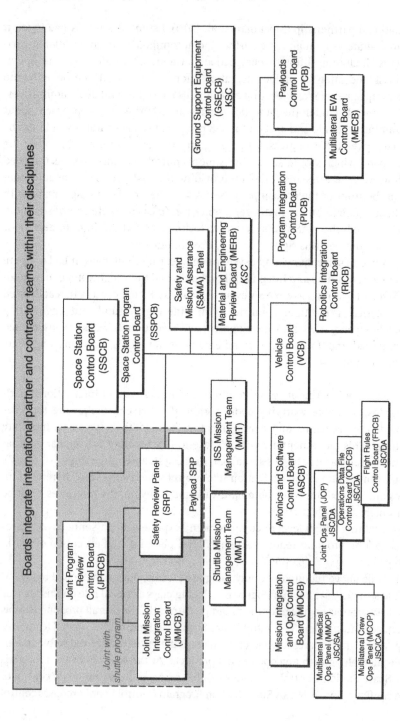

Figure 6.1 High-level board integration of the International Space Station program. *Source:* Stockman et al. (2010)/NASA/Public Domain.

over the several layers of more detailed ISS subsystem control boards associated with the U.S. elements. This process is also integrated with the Space Shuttle control boards. Each partner utilizes a similar control mechanism for their elements.

Coordinating the inclusion of the international partners whose system engineering approaches differed significantly was a major challenge of the ISS program. The problem was summed up well in a 1969 speech by Robert Frosch prior to the ISS and prior to becoming NASA Administrator:

> I believe that the fundamental difficulty is that we have all become so entranced with technique that we think entirely in terms of procedures, systems, milestone charts, PERT diagrams, reliability systems, configuration management, maintainability groups and the other minor paper tools of the "systems engineer" and manager. We have forgotten that someone must be in control and must exercise personal management, knowledge and understanding to create a system. As a result, we have developments that follow all of the rules, but fail. . . . Systems, even exceptionally large systems, are not developed by the tools of systems engineering, but only by engineers using tools.
>
> *(NASA 2010)*

The lesson here is that whether you are a systems engineer or a program manager you cannot rely on tools to understand how program performance and organizational value is measured and controlled, but instead must rely on leadership skills that encourage an environment of inclusivity, collective creativity, shared ownership, and large-scale transformation.

Integration in Program Delivery: Heathrow Terminal 5 (T5)

An important example of integration in program delivery that is often overlooked is the integration of the program team with the organization that will eventually operate the program. We can build the perfect program on schedule and budget, but if we fail to integrate the program with the operating organization, the program can ultimately be deemed a failure. As an example, Heathrow Terminal 5 (T5) was a megaproject that was on schedule and on budget since construction began, defying all the trends of previous megaprojects in the United Kingdom (Davies et al. 2009). T5 was seen as the first step in the regeneration of London's main airport in preparation for the 2012 Olympics. However, on the day the Terminal opened, what was to be a grand celebration instead turned into a national disaster due to baggage delays, temporary suspension of check-in, and the cancellation of 68 flights. This failure has been attributed to among other things a lack of systems integration and coordination between the program and the operating organization – each operating as separate systems. Once program management on Terminal 5 thought they surmounted the considerable issues related to building such a vast and technologically sophisticated terminal they suffered from technological hubris and forgot about the people issues related to the successful functioning of any large technical system (Brady and Davies 2010).

What really failed in the Heathrow Terminal Case Study was Integrated Project Delivery (IPD). The concept of IPD as defined by the American Institute of Architects (AIA), is a project delivery approach that integrates people, systems, business structures, and practices into a process that collaboratively harnesses the talents and insights of all participants to optimize program results, increase value to the owner, reduce waste, and maximize efficiency through all phases of design, fabrication, and construction (AIA 2008). If there is little integration between the program and operations the final outcome is almost certain to fail. Ironically, Heathrow Terminal 5 was a model in systems integration in many ways, yet the outcome was a failure due to the discontinuity in one of the most important areas of integration, program delivery.

Integration is important in all aspects of a program because it fosters collaboration, and collaboration fosters knowledge and trust – key elements of program success. In every megaproject, there is a need to integrate the processes and systems required to deliver the program with those involved in the operations of the program's end result (Davies et al. 2009). Recognizing the complexity of the delivery process in Heathrow Terminal 5, the CEO in his testimony before the House of Commons stated:

> ... with the benefit of hindsight, we might have adopted a more humble position, given the track record, (of other airport opening disasters) and it was unfortunate that we created an expectation of perfection in what was an extremely complicated programme.
>
> *Testimony of Colin Matthews, BAA CEO in House of Commons*
> *Transport Committee (HCTC) (EV 5) (2008)*

Key Takeaways from the Heathrow Terminal 5 Case Include:

- Systems engineering is the emerging paradigm in complex environments to transfer the governance from "project based" to "system based" and thereby increase the chance of holistic success (Locatelli et al. 2014).
- Integration between the systems engineer, the program management team, and the operations team should take place during the earliest phase of the program and continue throughout the life cycle.
- Integrated delivery is an approach that integrates people, systems, business structures, and practices into a process that fosters collaboration and trust.

- Systems integration is a form of governance that can detect problems and solve them long before they spiral out of control.
- The systems engineer and the program manager have overlapping roles that benefit from shared responsibility.

The London Olympics

The London Olympics is another example of integration of a large, complex system megaproject that had achieved some success through a system of systems model (Davies and Mackenzie 2014). The construction program for the 2012 London Olympics broke down into many individual projects and was managed as a program similar to most megaprojects. The construction program consisted of more than seventy projects (planned, approved, and managed by principal contractors), including fourteen temporary and permanent buildings, 20 km of roads, 26 bridges, 13 km of tunnels, and 80 hectares of parkland. At the level of system of systems, a structure and process were established with capabilities to understand the total system, manage external interfaces with multiple stakeholders, coordinate the progress of the overall program, and help manage individual system projects.

Davies et al. (2009) research underlines the importance of developing a system of systems integration capability to manage the most complex type of system megaproject. In addition to the technical and managerial capabilities required to design and integrate individual systems (Shenhar and Dvir 2007), a wider role of systems integration is required to manage the interfaces between multiple systems within a program, and between the system as a whole and external environment within which it is conceived, developed, and delivered (Davies and Mackenzie 2014).

Integration at the Big Dig

During the peak construction years on America's largest inner-city highway project, the Central Artery Third Harbor Tunnel Project, popularly known as the Big Dig, changed the project structure from a traditional program management model into an integrated project organization (IPO) as shown in Figure 6.2.

An IPO is an organization where both the owner's employees and the management consultant's employees work under one organization structure. This change was made because management decided that an integrated organization would enhance collaboration and reduce conflict at a time when it was desperately needed. The change involved challenges as the project had multiple employers each with different policies and procedures and work cultures. As shown in Figure 6.2, the integration meant that employees of one employer would be reporting to the senior staff of another employer. This transition to an IPO was made in 1999, just as construction activity was peaking. Ideally, an IPO should

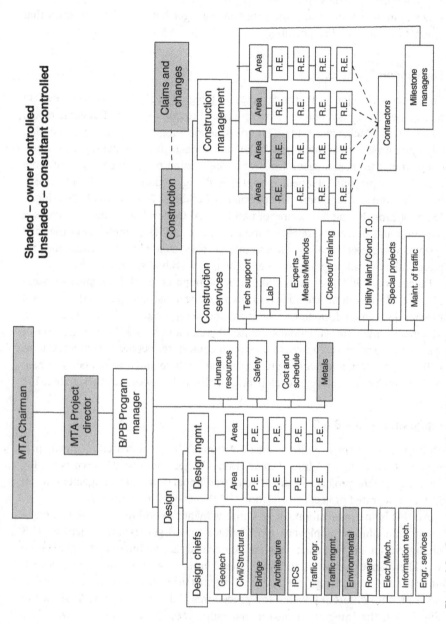

Shaded – owner controlled
Unshaded – consultant controlled

Figure 6.2 Central Artery Tunnel Project Integrated Project Organization. *Source*: Greiman (2013)/with permission of John Wiley & Sons.

have been established at the commencement of the project to ensure its full benefits (Greiman 2013). From the project perspective, the IPO solved many implementation problems and created greater efficiencies; however, this transition to the IPO was met with mixed reviews from the oversight perspective. In its Report on the Big Dig, the National Academies stated (BICE 2003): "The implementation of the IPO has complicated the control of expenses for the B/PB management-consultant team." One of the criticisms of the Big Dig during the process of integration was the loss of independence between the public- and private-sector managers of the project. Working as one integrated team has many advantages but the oversight function requires a separate independent structure (BICE 2003).

Integration at Crossrail

> The introduction of the integrated team simplified accountabilities and created a framework for the collaborative tackling of challenges throughout the project, whilst ensuring Crossrail could recruit the best person for any particular role.
> — *Bill Tucker, Central Section Delivery, Project Delivery Partner*

The Crossrail project is the biggest infrastructure project in Europe. It required the construction of 118 km of new railway from Shenfield and Abbey Wood in the East to Heathrow Airport and Reading in the West, including the construction of 10 new stations. Crossrail operates as an integrated team with staff from Crossrail Limited, the Program Delivery Partner, and the Project Partner. Each organization has its own recruitment processes and procedures as well as terms and conditions for staff.

Crossrail has defined its governance at two levels:

- Corporate Governance – established by the Crossrail Board which sets out delegated authority levels for the board, its committees and subcommittees as well as the scheme of delegated authorities for the executive directors of CRL.
- Program governance – which sits beneath this and constitutes all the forums which, in aggregate, control the Crossrail Project in accordance with the delivery strategy.

Crossrail has implemented processes and procedures to streamline the recruitment of the integrated team. A learning legacy paper on *Recruiting for an Integrated Team* reviews Crossrail's approach to working with multiple partners to secure the right person for the right role (Kosowski 2016). Crossrail also implemented processes and procedures to bring people from multiple organizations, with different terms and conditions together so that they can work in an integrated way as one team. This process is outlined in the *People Strategy paper* (Pascutto 2016). With staff from different employers, subject to different processes, the integration

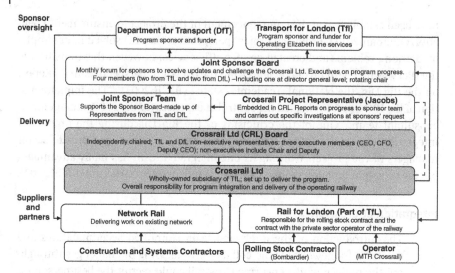

Figure 6.3 Crossrail integrated delivery structure. *Source:* Adapted from Crossrail Governance Documents (Meng 2018).

allowed for the development of processes to manage change and secure the right person for the right job. Crossrail was required to plan for and manage organizational change throughout the project lifecycle.

In contrast to the Big Dig one of the core principles of the governance structure for Crossrail is the clear separation of the sponsor group (or commissioning body), the delivery body (or executing body) and the users. In Crossrail's case, the sponsoring group comprised the two main funders, the delivery body was undertaken by the executive board of Crossrail Ltd (supported by the project management team and supply chain) and the users are the operational team (Figure 6.3). This separation is vital as some organizations, e.g., Transport for London, have a role in each of these elements so clarity of reporting and accountability line is essential. CRL has put prominence on core governance principles such as:

- Clear statement of objective and parameters – including arrangement for remedy
- Sufficient autonomy with a single "controlling mind"
- A clear system of delegation and process for timely decisions that fall outside the limits of delegation
- Process for controlling changes
- Process for reporting and other communications
- A collaborative culture and working relationships
- Boardroom understanding of project objectives, strategy, and approach of what needs to be achieved
- A defined system for assurance at all levels

Crossrail. *Source:* Nigel Young/Foster + Partners.

Part III. Public–Private Partnerships: The Sharing of Risk and Opportunity

An important form of integration over the past several decades in public megaprojects has been the public private partnership. A PPP is defined as "any contractual arrangement between a public entity or authority and a private entity, for providing a public asset or service, in which the private party bears significant risk and management responsibility." It includes brownfield and greenfield projects as well as performance-based management contracts (WBG 2016). Figure 6.4 shows the general responsibilities of the private sector and the government in the management of a PPP though roles can vary depending on the needs of the particular investment. Public private partnerships serve as a resource for much needed technical expertise but also to fill gaps in financing. A PPP will only work if the project environment is right for the development of a public private partnership. Figure 6.4 highlights the different roles of the private sector and the public entity though these can vary from one project to another.

Over the last several decades, the United States has experienced increased private involvement in infrastructure investment, development, and management – particularly in the transportation sector through PPPs. This contemporary activity has rekindled interest in public–private arrangements for infrastructure that were common in the nineteenth and early twentieth centuries but fell dormant until the 1990s (Clark and Hakim 2019). According to the World Bank,

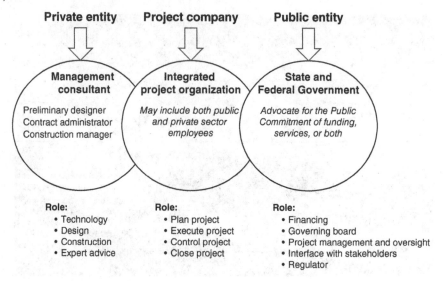

Figure 6.4 Public–private partnership integrated roles.

investments in PPPs have grown in absolute terms since 1991 with two notable periods of expansion and one period of contraction. Countries turned to PPPs throughout most of the 1990s, during which time there were massive commitments. PPPs grew steadily from $7B in 1991 to $91B in 1997, when governments felt the repercussions of the Asian financial crisis (1997–1998). When the global economy picked up steam in the mid-2000s, a second growth phase culminated in record investment of $158B in 2012 (WBG 2016). This second growth phase was unaffected by the global financial crisis of 2008 because many countries increased the public share in the financing of infrastructure projects to help boost investment. A significant decline of about 40% occurred in 2013. Since then, however, investment commitments in PPPs have grown, albeit slowly, reflecting the overall slowdown in key emerging markets, particularly Brazil and India (WBG 2016).

Fragility, conflict, violence, and weak institutions are critical development challenges that have affected many countries. In 2018, a WBG Report entitled, *The State of Infrastructure Public-Private Partnerships in Countries Affected by Fragility, Conflict or Weak Institutions* indicated that a total of 61 countries had been identified as an "Expanded list of Fragile and Conflict-Affected States" (EFCS). Economic and financial conditions, business environment, and rule of law are usually weaker in EFCS countries, raising private sector investment's risk in infrastructure projects. PPP markets in these countries therefore tend to be less developed (WGB 2018). Despite the less conducive environment, the *2018 WBG*

Report finds that for the period of analysis (2012–2016) some EFCS countries were able to bring projects to the market, and another few were able to create a PPP program.

The World Bank's *Private Participation in Infrastructure1 first half-year (H1) Report* shows clear signs of recovery in infrastructure investment since the first wave of the pandemic. In 2020, COVID-19 brought the infrastructure sector to a near standstill, with investment levels at historic lows since the inception of the PPI database. Although the pandemic continues to cause global disruption, investment levels are partially returning to pre-pandemic levels in many countries, indicating that the infrastructure business is slowly adopting to a new normal amidst the pandemic. As countries plan their recoveries, including the use of infrastructure spending to stimulate post COVID-19 economies, it is important for countries to ensure that proposed new infrastructure investments are green, resilient, and inclusive (WBG 2021).

As countries plan for increased megaproject activity, public private partnerships will fill a critical need. To ensure that a public private partnership can succeed, the structure and laws governing public private partnerships will need to be evaluated to make sure the necessary protections for the private sector partner(s) are in place.

Key Questions for Evaluating a Public Private Partnership:

1) Does the country of investment have laws and a governmental unit governing public private partnerships or is the structure embedded in policy and standards?
2) Is the procurement process open to all investors or are there restrictions on foreign investors?
3) Do the laws require that the host country law be applied or do the parties have a choice of law?
4) Is there a maximum or minimum amount for investment required by foreign investors?
5) What are the political conditions for investment in the country?
6) What are the forms for project delivery, i.e., BOT, BOOT, Other?
7) How can the value of a PPP be assessed?
8) Will both the public and the private sector receive their expected return on investment and benefits?
9) What are the grounds for termination of a PPP?
10) What are the critical success factors in PPP development and what are the factors that trigger failure in these projects?

Valuation of a PPP Opportunity

PPP (P3) opportunities are assessed in different ways by public procurement processes, intergovernmental organizations, multilateral banks, and private investors. Organizations use different analytical tools and data to better understand the concepts, inputs, key assumptions and outputs from evaluations of risk, financial feasibility, benefit-cost, and "value for money" analyses to compare the aggregate financial benefits and costs of a P3 alternative with conventional procurement. The following tools are commonly used in the evaluation of a P3.

1) The risk assessment component assists the user in understanding the process used in identifying, defining, valuing, and allocating risks. The outputs from this component are used as inputs into the value for money, benefit-cost, and financial viability components.
2) The value for money analysis component assists the user in understanding the process used in conducting an evaluation of the financial impacts of P3 delivery in comparison with conventional delivery.
3) The benefit-cost analysis component assists the user in understanding the process used in conducting an evaluation of the societal impacts of P3 delivery in comparison with conventional delivery.
4) The financial viability analysis component assists the user in understanding the process used in conducting an evaluation of the affordability to the public agency of the P3 delivery option and the conventional delivery option.

Multilateral banks such as the World Bank, the EBRD and the Inter-American Development Bank use a number of financial ratios to test the financial viability including:

- Debt-to-equity Ratio – Compares the amount of debt in the project against the amount of equity invested.
- Debt-service cover ratio (DSCR) measures the income of the project available to meet debt service (after deducting operating expenses) against the amount of debt service due in the same period. This ratio can be either backward- or forward-looking.
- The loan life cover ratio (LLCR) is the net present value of future project income, available to meet debt service, over the maturity of the loan against the amount of debt.

The typical structure of a PPP is shown in Table 6.1.

Typical Steps in the PPP Process

The development of a PPP process can be complex and time consuming and must comply with local and national laws to ensure that the agreement will be legally sound and enforceable if disputed by the parties themselves or a third party that has an interest in the project. Box 6.3 provides an example of the PPP process in Ethiopia, yet it is similar to many of the requirements for approval of PPPs in both developed and developing countries (Ethiopia 2018). One of the key aspects in

Table 6.1   Structure of a public–private partnership (PPP).

Typical structure of a PPP
Involves a contract, or concession agreement, between a governmental agency and a private entity to design, build, finance, operate, and/or maintain a facility;
Where the private entity is often a special purpose company (SPC) established exclusively for the intended functions and a number of private firms provide funds or services to the company;
Typically have contract durations of 10 years or more;
Include a financing package that the SPC puts together comprised of equity from the company's sponsors and debt provided by bonds or commercial loans; and
Equity and debt are secured solely by the revenue stream that the SPC receives from the facility/project.

Box 6.3   PPP Process in Ethiopia

1) Once a potential PPP project has been identified, a public sector comparator will need to be developed for initial approval by the PPP Board.
2) Next, a feasibility study will need to be undertaken in order to seek authorization to tender from the PPP Directorate.
3) Projects will be procured through an open bidding process.
4) The private sector will be invited to prequalify.
5) Following identification of suitably qualified bidders, they will be invited to submit bids pursuant to a request for proposals issued by the PPP Directorate setting out the technical and financial conditions required to be met.
6) A preference margin may be granted to proposals reflecting local participation.
7) Following any necessary clarifications, technical bids will be opened first and then, for those bids which are responsive, financial bids will be opened. An evaluation report will then be prepared to establish bidder rankings and the results published.
8) The Proclamation envisages various different methodologies for bidding: normally, either a two-stage process or competitive dialogue. However, direct negotiation may be allowed where there is an urgent need and

either (i) the former two processes are considered impractical; (ii) the project is of short duration; or (iii) the project relates to national defense or national security. The proclamation sets out the process for each of these options. The proclamation also contemplates PPPs proceeding by way of unsolicited proposals, provided these do not relate to a project which has already received approval or is being studied.

9) Successful bidders must establish an Ethiopian company as the project vehicle, which may include a public entity as a minority shareholder.

10) The PPP project agreement will set out the terms of the PPP arrangement, respective obligations and fees payable to, or tariffs permitted to be levied by, the private sector party.

11) If any Government support is justified and agreed on a value for money basis, direct payments, contributions in kind, or guarantees may be provided.

12) The public sector will assist the private sector party with any necessary land rights. Subject to approval, the private sector may create security interests over assets, rights, or interests required to secure financing.

13) Dispute resolution mechanisms in whatever forum may be agreed by the parties.

Source: Proclamation, Government of Ethiopia.

ensuring that a public private partnership will be viable is the openness and transparency of the procurement process for the selection of the private sector partner. Any failure to follow the tender procedures may result in revocation of the PPP agreement at a later date.

Program and Project PPP Development

The structure of a PPP program can vary widely depending on the financial, economic, social, technical, and legal analysis. In some countries, the PPP development may involve a two-step process. As shown in Figure 6.5, the PPP should be structured first at the program level which usually requires establishment of a statutory and regulatory authority which defines the public and private obligations and returns, develops capabilities and determines a procurement approach. The PPP at the project level requires preliminary project planning, eligibility of PPP, solicitation of proposals, and procurement and team integration and development. This is generally a negotiated process whereby the government entity will have certain requirements that may be established by law or regulation and the private sector will have certain requirements based on an analysis of their business case.

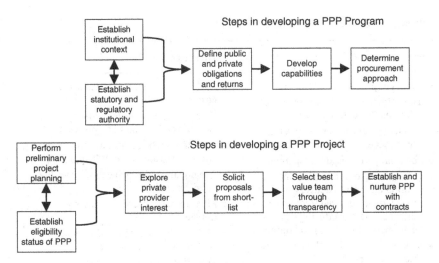

Figure 6.5 PPP program and project development flowchart.

Advantages and Challenges of Implementing P3s

Although PPPs are popular in many European countries and in many developing countries where the infrastructure needs are greatest and financing is scarce, P3s have gotten off to a relatively slow start in the United States. However, they are increasingly used for large-scale infrastructure and public works projects, particularly highway projects. Many P3 projects in recent decades have been extremely successful. For example, the high-occupancy toll lanes project in Virginia involving several private sector firms, resulted in cost savings in the millions of dollars. In addition, the collaboration between government and private partners brought expanded highway capacity online years earlier than a traditional government-does-all approach might have done. A key element of these contracts is that the private party must take on a significant portion of the risk because the contractually specified remuneration – how much the private party receives for its participation – typically depends on performance with incentives sometimes given for better than expected performance. Table 6.2 provides a list of some of the typical advantages of a PPP along with the challenges though every project has some unique advantages and challenges depending on the complexity and uncertainty that exists, particularly in megaprojects.

The Global Implementation of P3s: The Demand and Supply Side

The number of projects using PPP as a model of service delivery has increased in Emerging Markets and Developing Economies (EMDE). Between 2004 and 2013,

Table 6.2 Advantages and challenges of a PPP.

Advantages of a PPP	Challenges of a PPP
• Access to private sector finance	• PPP procurement can be lengthy and costly
• They provide better infrastructure solutions through innovation	• Does sufficient private sector expertise exist to warrant the PPP approach?
• Efficiency advantages from using private sector skills and risk transfer	• Does the public sector have the capacity and skills to adopt the PPP approach?
• Potentially increased transparency	• It is not always possible to transfer life cycle cost risk
• Enlargement of focus from only creating an asset to delivery of a service, including maintenance of the infrastructure asset during its operating lifetime (focus on outputs)	• PPPs do not achieve absolute risk transfer
• Risks are allocated to the party best able to manage the risk	• PPPs imply a loss of management control by the public sector
• Creates incentives to reduce the full life-cycle costs (i.e., construction costs and operating costs)	• The private sector has a higher cost of finance
• Better value for money over the lifetime of the project	• PPPs are long-term relatively inflexible structures
• Value for Money means that the undertaking of the public service activity by a Private Party under a PPP results in a net benefit accruing to that contracting authority or consumer defined in terms of cost, price, quality, quantity, timeliness of implementation and other factors which influence the determination of the best economic value compared to other options of delivering this public service	• In cases where the PPP site has been purchased from community members, formal, and informal tenants may need to be resettled, some community members may have their livelihoods disrupted, while others may potentially be affected by noise, congestion, and air emissions caused by the new facility
• Activity or use of government property	• PPPs must be paid for, and payments bring with them a range of other potential impacts on communities
	• Users pay PPPs that levy fees, tolls, tariffs or other charges on the community have a further direct impact on communities

G-20 (Group of 20) countries accounted for 14 of the top 20 countries in terms of PPP project numbers, based on PricewaterhouseCoopers (PwC) analyses. Of those, three countries (China, India, and the United Kingdom) accounted for almost half of the total number of projects carried out by the top 20 countries. EMDE have tended to focus on more-fundamental economic infrastructure such as energy and transport, while high-income countries have implemented social infrastructure PPP projects, such as schools, hospitals, and leisure facilities. In EMDE, there is a steady and recognizable demand for investment in PPP infrastructure (WBG 2021).

Though the demand for PPPs can be great to make otherwise unviable projects bankable, the supply side is equally important. In other words, PPPs require the right projects to make the financing structure viable, they must have a regulatory environment with sustainable goals and they must also be able to attract the right kind of technical expertise combined with the ability to have a reliable revenue source that will be available throughout the expected life cycle of the project and beyond (WBG 2016). Box 6.4 highlights the PPP challenges in constructing Delhi Metro Rail.

Proposed New Delhi Railway Station. *Source:* Bennett, Coleman & Co. Ltd.

Box 6.4   Airport Express Metro Line: Delhi Metro Rail Corporation, India

As part of the preparation for Delhi to host the XIX Commonwealth Games in 2010, India embarked on a major project to link the New Delhi Railway Station with Indira Gandhi International (IGI) Airport via a metro system. However, the project was never completed in time for the Games. The project was implemented through a public private partnership (P3) which was highly leveraged with debt and equity sponsorship. The project has been successful as it stayed within budget, but also had some failure as the concession agreement was terminated and ridership was less than expected. The megaproject raises many challenges that (P3) projects face globally. The important questions and lessons from this megaproject include:

1) Is the P3 model the best model for public transit?
2) Will the P3 model be able to provide the return on investment needed for the private partner?
3) Who will be responsible for the construction risk where delays in the completion of projects can adversely affect the business case?
4) How will public support for the project be maintained throughout implementation and during operations?
5) How will the relationship between the contractor, the local and national government, the operator and concessionaire be governed?
6) Is it reasonable to make a commitment for a project that will operate for 60 years without regular negotiations due to the changing financial, economic and social environment?

Source: Adapted from Jacob and Raghuram (2013) and Pratap (2013).

The Economic Case for a PPP: Power Purchase Agreements

In order to make a PPP workable, the potential cash flow and revenue stream must be economically sound and predictable to the fullest extent possible (CRS 2020). Without a reliable revenue stream, a PPP will not be viable. Some PPP projects are funded wholly or primarily through user payments. This is most common in economic infrastructure sectors such as toll roads and electricity payments. Other PPP projects are funded wholly or primarily through government sources. This is the case in most social infrastructure projects, but government payments also occur in many economic infrastructure projects. For some forms of economic infrastructure (such as rail transport or water), a PPP project may be just one component of a broader network or service that is operated by another entity (the incumbent operator) and the user is paying that operator for the final service (transportation, water supply to homes). Examples are HSR projects in Europe (mainly in France and Spain), or wastewater and treatment projects (WWTP) and purification plants where the water is taken off by a public water authority that operates the service. For other projects, where it would be possible to charge the user for the use of the infrastructure contracted under the PPP (e.g., a road), it may be decided that no charge will be applied (that is, it would be a toll-free highway). That is a public finance decision as to whether the project should be funded through user charges to the specific users or through utilizing tax revenues to make government payments (the tax revenues could be derived from usage-related taxes, such as a fuel tax, or from general tax revenues). Examples are found worldwide, including in Canada, Hungary, Mexico, Spain, and Portugal.

A common form of revenue sourcing in the energy industry is the power purchase agreement between a producer of energy and an offtake purchaser know as a take-or-pay contract. Offtake Agreements significantly increase the likelihood of project loan approval by reducing long-term project risk and offering stable cash flow for many years. To attract PPP investors and lenders a project must have a viable economic structure. This means there must be sufficient funds to cover the debt ratio as well as to provide a reasonable return on investment to project equity holders. Depending on the type of project, generally revenue is secured from user fees, government guarantees or publicly funded sources, or in the case of energy projects possibly through an offtake purchaser such as an electric or telecommunications company. Whatever, the source of financing it must be secured before project financing documents can be completed. Key features of a power purchase agreement in Vietnam are highlighted in Box 6.5. This document provides a form of guarantee for the project sponsors and lenders, but also provides for allocation of risk and transfer of liabilities under certain agreed upon circumstances. The power purchase agreement is an essential document to make a project bankable when user fees contain more uncertainty. Box 6.6 highlights the characteristics of some successful PPP projects in Hong Kong.

Box 6.5   Power Purchase Agreement

Key Features of Sample BOT Contract for a PPP Power Purchase Agreement in Vietnam:

1) Ministry of industry and trade of the socialist republic of Vietnam (MOIT) responsibilities – the MOIT is responsible for a number of deliverables in respect of the project, including:
    - Procuring a land lease with the local provincial government for the project site.
    - Procuring a number of key regulatory approvals (e.g. investment certificate for the establishment of the project company, electricity license for the plant, and import/export permits).
    - Passports, visas, and work permits for some of project company's personnel; and
    - Negotiations between the project company (and sponsors) and other government departments.
2) Financing – the sponsors are required to fund the construction and operation of the power plant through a combination of equity injection into the project company and loans (clause 7).

3) Remittance and foreign exchange rights (clauses 16.2 and 16.3) – the project company and sponsors are given assurances by the MOIT that they have rights to make remittances according to the financing and equity documents and that they can convert amounts received in Dong (VND) into USD on a monthly basis in accordance with the financing documents. This is to address risks arising from the regulated nature of foreign exchange and overseas remittances in Vietnam (clauses 16.2 and 16.3).

4) Tax incentives (clause 17) – the project company is given a tax holiday in respect of corporate income tax for the initial part of the project (100% exemption for the first four profit making years, 50% reduction for the next nine years).

5) Government events – the project company is relieved from complying with its obligations to the extent affected by a Government Event and the MOIT must pay the project company the Capacity Charges under the power purchase agreement to the extent the project company is unable to deliver energy due to the Government Event. Government Event is defined to include a range of political and regulatory events such as civil unrest in Vietnam, nationalization, failure to obtain or maintain approvals other than due to a breach by the project company.

6) Early termination payments – MOIT is required to buy the power plant and pay the project company an early termination amount in the event of early termination of the BOT contract. Where the early termination is due to MOIT default, the amount is calculated with reference to the project company's project debts, financing costs, equity investment and an agreed rate of return on equity. For other termination events (e.g. prolonged force majeure and project company default), the amount *payable is scaled* down.

Source: Adapted from The World Bank (PPPLRC) (2021).

Case Studies in Public–Private Partnership Development

Box 6.6   Hong Kong PPP Development

PPPs in Hong Kong have been applied to infrastructure development, hospital services, and tourism-related preservation. Although there are several successful PPP infrastructure projects in Hong Kong, including the Tsing Ma Control Area, the Chemical Waste Treatment Plant, and the Asia World-Expo (AWE), other projects are quite controversial. These include the Western Harbor Crossing and the West Kowloon Cultural District. The PPP has also been used to facilitate public housing projects in Hong Kong. Hong Kong is one of the world's most expensive housing markets. To deal with the climbing rents in Hong Kong's private property market, the government has attempted to increase the provision of subsidized housing.

The P3 concept has been applied to many megaprojects in Hong Kong including the Cross Harbor Tunnel (CHT) and the Hong Kong Disney Theme Park. The success of Hong Kong's first BOT tunnel the CHT could be attributed to the strong commitment and support shown by the government toward the development of the tunnel project. In essence, the government demonstrated commitment toward the project by retaining a percentage of the project company's share. Considering the successful progress and significant contribution of the Disney PPP project to Hong Kong's economy and international reputation, it is regarded as one of Hong Kong's landmark PPP projects particularly in the tourism sector (HKDL 2022). Contributing factors towards the success of the Theme Park is the commitment demonstrated by the government, the proper management techniques, and innovations employed (Osei-Kyei and Chan (2021).

Privatization of the World's Water Supply

Privatization of the world's water supply has been subject to controversy in Europe and other parts of the world for some time. At the European level, consistent with the definition of water as a public good of fundamental value for all Union citizens (EC 2014), water is lately at the center of a debate on a return to public management of water (Novaro and Bercelli 2017). The new European regulatory framework seems to widen the range of opportunities for the adoption of public management models for water control: on one hand, by excluding water services from the application of the market-oriented rules governing the concession contract; on the other, by extending the application of the in-house providing scheme (p. 2385).

France could be described as the birthplace of water privatization. Private companies have run French waterworks to one degree or another since the Napoleonic era (Godoy 2003).

Private water firms have had a dominant role in France for more than a century and in England and Wales since the 1989 Thatcher privatizations. Private utilities also have significant market shares in Spain, Italy, and Denmark, but attempts to privatize water elsewhere have met major resistance and support for public ownership has been growing across Europe for more than a decade.

The risks of privatization, of water resources according to a report by Léo Heller, former United Nations special rapporteur on the human rights to safe drinking water and sanitation, include unaffordability and unsustainability, possible deterioration of services, and increased opportunities for corruption (Heller 2020).

With water-related hazards on the rise and the number of people experiencing water stress increasing, experts say the need for better water management is key. Yet, access to clean water still eludes 76% of people in sub-Saharan Africa. Sometimes, according to Aminata Touré, former prime minister of Senegal and a member of Sanitation and Water for All's Global Leadership Council, the private sector is the one with the tools and capabilities to access water that may be hard to reach (Root 2021). According to the World Bank (2016, p. 33) project cancellation rates vary greatly among sectors – for instance, deals in transport (5.1%) and water (5.7%) have a much higher cancellation rate than those in the energy sector (2%). With the lowest cost recovery among all sectors, it should be no surprise that water has the highest rate of cancelled investment commitments at 28.3%. In addition, the water sector has been plagued by the negative perception among stakeholders that water supply should not be provided by the private sector (WBG 2016). The impact of the pandemic has underscored the importance of strengthening key provisions in PPP legal frameworks to ensure that resilience to global disasters, such as pandemics, and other kinds of external shocks is strongly integrated in PPP projects and programs' (WBG 2021).

Public–Private Partnerships: Goals and Advantages

The realities of the private sector marketplace exert a powerful discipline on businesses to maximize efficiency and take full advantage of business opportunities. Successful PPPs enable the public sector to access the discipline, skills and expertise of the private sector.

The Luis Muñoz Marín International Airport in San Juan, Puerto Rico, was privatized in 2013 through the U.S. Federal Aviation Administration (FAA) Airport Privatization Pilot Program (APPP) (FAA 2022). The partners involved were the public airport owner, Puerto Rico Ports Authority, the Puerto Rico P3 Authority, and Aerostar – a 50–50 venture between Highstar Capital, an infrastructure investor, and Grupo Aeroportuario del Sureste SAB de DV, which operates nine airports in Mexico. The process took four years to complete and resulted in a 40-year lease under the Aerostar name. The privatization of San Juan's Luis Munoz Marin International Airport demonstrates that privatization can occur in the United States if both the public and private sectors see that the benefits outweigh the costs. Meanwhile, instead of privatization under the APPP, many public sector airport owners have engaged the private sector through a variety of partnerships ranging from management contracts to development agreements to help reduce costs, improve services, and obtain capital investment without transferring airport control. While privatization may be an option for those public-sector airport owners that determine the potential benefits outweigh the costs, the current structure of the U.S. airport system provides for a broad range of private-sector participation making the need for full privatization less likely (CRS 2021).

Transit Services Public–Private Partnerships

The Reason Foundation in a study for the World Bank described private provision of transit services in three forms (Feigenbaum 2019). The purest provision – privately financed, operated, and maintained services – is limited to locations in which operating transit services can turn a profit. Thus far that has been limited to cities with a population density of 10 000 people or more per square mile. These include dense Asian cities such as Hong Kong and Tokyo, as well as other densely populated cities such as Mexico City and Cairo. As these are the only places in the world where transit systems can be operated profitably, there are a limited number of truly private services.

In a second type of private service, the public sector procures a contract with the private sector to design, build, finance, operate, and maintain service. These public–private partnerships (P3) are becoming an increasingly popular method of building new rail lines. Some experts believe that they could be used for bus rapid transit lines as well. P3s have a track record of producing a higher-quality, lower-cost transit solution compared with public management. Most popular in parts of Europe and Latin America, P3s for transit are now being used in Canada, the United States, and the Middle East. Most transit P3s use some amount of public funding to help build and/or operate the service.

A third option is for the private sector to plan and operate transit service via a competitive service contract, often called tendered service. Public entities will choose to contract different service provisions including operations, service planning, and maintenance to a private entity. These public agencies seek requests for qualifications and requests for proposals from several consortiums. They select between three and five finalists and award service to the team that offers taxpayers the best overall value (Feigenbaum 2019).

Public Private Procurement in the United States

The U.S. Department of Transportation sets forth detailed guidance in its *Public Private Partnership (P3) Procurement Guide for Public Owners* (USDOT 2019). There are rare exceptions to the open bidding process in a public project and they are usually set out in law or regulations. For example, if a project requires unique technical skills and engineering knowledge not readily available in the market, the public sponsor may seek a waiver or may pursuant to statute seek a sole source for the project without having to go through a public bidding process.

Box 6.7   Unsolicited Proposal, Capital Beltway Project, Fluor Enterprises

The procurement of the $2B I-495 Capital Beltway High-Occupancy Toll (HOT) Lanes project started with an unsolicited proposal. In June 2002, Fluor Daniel (now Fluor Enterprises), a private engineering, procurement, construction, maintenance, and project management company based in Irving, Texas, submitted an unsolicited proposal to Virginia Department of Transportation (VDOT) to design, build, finance, operate and maintain HOT lanes on the Capital Beltway on a P3 basis.

After issuing a request for competing proposals (and receiving none), VDOT negotiated the P3 agreement with Fluor. The project ultimately expanded and improved a 14-mile section of the Capital Beltway (I-495) in Fairfax County, Virginia. As well as adding four new HOT lanes (two in each direction) and reconstructing the existing general-purpose lanes, the project replaced over 50 bridges and instituted vehicle occupancy requirements on the managed lanes.

Motorists that declare high-occupancy vehicle (HOV) status with their transponder are required to have three or more people in the vehicle. Initially, the plan was for Fluor to provide 100% of the financing. However, Fluor eventually partnered with Transurban to improve its position relative to toll road operation and its ability to finance the project. The I-495 procurement presents an example of how market conditions affect P3 procurements. The project faced significant challenges to its financing as a result of the global financial crisis.

Source: Adapted from U.S. Department of Transportation (USDOT) (2019).

Or, as illustrated in the public procurement for the I-495 Capital Beltway High-Occupancy Toll (HOT) Lanes project in the Washington, DC area in Box 6.7, an exception may come about by an initial solicitation of a proposal by the private sector and no other bids are forthcoming illustrating the process for an unsolicited proposal.

The Seoul Metro System Private Procurement

Another example of private procurement for public transportation projects, is the Seoul Metro System, considered one of the best in the world (Railway Technology 2020). The subway is funded by the Macquarie Korea Infrastructure Fund (MKIF). It is the first private investment in the subway system. The company has made investments in the form of equity and subordinated loans. MKIF has a concession term of 30 years, starting from July 2009. On a daily basis, Seoul's subway system serves more than 7 million passengers, compared to Beijing's 6.74 million, Tokyo's 6.22 million, and New York City's 5.47 million. Seoul also has the highest number of subway stations in operation, comparable to New York City (468 stations) and Paris (301 stations); and a significantly longer track length than Beijing, London (415 km), and New York City (368 km) (Warsen et al. 2018).

Emerging Trends and Social Considerations for PPP Development

1) Community Engagement by the PPP

An essential aspect of all PPP projects is addressing the concerns of the local community. It is rare when a community project does not have an impact in some way on people's lives, property or way of life. They must be constantly analyzed and observed for potential risks that may emerge to mitigate the risk at the earliest possible time (IFC 2019). Some of these concerns include:

- Expectations for jobs from the local community or, in the case of an existing service, fears about losing employment or being transferred from the public to the private sector.
- Construction and how it disrupts daily activities for citizens (e.g. related dust, noise, traffic patterns, utility services, etc.).
- In-migration, as people come to the project area in search of work. This can place additional demand on local services and infrastructure and create issues with the local population. The social impact of migration/temporary workers, particularly of the potential for gender-based violence.
- Relocation or resettlement impacts on households and on small businesses.
- The project will not meet the needs of the community that will use the PPP because it has not been designed with community input.
- The construction or operation of a PPP project will affect their ability to access local services or infrastructure. A new road may have a toll charge that local residents cannot afford to pay.
- The environmental impact of the PPP.
- User charges, or the taxes required to pay for a PPP.
- Changes in the availability or nature of public services. Some services may be discontinued to make way for the PPP.
- How the project will be gender friendly, responsive to climate change and protect the poor.
- Cultural heritage that the project may affect or would incorporate.

2) Building Social Infrastructure Through a P3

Social infrastructure P3s, like transportation and environmental P3s, present unique procurement and contracting challenges for state and local governments.

Social infrastructure P3s are strictly speaking neither construction contracts nor service contracts; they are both. This dual construction/service nature is frequently at odds with the procurement and contracting statutes, ordinances and regulations of many state and local governments. For this reason, the Government Finance Officers Association (GFOA) recommends that states pass P3 enabling legislation. As of 1 September 2018, 37 states had passed some form of P3 enabling legislation (NCSL 2019). Unfortunately, the majority of this legislation restricts the use of P3s to transportation projects only or makes no specific provisions for social infrastructure. Consequently, only a few states have P3 enabling legislation that applies broadly to social infrastructure, but there is increasing interest in moving toward a more socially friendly PPP structure (Clark and Hakim 2019).

3) Build Sustainability into the Public–Private Partnership

Importantly, one critical issue that is emerging with the rapid development of PPPs is how sustainability could be enhanced through the PPP concept (Osei-Kyei et al. 2019). Sustainability is a global issue and its use in PPPs has become essential considering the impact that PPPs have on society. Recently, the adoption of social responsibility (SR) initiatives in PPPs has become critical to avoid resistance and opposition to major megaprojects. They also add economic value that is often not understood by the partnership and therefore not incorporated in benefit-cost analysis in the early stages of the project. Osei-Kyei et al. (2019) through their study of social responsibility in PPPs provide the following useful recommendations for enhancing sustainability in practice:

---

1) Focus on improved quality and quality control measures to increase the overall satisfaction of users and project stakeholders.
2) Utilize resources including land, capital, and employees and tools such as building information modeling (BIM) to estimate costs at early stages.
3) Enhance profitability through the adoption of smart and digitalized technology.
4) To create more job opportunities, construction and operational works should be subcontracted to local businesses and suppliers.
5) A detailed environmental plan should be prepared and reviewed by environmental protection agencies in the host country. In addition, adequate measures should be adopted to avoid the pollution of water bodies and the emission of harmful gases during the construction and operation of the facility and beyond.

4) PPPs for the Fourth Industrial Revolution (4IR)

The PPP model is based on three principles: risk allocation and transfer, affordability, and value for money. Traditionally, PPPs have been leveraged for hard service development such as infrastructure development. The advancement of technology within the context of the Fourth Industrial Revolution (4IR) has created new opportunities and risks for PPPs as important mechanisms for the promotion of development. The 4IR has implications for government service delivery, which have brought about an increased demand for service delivery innovation and the development of information and communications technology (ICT).

5) Public–Private Partnership Legislation and Regulation

Globally, countries are at different stages of enacting public private legal frameworks and public private partnership laws and regulations with some having no formal framework while others have dedicated PPP units. As an example, in Australia, only three state/territory governments (New South Wales, South Australia, and Victoria) have established a dedicated unit. Other states and territory governments (i.e. Tasmania, Western Australia, Australian Capital Territory, Northern Territory) do not have a dedicated PPP units and place responsibility within their finance ministry more generally. In the United States responsibility for public private partnerships lies with three entities – The U.S. Department of Transportation, Federal Highway Administration (FHWA), the Department of State, Global Partnership Initiative and the National Council for PPP.

Table 6.3 provides an overview of public private partnership legislation in six countries demonstrating the distinctions and similarities among these laws. PPP frameworks and delivery methods come in a variety of forms ranging from concession agreements to models of delivery that include Design Build, Design Bid Build, and Build Own Operate and in some countries PPPs are industry specific. Examples of PPP projects are included along with some of the special provisions of the legislation that may create challenges for the private sector. The context of the PPP environment must be carefully analyzed in a given country to ensure the viability of privately financed projects in that country.

Summary

In this chapter, an overview of the successes and the challenges of building an integrated project organization, an integrated systems approach, and integrating project disciplines including systems engineering and program management and building global PPPs was presented. The PPP structure is a commonly recognized approach to financing infrastructure where public funds are insufficient to fund these costly undertakings and greater technical skills are needed that the

Table 6.3 Public private partnership legislation.

Country/PPP law	Form of PPP	Special provisions	PPP projects
Bosnia and Herzegovina Law on Concessions of Bosnia and Herzegovina (Concession Law) Book of Rules on Request Submitting Procedure and Concession Granting Procedure (Procedure Rules) Sectoral Law: Law on Public-Private Partnership in the Republic of Srpska	Concession	Public invitation for prospective bidders. The Conceding Party shall submit to the Commission on Concessions (Commission) an RFP for approval. (Concession Law, art.22, art.23; Procedure Rules, art.8, art.9, art.10). The Commission may decide to open an international tender, depending on the value of project, the need to engage specific and modern know-how and technology, project financing structure and other cases when the Commission considers a project of particular importance. (Procedure Rules, art.13)	**Infrastructure sector** - Telecom Telekom Srpske Infrastructure sector – Power EFT Stanari Coal Plant Infrastructure sector – Telecom HT Mobilne Komunikacije d.o.o. Mostar
Ethiopia Ethiopia enacted a new Proclamation No. 1076/2018 facilitating Public-Private Partnership (PPP). PPP Board, and a PPP Unit within the Ministry of Finance, PPP Directorate General, PPPDG	Concession DBOT DBFM BF OM DB	The goals of the law are to enhance transparency, fairness, value for money, and efficiency through the establishment of specific procedures. The private sector will submit bids through an open bidding process	**Expressway Project, Mieso-Dire Dawa**, 445 million Oromiya and Somali **Hydro Power Project** Tams, 3.36B, Gambella Regional State **Solar Projects**, Wolenchiti, 165 million, 150 MW, Oromiya Regional State

India There is no specific legal or statutory framework for PPPs at the central level. However, a few states (for example, Punjab) have State Acts that govern construction contracts and have their own statutory tribunal where disputes are resolved. Government of India has set up Public Private Partnership Appraisal Committee to streamline appraisal and approval of projects. Transparent and competitive bidding processes have been established	There is no typical procurement/tender process in PPP transactions in India. All government tendering is governed by the guidelines laid down by the Central Vigilance Commission to ensure transparency and avoid corrupt practices, and the leading lending agencies such as the World Bank and other multilaterals	There are no standard forms for PPP projects. However, various authorities do have approved formats, which are regularly amended. For example, the National Highways Authority of India (NHAI) has its own format which is regularly updated and amended	**Mundra Ultra Mega Power Plant** **L&T Hyderabad Metro Rail Private Limited** **GMR Kishangarh Udaipur Ahmedabad Expressway Limited**
The Netherlands PPP projects are usually tendered, making use of existing (EU-based) procurement legislation. There is no formal PPP Law in the Netherlands. (PPPLRC 2022)	The vast majority of PPP projects in the Netherlands including private financing (DBFM or DBFMO) are tendered by the central government	The Dutch government does not provide financing or guarantees for PPP projects. It does accommodate private financing, for instance, by agreeing to enter into a direct agreement with the financiers of the project. In addition, the standard DBFM(O) contracts contain detailed provisions dealing with the repayment of financing costs in case of early termination of the project due to, for instance, force majeure or a breach of contract by the contracting authority	Construction and reconstruction of the **Second Coen Tunnel** (DBFM). Construction and reconstruction of the **A15 Maasvlakte-Vaanplein motorway** (DBFM). Reconstruction of the **A2 Hooggelegen** (alliance).

(*Continued*)

Table 6.3 (Continued)

Country/PPP law	Form of PPP	Special provisions	PPP projects
Thailand 2562 (2019) (PPP Act) was enacted to govern public private partnership (PPP) projects in Thailand and provide a transparent framework for the development and delivery of such projects	Concession Joint Venture Build-Operate Transfer Build-Transfer-Operate A joint venture (JV) led by Gulf Energy Development Plc will develop Phase III of Thailand's Laem Chabang port following a new deal signed with the national authority	International Standards Transparent Procurement Procedures, including negotiation. A direct agreement entered into by and between the private entity, the financiers and the procuring government agency to increase the bankability of the project – is allowed	**U-Tapao Airport and Eastern Airport City** Part of the combined U-Tapao and Eastern Airport City project. The aim is to establish a third main international airport in Thailand **High Speed Rail Connection** In October 2019, the State Railway of Thailand signed a public–private partnership agreement with a consortium led by Thailand's Charoen Pokphand Group, which also included investors from China and Japan **Laem Chabang Port Phase III** Creation of a deep-sea port and other facilities. **The Bang Pa-in-Nakhon Ratchasima Intercity Motorway (M6)**
Vietnam The Law on Public-Private Partnership (the PPP Law) effective from 1 January 2021. (Steelberg 2021)	The following delivery methods are accepted: BOT BTO, BOO, O&M BTL, and BLT, but no longer include BT. Can issue bonds without waiting period	Vietnamese Law is mandatory and only standard form contracts are permitted and international law and the commonly used English or Singaporean contract law including liquidated and consequential damages is no longer permitted	The first two BOT power plants built comprised one (**Phu My 2.2**) that resulted from a World Bank-assisted tender; and one by investor proposal (**Phu My 3**). The next built was Mong Duong 2, also investor proposal. Currently under construction is **Nghi Son 2**, which resulted from an IFC-assisted tender

Source: Adapted from Country PPP legislation.

government owner can provide. The framework for PPPs must be reviewed in each country so that the political, technical, financial, social, and economic risks can be properly assessed. The viability of a PPP project depends on the experience of a particular country the project selection process, preparation for the partnership, and the reliability of the project revenue stream among other factors.

Lessons and Best Practices

1) Integrating people, processes, and systems can create greater efficiencies and improve project performance.
2) Integrating disciplines such as systems engineering and program management through shared values, standards, policies and procedures can reduce conflict and enhance the likelihood of success.
3) Public–private partnerships are typically developed to provide additional financing when public sources are insufficient. They also provide needed technical expertise for complex projects where local talent is unavailable.
4) The legal and regulatory process is key to attracting private investors so that risks and liabilities are reduced to the fullest extent possible.
5) Risk allocation is a primary benefit for the megaprojects owner and sponsor(s).
6) Finding an offtake purchaser for the product or project produced can be key to securing project financing.
7) Transparency and an open competitive process is essential to attracting private investment with rare exceptions.
8) An environmental and social assessment is required for all projects to ensure the goals of inclusivity, sustainability, and feasibility are met.
9) The viability of a public private partnership depends on both the demand and supply side. Particular attention should be paid during the selection of the private partner to make sure the expertise is available for the entire life cycle of the project.
10) Failure of governments to continuously develop new projects can result in a significant lag in the PPP project pipeline which is an ongoing process.
11) Integration is critical to ensure value from the partnership both at the organizational and the project level.
12) Renegotiation is an opportunity to improve the PPP arrangements particularly when unexpected events impact the sustainability of the project.
13) Evaluate supply chain reliability throughout the project life so that unnecessary delay and increased cost is avoided.
14) Build a collaborative environment where trust, transparency, and openness are key attributes.

Discussion Questions

1 Why is integration an important tool in developing and implementing global megaprojects?

2 How can the roles and responsibilities of program managers and systems engineers be integrated in the context of a megaproject?

3 How can integration improve the governance of megaprojects?

4 What are the key policy issues that must be decided prior to initiating a public private partnership?

5 What are the essential criteria a government must consider in selecting a private sector partner?

6 What is systems thinking and how can it be utilized to add value to the partnership?

7 How would you go about integrating the project organization and the numerous disciplines that exist on a project?

8 Why are public private partnerships important to developing countries?

9 How should public private partnerships be structured and governed?

10 What are the cultural considerations in establishing a PPP between the host government and foreign investors?

11 How can the PPP ensure private investors that the return on investment is realistic?

12 Who should bear the financial, economic, political, construction, operations and maintenance risks of a PPP?

13 What are the essential phases in building a PPP?

14 What are the considerations in transferring a PPP from a project environment to operations? How would you mitigate risk during and after the transition?

15 Why would a project seek out an offtake purchaser and in the case of an energy project what are the advantages of a power purchase agreement and who would the agreement benefit the most?

16 How do you ensure the integrity of the PPP and its partners?

References

American Institute of Architects (AIA) (2008). *Integrated Project Delivery: A Guide.* Washington, DC and Sacramento, CA: AIA National Council, and AIA California Council.

Beasley, R., and Partridge, R. (2011). *The Three T's of Systems Engineering—Trading, Tailoring, and Thinking.* [Paper presentation] The 21st Annual Symposium of the International Council on Systems Engineering (INCOSE), Denver, CO, June 20–23.

Board on Infrastructure and Constructed Environment (BICE) (2003). *Completing the Big Dig: Managing the Final Stages of Boston's Central Artery/Tunnel Project.* National Academy of Engineering, National Research Council, Transportation Research Board of the National Academies.

Brady, T. and Davies, A. (2010). From hero to hubris: Reconsidering the project management of Heathrow's Terminal 5. *International Journal of Project Management* 28 (2): 151–157. https://doi.org/10.1016/j.ijproman.2009.11.011.

Campagnolo, D. and Camuffo, A. (2010). The concept of modularity in management studies: A literature review. *International Journal of Management Reviews* 12 (3): 259–283. https://doi-org.ezproxy.bu.edu/10.1111/j.1468-2370.2009.00260.x.

Chia, A. (2008). *A Large-Scale System Engineering Perspective of Water Management in Singapore.* Proceedings of the 18th Annual INCOSE International Symposium, Utrech, The Netherlands, 15–19 June 2008.

Clark, R.M. and Hakim, S. (ed.) (2019). *Public Private Partnerships: Construction, Protection, and Rehabilitation of Critical Infrastructure.* Switzerland, AG: Springer Nature.

Congressional Research Service (CRS) (2020). *Funding and Financing Highways and Public Transportation.* Updated 11 May 2020. (pp. 19–20) CRS Report, R45350. https://sgp.fas.org/crs/misc/R45350.pdf.

Congressional Research Service (CRS) (2021). *Airport Privatization: Issues and Options for Congress.* Washington, DC: CRS.

Davies, A. and Mackenzie, I. (2014). Project complexity and systems integration: Constructing the London 2012 Olympics and Paralympics games. *International Journal of Project Management* 32 (5): 773–790.

Davies, A., Gann, D., and Douglas, T. (2009). Innovation in megaprojects: Systems integration at London Heathrow Terminal 5. *California Management Review* 51 (2): 101–125.

Ethiopia (2018). *Proclamation No. 1076/2018 Facilitating Public-Private Partnership (PPP)*. Ministry of Finance, PPP Directorate General, PPPDG. https://www. informea.org/en/legislation/ public-private-partnership-proclamation-no-10762018.

European Commission (EC) (2014). Communication on the European Citizens' Initiative. *Water and sanitation are a human right! Water is a public good, not a commodity!* (COM/2014/177 final).

European Court of Auditors (EU) (2018)/Special Report: Public Private Partnerships in the EU: Widespread shortcomings and limited benefits. *European Procurement and Public Private Law Review* (EPPPLR) (2021).

Federal Aviation Administration (2022). *Airport Investment Partnership Program (AIPP) – Formerly Airport Privatization Pilot Program*. https://www.faa.gov/newsroom/ airport-investment-partnership-program-aipp-formerly-airport-privatization-pilot-program-0.

Federal Highway Administration (FHWA) (2011). *Lessons Learned: Summary of Lessons Learned from Recent Major Projects*. Washington, DC: Office of Innovative Program Delivery (IPD), Department of Transportation.

Feigenbaum, B. (2019). Summary of Transit Public–Private Partnerships. In: *Public Private Partnerships. Competitive Government: Public Private Partnerships* (ed. R.M. Clark and S. Hakim). Cham: Springer https://doi. org/10.1007/978-3-030-24600-6_3.

Godoy, J. (2003). *The Water Barons: Water and Power: The French Connection*. Washington, DC: The Center for Public Integrity.

Greiman, V.A. (2013). *Megaproject Management: Lessons on Risk and Project Management from the Big Dig*. Hoboken, NJ: John Wiley & Sons.

Heller, L. (2020). *Privatization and the Human Rights to Water and Sanitation*. Report by the Special Rapporteur on the human rights to water and sanitation, United Nations Human Rights Office of the High Commissioner, Geneva.

Hong Kong Disneyland (HKDL) (2022, March 21). *Hong Kong Disneyland Sees Record High Local Attendance and Annual Pass Membership with the launch of New Castle of Magical Dreams and Daytime Show*. https://news.hongkongdisneyland.com/en/ press/2022-03-21/.

House of Commons Transport Committee (HCTC) (2008). (EV 5) The opening of Heathrow Terminal 5 Twelfth Report of Session 2007–08 Report, together with formal minutes, oral and written evidence. Ordered by The House of Commons to be printed 22 October 2008.

Hughes, T.P. (1998). *Rescuing Prometheus: Four Monumental Projects that Changed the Modern World*. New York, NY: Vintage Books, Random House.

Hughes, T.P. (2004). *Human-Built World: How to Think About Technology and Culture*. Chicago: The University of Chicago Press.

Huxham, C. and Vangen, S. (2005). *Managing to Collaborate; The Theory and Practice of Collaborative Advantage*. London, UK: Routledge.

IEEE (2005). *Application and Management of the Systems Engineering Process*, (5th ed.). New York, NY: The Institute for Electrical and Electronics Engineers (IEEE).

International Finance Corporation (IFC) (2019). *A Guide to Community Engagement for Public-Private Partnerships* (Draft for discussion). Washington, DC: The World Bank Group.

Jacob, J. and Raghuram, G. (2013). *Delhi Metro: Airport Express Line*. Indian Institute of Management-Ahmedabad, Ivey Publishing: Ontario, Canada.

Koppenjan, J.F.M. (2005). The formation of public-private partnerships: Lessons from nine transport infrastructure projects in The Netherlands. *Public Administration*, 83 (1) 135–157. https://doi.org/10.1111/j.0033-3298.2005.00441.x. Wiley Online.

Kosowski, L. (2016, September 27). *Recruiting for an Integrated Team, Crossrail Learning Legacy*. London, UK: Crossrail Ltd. https://learninglegacy.crossrail.co.uk/documents/recruiting-integrated-team/.

Langley, M., Robitaille, S., and Thomas, J. (2011). *Toward a New Mindset: Bridging the Gap Between Program Management and Systems Engineering*. PM Network, September. http://www.pmi.org/Business-Solutions/~/media/PDF/Business-Solutions/Lean-Enablers/PMN0911-INCOSE.ashx.

Locatelli, G., Mancini, M., and Romano, E. (2014). Systems engineering to improve the governance in complex project environments. *International Journal of Project Management* 32 (8): 1395–1410.

Mahnken, T.G. (2008). *Technology and the American Way of War Since 1945*. New York, NY: Columbia University Press.

Meng, J.M. (2018). *Crossrail Programme Governance*. London, UK: Crossrail, Ltd.

Miller, R. and Hobbs, B. (2005). Governance regimes for large complex projects. *Project Management Journal* 36 (3): 42–50.

Miller, R. and Lessard, D. (2001). *The Strategic Management of Large Engineering Projects: Shaping Risks, Institutions and Governance*. Cambridge, MA: MIT Press.

Morris, P.W.G. (2000). Researching the Unanswered Questions of Project Management. In: *Project Management Research at the Turn of the Millennium: Proceedings of PMI Research Conference 2000* (ed. D.P. Slevin, D.J. Cleland and J.K. Pinto), 87–102. Newtown Square, PA: Project Management Institute.

Müller, R. (2009). *Project Governance*. Aldershot, UK: Gower Publishing.

NASA (2021). *Visitors to the Station by Country*. Washington, DC: National Aeronautics and Space Administration.

National Aeronautics and Space Administration (NASA) (2010, February 21). Robert Frosch on systems engineering. *NASA APPEL Knowledge Services* 1 (5).

National Commission on State Legislators (NCSL) (2019). *NCSL P3 State Legislative Update: 2016–2018.* Denver, CO and Washington, DC: NCSL.

Nicholas, J.M. and Steyn, H. (2008). *Project Management for Business, Engineering, and Technology*, 3rde. Burlington, MA: Butterworth-Heinemann.

Novaro, P. and Bercelli, B. (2017). Water services are the bridgehead for a return to publicly owned utilities in Europe. A comparative analysis. *Water Resources Management* 31 (8): 2375–2387. https://doi.org/10.1007/s11269-016-1535-z.

Oehmen, J., Thuesen, C., Ruiz, P.P., and Geraldi, J. (2015). *Complexity Management for Projects, Programs, and Portfolios: An Engineering Systems Perspective.* [PMI White Paper]. Newtown Square, PA: Project Management Institute.

Osei-Kyei, R., and Chan, A.P.C. (2015). Review of studies on the critical success factors for public–private partnership (PPP) projects from 1990 to 2013. *International Journal of Project Management*, 33 (6) 1335–1346. https://doi.org/10.1016/j.ijproman.

Osei-Kyei, R. and Chan, A.P.C. (2021). *International Best Practices of Public Private Partnership: Insights from Developed and Developing Economies.* Singapore Pvt. Ltd: Springer Nature.

Osei-Kyei, R., Chan, A.P.C., Yu, Y. et al. (2019). Social responsibility initiatives for public-private partnership projects: A COMPARATIVE STUDY between China and Ghana. *Sustainability* 11: 1338. https://doi.org/10.3390/su11051338.

Pascutto, N. (2016, Sept 27). *People Strategy.* London, UK: Crossrail Learning Legacy. Crossrail Ltd.

Patanakul, P. and Shenhar, A. (2009). Exploring the concept of value creation in program planning and systems engineering processes. *Systems Engineering* 13 (4): 340–352. Wiley online.

Pollitt, M.G. (2002). The Declining Role of the State in Infrastructure Investments in the UK. In: *Private Initiatives in Infrastructure: Priorities, Incentives and Performance* (ed. S.V. Berg, M.G. Pollitt and M. Tsuji), 67–100. Cheltenham, UK: Edward Elgar.

Poppo, L. and Zenger, T. (2002). Do formal contracts and relational governance function as substitutes or complements? *Strategic Management Journal* 23 (8): 707–725.

Pratap, K.V. (2013). Delhi Airport Metro Fiasco: What can be done to redeem the project? *Economic and Political Weekly* 48 (49): 18–20. https://www.jstor.org/stable/24478367.

Public Private Partnership Legal Resource Center (PPPLRC) (2022). *Procurement Processes and Bidding Documents.* Washington, DC: The World Bank, IBRD, IDA, World Bank Group.

Razzetti, E.A. (2020) System Integration - Enabling Capability Through Connectivity. *Defense Acquisition Magazine*, Defense Acquisition University, Ft. Belvoir, VA.

Rebentisch, E.S., Nelson, R.M., Townsend, S.A. et al. (ed.) (2017). *Integrating Program Management and Systems Engineering: Methods, Tools, and Organizational Systems for Improving Performance*. Hoboken, NJ: John Wiley & Sons.

Root, R. (2021). *African Water Activists Push Back on Corporate Privatization*. London, Washington, DC: Devex.

Seoul Metropolitan Subway (2020, July 30). *Railway Technology*. https://www.railway-technology.com/projects/seoul-metro/.

Shenhar, A.J. and Dvir, D. (2007). Project management research: The challenge and the opportunity. *Project Management Journal* 38 (2): 93–99. https://doi.org/10.1177/875697280703800210.

Singapore PUB (2022). *Singapore's National Water Agency. Singapore Water Story*. https://www.pub.gov.sg/about/organisationalchart.

Steelberg, V.C. (2021). Late Development and the Private Sector: A Perspective on Public-Private Partnerships in Vietnam, 59 *Columbia Journal of Transnational Law* 726.

Stockman, B., Boyle, J., and Bacon, J. (2010). *International Space Station Systems Engineering Case Study*. Air Force Center for Systems engineering and the National Aeronautics Space Administration.

U.S. Department of Transportation (USDOT) (2019). *Public-Private Partnership (P3) Procurement: A Guide for Public Owners*. Washington, DC: Federal Highway Administration.

Walden, D.D., Roedler, G.J., Forsberg, K.J. et al. (ed.) (2015). *Systems Engineering Handbook: A Guide for System Life Cycle Processes and Activities*, (4th ed.)e. San Diego, CA: International Council on Systems Engineering.

Warsen, R., Nederhand, J., Klijn, E.H. et al. (2018). What makes public-private partnerships work? Survey research into the outcomes and the quality of cooperation in PPPs. *Public Management Review* 20 (8): 1165–1185.

Warsen, R. Klijn, E.H., amd Kippenjan, J. (2019). *Mix and Match: How Contractual and Relational Conditions Are Combined in Successful Public-Private Partnerships*. Oxford University Press on Behalf of the Public Management Research Association.

World Bank Group (WBG) (2016). The State of PPPs Infrastructure Public-Private Partnerships. In: *Georgia Institute of Technology Markets and Developing Economies 1991–2015* (ed. B. Feigenbaum). Washington, DC: Reason Foundation, M.C.R.P. Transportation Planning, PPIAF, WBG.

World Bank Group (WBG) (2018). *The State of Infrastructure Public-Private Partnerships in Countries Affected by Fragility, Conflict or Weak Institutions*. Washington, DC: WBG.

World Bank Group (WBG) (2021). *Private Participation in Infrastructure (PPI) as Defined by the Private Participation in Infrastructure Database*, 2021 Half Year Report (H1). Washington, DC: WBG http://ppi.worldbank.org/methodology/ppi-methodology.

7

Managing the Megaproject Implementation and Delivery

Let our advance worrying become advance thinking and planning.
— Winston Churchill

Introduction

The future is intelligent: By 2030, artificial intelligence (AI) will add $15.7 trillion to the global GDP, with $6.6 trillion projected to be from increased productivity and $9.1 trillion from consumption effects (PwC 2017). With the advancement of AI, it will be important to prepare for a hybrid workforce in which AI and human beings work side-by-side. AI will play a major role in the delivery of the megaprojects of the future and will improve performance dramatically. Thus, the implementation of megaprojects in 5–10 years will look very different than it does today.

In this chapter, we introduce the steps in implementation of two of the world's major megaprojects – London's Crossrail Project (Europe's largest megaproject) and Boston's Third Artery/Harbor Tunnel Project popularly known as the Big Dig (America's largest inner-city project) that used some of the same resources, methodologies and policies though there were also some important differences. We contrast these two case studies to better understand the various challenges in the implementation of megaprojects which are characterized by:

- Long duration and over $1 billion in costs
- Large-scale public policy making
- Complex systems and interorganizational relations
- Emergent risk and catastrophic potential
- Technological and procedural complexity
- Collaborative contracting, integration, and partnering

Global Megaprojects: Lessons, Case Studies, and Expert Advice on International Megaproject Management, First Edition. Virginia A. Greiman.
© 2023 John Wiley & Sons, Inc. Published 2023 by John Wiley & Sons, Inc.

- Dynamic governance structures
- Extensive social and environmental frameworks
- Major events and political adaptation

Development of a Megaproject

In this section, we are introduced to the major phases in the life cycle of megaproject development. As illustrated in Figure 7.1, the five major phases of megaproject development include: (i) initiate: development of the concept, securing the financing, and seeking public support; (ii) procure: scope development and procurement; (iii) execute: implementation of the megaproject; (iv) control: controlling and monitoring the megaproject; and (v) transition: The transition from building a megaproject to operating the structure and services provided by the megaproject. These phases are not always linear and may overlap or stop and start again at various points in the long lifecycle. Projects are constantly altering existing scope and adding new scope as decisions are made regarding project design and construction. For example, in a fast-track project where construction is started before design has finished there will be continuous planning and design until all elements of the project, i.e., roads, bridges, and tunnels are complete.

Similarly, stakeholder demands throughout the life cycle such as environmental concerns, sustainability, and megaproject social frameworks will continue to evolve requiring changes in plans, scope, schedules, and budgets. The shaping of megaprojects results from interactive decision making among numerous stakeholders over many years. It is a complex process and results from progressive

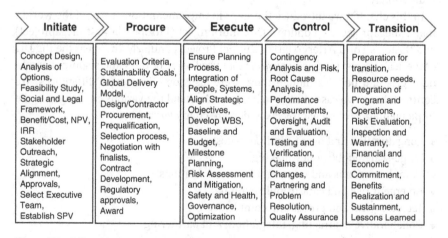

Initiate	Procure	Execute	Control	Transition
Concept Design, Analysis of Options, Feasibility Study, Social and Legal Framework, Benefit/Cost, NPV, IRR Stakeholder Outreach, Strategic Alignment, Approvals, Select Executive Team, Establish SPV	Evaluation Criteria, Sustainability Goals, Global Delivery Model, Design/Contractor Procurement, Prequalification, Selection process, Negotiation with finalists, Contract Development, Regulatory approvals, Award	Ensure Planning Process, Integration of People, Systems, Align Strategic Objectives, Develop WBS, Baseline and Budget, Milestone Planning, Risk Assessment and Mitigation, Safety and Health, Governance, Optimization	Contingency Analysis and Risk, Root Cause Analysis, Performance Measurements, Oversight, Audit and Evaluation, Testing and Verification, Claims and Changes, Partnering and Problem Resolution, Quality Assurance	Preparation for transition, Resource needs, Integration of Program and Operations, Risk Evaluation, Inspection and Warranty, Financial and Economic Commitment, Benefits Realization and Sustainment, Lessons Learned

Figure 7.1 Life cycle in megaproject development.

problem solving over multiple events. Shaping is a much under studied area of project management and there has been a call by scholars for more research in this area (Pinto and Winch 2016).

The five commonly recognized phases of a megaproject are highlighted in Figure 7.1. Each phase brings its own challenges and each succeeding phase depends on the success of the preceding phase. In this section, we will look at the problems and the actual solutions to the cases presented with an aim toward understanding the important issues that must be addressed proactively for a project to succeed in meeting its goals and objectives.

Phase 1: Initiation of the Megaproject

The lifecycle of a megaproject can last decades when compared to smaller projects, requiring a critical upfront planning phase to address the multiple issues that arise in these mammoth undertakings (Miller and Hobbs 2005). Public investment projects require knowledge-based decisions at all process steps. However, decisions at the front end of the project are crucial. This is because front-end decisions determine the fate of the project either to continue or halt the process from going forward to the next step (Belay and Torp 2021) based on an evaluation of options.

The development of the Norwegian project approval process which requires compliance with applicable national and European Union law illustrate the multitude of requirements that must be met to ensure successful outcomes years later. Norway is also a member of the World Trade Organization and has ratified the Plurilateral Agreement on Government Procurement (GPA) (Norway 2016). The GPA is intended to make laws, regulations, procedures, and practices regarding government procurement more transparent, and to prevent the protection of domestic products or suppliers, or the discrimination against foreign products or suppliers. According to §5 in the Norwegian public procurement law (§5 LOA), public organizations shall promote human rights when procuring products with a high risk of human rights violations in the supply chain.

As an example of the complexity of these contracts, Norway (2021) in its projects funded through the European Economic Area (EEA) and Norway Grants provides for the following standard terms and conditions:

> The Project shall contribute to good governance, sustainable development, and gender equality. It is a condition that the business shall be based on good business practice and stringent ethical requirements, and that it does not contribute to corruption, the violation of human rights or poor working conditions, or have a harmful impact on local communities

and the environment. Innovation Norway expects customers to have and use ethical requirements (policy) for its own operations and anti-corruption measures adapted to size, business type and risk. If serious matters are uncovered that are in breach of the above-mentioned principles, this may constitute grounds for Innovation Norway to terminate the contract.

During the initiation phase is a time when the project sponsors and the executive project team must explore concept designs, alternatives to the proposed project, conduct comparative cost/benefit analysis, and a feasibility study. They also conduct stakeholder outreach to numerous and diverse interests including conservationists, community members, local businesses, labor, and industry groups. In addition, this phase addresses how project governance may fit into the overall host's organizational framework where project delivery is either the business of the organization or the organization conducts its business through a series of projects with a program and/or portfolio of projects. The literature tends to treat governance issues as being static, but project development processes and environments are dynamic. The governance regimes must adapt to the specific project and context, deal with emergent complexity, and change as the project development process unfolds (Miller and Hobbs 2005; Miller and Lessard 2001).

Three Gorges Dam. *Source:* Yao YilongImaginechina/AP Images.

Opposition and Public Outreach to Stakeholders

Megaprojects attract a lot of attention because of their sheer size and cost and because they are building something that has never been done before. Preparing for opposition to the project is critical during the initiation phase. Projects that have failed or been delayed because of public opposition are legendary. Opposition has ranged from impacts on the environment, health and safety, cost to the taxpayers, resettlement and quality of life, construction blasting, vibration, noise, dirt and transportation disruption, and social concerns. Examples of extensive opposition to megaprojects include:

- The Three Gorges Dam in China which had serious environmental concerns, labor rights violations, and involved the relocation of more than one million people.
- Another example is the environmentally opposed Interstate H-3, which cuts through the Hawaiian island of O'ahu. It took 37 years, and $1.3 billion in funds and an immense amount of controversy. The Interstate H-3 project did not commence until the late 1980s with protests, environmental complaints, and legal challenges significantly delaying its completion. The highway eventually opened in 1997, and though many Hawaiians have made peace with the highway to this day some local opposition remains strong.

The best cure for public opposition is to partner early on with the local community and all impacted or interested stakeholders. Reach out to homeowners, businesses and in the case of the developing world those who live and work in the community. Sessions should be held with all stakeholders to find out their concerns and to address the risk and problems at an early stage. The concerns and complaints will continue throughout the life of the project so regular meetings should be established so there will be continuous communication. Stakeholders in the developing or emerging economies can be just as sophisticated as those in the developed world so one should never make assumptions about the interests of the stakeholders. It is also necessary to have regular press briefings sometimes weekly to make sure the press understands the issues and can communicate effectively to all stakeholders. Researchers and public interest groups such as environmentalists, conservationists, and economic groups particularly those that represent big business can be of great assistance in supporting the project and helping to diffuse any opposition surrounding the project.

The United Nations UNCTAD emblematic public building project in Santiago, Chile is a great example of urban development and collaboration. Dedicated in April 1972, UNCTAD III was a megaproject whose public–private dynamics helped bring together social actors possessed of a singular work ethic (Schlack and Varas 2019). This major complex was completed in a record 275 days by an accomplished team that made highly efficient use of technology. The actors involved in the construction were the government together with the Corporation for Urban Improvement (the CORMU) and various services and productive enterprises dependent on the state, organized workers in unions and artists, who worked in three shifts self-organized by a "committee of workers" that allowed to maintain coordination among the workers, technicians, and professionals in order to quickly solve any difficulty or problem that may arise in the development of all design, procurement and construction activities (Wong 2011, p. 67).

There are many examples of successful urban renewal projects including India's smart cities, the River Project in Portland Oregon, the Marina Bay in Singapore and the Hafen City Project in Hamburg on the Elbe River, which was one of the most ambitious urban construction sites in Europe.

The Megaproject Charter or Concession Agreement

Megaprojects require long up-front planning to clarify political, socioeconomic, and environmental concerns. For projects as complex as infrastructure projects, it is not unusual for several months to elapse in the final negotiations before the parties are ready to sign the project agreement. Additional time may also be needed in order to accomplish certain formalities that are often prescribed by law, such as approval of the project agreement by a higher authority. The entry into force of the project agreement or of certain categories of project agreements is in some countries subject to an act of parliament or even the adoption of special legislation. Given the cost entailed by delay in the implementation of the project agreement, it is advisable to find ways of expediting the final negotiations in order to avoid unnecessary delay in the conclusion of the project agreement.

A number of factors have been found to cause delay in negotiations, such as inexperience of the parties, poor coordination between different public authorities, uncertainty as to the extent of governmental support and difficulties in establishing security arrangements acceptable to the lenders. Also, as discussed in previous chapters, the role of private investors, for example in a public private partnership, may take time to recruit and the government may make a significant contribution by providing adequate guidance to negotiators acting on behalf of the contracting authority in the country. The clearer the understanding of the parties as to the provisions to be made in the project agreement, the greater the chances that the negotiation of the project agreement will be conducted successfully. Conversely, where important issues remain open after the selection process and little guidance is provided to the negotiators as to the substance of the project agreement, there may be considerable risk of costly and protracted negotiations as well as of justified complaints that the selection process was not sufficiently transparent and competitive.

Phase 2: Global Project Delivery Methodologies and Procurement

Although the implementation phase begins after the project has been selected and financing has been secured everything that happened during those phases will have an impact on the future viability of the project. It is rare when the project director, the executive team and the project teams have been involved in the actual selection and financing of a megaproject. This is because it requires different skills and know how to select, structure, and finance a project than it does to implement multiple projects under the umbrella of the program organization. Projects are selected by the project owner and in the case of megaprojects, this is often a public owner represented by a government entity at the local or national level but are often delivered by a private sector entity with support and oversight from the public owner.

As discussed in earlier chapters, the success of a megaproject depends upon the proper structuring of a project in its early upfront planning phase where the megaproject can last for decades. In some megaprojects, the planning phase can be as long or even longer than the construction phase. For example, the South North Water Transfer project in China has one of the longest time horizons for completion as the project took 50 years to plan from conception to commencement in 2002 and will take another 50 years for the construction phase as it is not expected to complete until 2050. Once finished, the nearly $80 billion project will supply residents of the north with 44.8 billion cubic meters of water every year.

When California voters approved a bond in 2008 for a high-speed rail system from Los Angeles to San Francisco, the project was projected to cost $33 billion and be completed by 2020. The project is now projected to finish in 2033 at a cost of $100 billion. The Honolulu Rail Transit System has grown from $5.122 to 12.45 billion since 2012 and it is now estimated that the full system would not be operational until March 2031 – 11 years later than promised.

As a response to an increasingly demanding marketplace, project delivery methods have evolved from the more traditional methods of design-bid-build, design-build, and construction-manager-at-risk to what is now known as integrated project delivery (IPD). In the typical construction contract, each party seeks to avoid and transfer risk to other parties. The IPD approach is unique because it allows the project participants to accept and manage design and construction risks as a team reducing the likelihood of disputes. The pure IPD method often does this with a single, multi-party contract that is executed by the owner, general contractor, and designer. The team members agree to a multi-party contract, share financial risks and rewards using a profit/incentive pool that is based upon measurable project-outcomes. Team members collaborate on how the profit and incentive pool is structured to ensure that each member is accountable for its contribution to the project outcome. The goal is to motivate each member in a way that encourages candid and transparent communication, teamwork, and shared accountability for overall design and construction.

When contractual relationships are properly structured, the IPD approach can assist in reducing change orders, delays, disputes, and claims (Xie and Liu 2017). The risk of those contingencies is collectively managed by the project participants, who prioritize the owner's objectives over their own interests. The IPD approach facilitates this end by directly linking project participants' financial incentives to the achievement of negotiated, agreed-upon, shared, and measurable goals that tie back to the owner's objectives.

There are a number of ways to structure the IPD relationships to achieve that goal. The factors to be considered in selecting an approach will depend on the nature of the project, structure of the teams, and how the sponsor(s) want to allocate project risks. The IPD process and contracting structure can and should be tailored to the specific needs of a given project. The structure can be designed based on the American Institute of Architects ("AIA") C191 contract documents (AIA 2009).

Complex megaprojects require innovation in contracting and procurement to address the allocation of risk during the early planning stages. Megaprojects are known for varied and unique delivery methods. Project delivery is a description of the contracting methods and relationships between the owner, designer, and contractor required to design and build a construction project and includes planning, budgeting, financing, design, construction, and operations. In the United States, individual state departments of transportation (DOTs) typically manage and control the full-cycle of project delivery, from inception through construction. They may elect to contract with engineering consultants or construction contractors to perform various services related to the project development process. Some of the more common project delivery methods include design-bid build (DBB), design-build (DB), construction manager/general contractor at risk (CMR/GC), and build-operate-transfer (BOT). It is contended that DB projects provide greater opportunities for small business (as subcontractors) to perform substantial portions of such projects.

Innovative Delivery Methods

New methods for both delivering and financing infrastructure have experienced a rebirth over roughly the past two decades. Integrated delivery strategies such as design-build operate (DBO) and design-build-finance-operate (DBFO) have increased the emphasis upon viewing constructed facilities as systems and have forced the recognition that sustaining these facilities requires more than just the initial capital investment.

Box 7.1   Best Value Delivery

Best value is another way of ensuring that the best team is employed. Under best-value selection, the owner considers both qualifications and cost when selecting the design-build team. Best value can be thought of as the measure (value points) of the quantity and quality that the owner may receive on a "cost per value point" basis. The relative value is clearly expressed and understood for technical innovation, design excellence, management capability, past performance record, and overall proposal value, including cost. The design-phase manager must understand the nuances of this selection method and how they will influence the project.

As illustrated, there are a multitude of ways in which contracts can be developed. Many megaprojects are built on a turnkey basis, implemented by way of an engineering, procurement, and construction (EPC) contract. Commonly used in power and water projects but less so in oil and gas projects. Under an EPC contract, the project sponsors will seek to shift as many of the risks as possible on to the turnkey or EPC contractor such as cost and schedule risk. An EPC contract also creates a single point of responsibility which is important on megaprojects where there are complex interactions and interdependencies.

Best-value procurement (BVP) (and its variations) as illustrated in Box 7.1 is an innovative means by which owners select their design-builders and is presently being used by the government of New Zealand in their procurement process (Ying et al. 2022).

Phase 3: Implementation

Implementation of a megaproject is a complex process requiring lots of data, documentation, contracts, reports, and other paperwork. During implementation, issues concerning scope, rising costs, schedule delays, quality assurance, risk management, supply chain reliance, and other aspects of the project will evolve. Projects should prepare for this evolution by incorporating best practices in all areas of project implementation and then evaluate those practices throughout the project life cycle. There are many lessons to be gleaned from some of the world's major projects. Due to the many activities that occur during the "Implementation Phase" on megaprojects, this phase is divided into the following four sections: (i) the project agreement; (ii) scope development; (iii) megaproject schedules; and (iv) megaproject cost, budgets, baselines, and contingencies.

1. Implementation: The Project Agreement

> We can lick gravity, but sometimes the paperwork is overwhelming.
> — *Wernher von Braun, Chief Architect of Apollo's Saturn V*

The "project agreement" between the contracting authority and the concessionaire is the central contractual document in all megaprojects. The project agreement defines the scope and purpose of the project as well as the rights and obligations of the parties; it provides details on the execution of the project and sets forth the conditions for the operation of the infrastructure or the delivery of the relevant services. Project agreements may be contained in a single document or may consist of more than one separate agreement between the contracting authority and the concessionaire. This section discusses the relation between the project agreement and the host country's legislation on public and privately financed infrastructure projects. It also discusses procedures and formalities for the conclusion and entry into force of the project agreement.

The Project Site, Assets, and Easements

Provisions relating to the site of the project are an essential part of most project agreements. They typically deal with issues such as title to land and project assets, acquisition of land, and easements required by the concessionaire to carry out works or to operate the infrastructure. To the extent that the project agreement contemplates transfer of public property to the concessionaire or the creation of a right of use regarding public property, prior legislative authority may be required. Legislation and acquirement via eminent domain may also be needed to facilitate the acquisition of the required property or easements when the project site is located on private property.

Ownership of Project Assets

The need for clarity in respect of ownership of project assets is not limited to legal systems where physical infrastructure required for the provision of public services is regarded as public property. Generally, where the contracting authority provides the land or facility required to execute the project, it is advisable for the project agreement to specify, as appropriate, which assets will remain public property and which will become the private property of the concessionaire. The concessionaire may either receive title to such land or facilities or be granted only a leasehold interest or the right to use the land or facilities and build upon it, in particular where the land remains public property. In either case, the nature of the concessionaire's rights should be clearly established, as this will directly affect the concessionaire's ability to create security interests in project assets for the purpose of raising financing for the project.

In addition to the ownership of assets during the duration of the concession period, it is important to consider the ownership rights upon termination of the project agreement. In some countries, the law places particular emphasis on the contracting authority's interest in the physical assets related to the project and generally require the handover to the contracting authority of all of the assets, whereas in other countries privately financed infrastructure projects are regarded primarily as a means of procuring services over a specified period, rather than of constructing assets. Thus, the laws of the latter countries limit the concessionaire's handover obligations to public assets and property originally made available to the concessionaire or certain other assets deemed to be necessary to ensure provision of the service. Sometimes, such property is transferred directly from the concessionaire to another concessionaire who succeeds it in the provision of the service.

Ownership of Land

Where a new infrastructure facility is to be built on public land (that is, land owned by the contracting authority or another public authority) or an existing infrastructure facility is to be modernized or rehabilitated, it will normally be for the owner of such land or facility to make it available to the concessionaire. The situation is more complex when the land is not already owned by the contracting authority and needs to be purchased from its owners. In most cases, the concessionaire would not be in the best position to assume responsibility for purchasing the land needed for the project, in view of the construction and operation of infrastructure.

Eminent Domain, Right of Way Agreements and Easements

Eminent domain, compulsory purchase, expropriation, or simply a "taking," are different names for one and the same legal concept. It refers to the power of the government to take private property and convert it into public use. This is a practice that occurs in many different countries under different names and different legal structures. In the United States, the Fifth Amendment provides that the government may only exercise this power if they provide just compensation to the property owners. In large megaprojects, this sometimes means the right to take another's property in exchange for giving the private owner or business the fair market value of their property or to agree to relocate the property if it is physically feasible to do so.

An easement is a right to use someone else's property in the ways described in the easement. This can include a shared driveway, a private road, a right to convey water through pipes running under someone else's property, a right to flood someone else's property or to keep the vegetation cut to allow flying over the property.

Urban megaprojects that are built in or near local communities and residential areas must consider these legal rights of governments before committing financing to a project that may not be feasible without government support and a legal system that will enforce these laws.

Construction Consultant Agreements

Contracting authorities purchasing construction and technical services typically act as the employer under a construction contract and retain extensive monitoring and inspection rights. This includes the right to review the construction project and request modifications to it, to follow closely the construction work and schedule, to inspect and formally accept the completed work and to give final authorization for the operation of the infrastructure. Instead of assuming direct responsibility for managing the details of the project, the contracting authorities may prefer to transfer that responsibility to the concessionaire by requiring the latter to assume full responsibility for the timely completion of the construction. The concessionaire, too, will be interested in ensuring that the project is completed on time and that the cost estimate is not exceeded, and will typically negotiate fixed-price, fixed-time turnkey contracts that include guarantees of performance by the construction team. Therefore, in privately financed infrastructure projects it is the concessionaire who, for most purposes. performs the role that the employer would normally play under a construction contract.

Choice of Subcontractors

The concessionaire's freedom to hire subcontractors is in some countries restricted by rules that prescribe the use of tendering and similar procedures for the award of subcontracts by public service providers. Such statutory rules have often been adopted when infrastructure facilities were primarily or exclusively operated by the government, with little or only marginal private sector investment. The purpose of such statutory rules is to ensure economy, efficiency, integrity, and transparency in the use of public funds. However, in the case of privately financed infrastructure projects, such provisions may discourage the participation of potential investors. This is because the project sponsors typically include engineering and construction companies that participate in the project in the expectation that they will be given the main contracts for the execution of the construction and other work.

The Operations Manuals

Throughout the project life cycle the lead engineers commonly referred to as resident engineers (REs) also known as civil engineers, oversee the construction process and staff for various projects, including buildings, roads, and bridges, and often work on behalf of a contractor. They make sure all of the construction is up to code, and they often work out on the construction site itself. An operations manual is essential to chart the implementation for all participants in the project.

Box 7.2   Resident Engineers (RE) Operations Manual: Typical Provisions

The RE manual describes procedures to be used for construction site activities and administration. They are the activities that involve primarily or exclusively field personnel and the activities for which the RE is primarily responsible. They do not describe fully the responsibilities of the RE. The primary functions of the RE shall be to monitor the construction in accordance with the construction contract documents, project procedures and standards, and the RE manual so that:

- Construction proceeds and is completed in accordance with the terms of the contract documents, and good engineering and construction practices.
- The work is performed in compliance with applicable safety regulations.
- A list of required licenses and permits and key steps and processes to obtain them.
- All required testing and commissioning frameworks indicating which party is responsible and the framework for supervision and approval.
- Sample appointment and approval letters, change orders, and other key communications needed for smooth project implementation.
- Model process schedules, budgets, and action plans for performance assessments and reporting.
- A summary of rights and obligations under the project agreements, with due dates and milestones.
- A dispute resolution process that becomes part of the contract obligations.
- A risk matrix and management plan, addressing each risk and how it is to be allocated, monitored, mitigated, and controlled. Ownership of each risk should be clearly identified on a risk matrix, setting out a clear strategy.
- Other practical assistance for project implementation and contract administration.

These manuals will provide guidance throughout the project life cycle as new teams enter the project and other teams exit the project upon completion of the work. In Box 7.2, we provide an overview of the general requirements that should be included in these manuals. These will not supersede design or construction contracts but will provide supporting clarification of the contract language.

Issues of Concern to the Project/Resident Engineers

Because of their daily presence on the assigned project sites, they keep rigorous notes and logs of the daily activities as well as action items that will be discussed in weekly meetings with the project teams. They calculate cost of everything from construction materials to equipment, as well as ensure these supplies are available

---

Box 7.3   Typical Responsibilities of the Resident Engineers

- Status of new or revised construction or policy and current work activities
- Issues impacting safety and health of the workers and members of the local community
- Project procedural changes
- Integration of project teams
- Active modifications and change orders
- Risk assessment, control, and mitigation
- Contractor performance and problem resolution
- Quality control and assurance
- Environmental and social issues
- Ensuring stakeholder satisfaction in the local communities
- Cost/schedule discrepancies
- Testing and inspection items
- Abutter and adjacent contract issues
- Personnel administrative issues
- Training schedules
- Client and employee feedback
- Compliance with all standards, laws, and regulations

---

throughout the process. They also check the quality of materials and equipment used during construction. They create a project schedule and look for obstacles that may delay the project. They also help plan the construction by looking at maps and blueprints, as well as going to the field to survey the site. REs are essential to the success of every project regardless of the nature of the project, the location of the project or the technological difficulty. Box 7.3 highlights the typical responsibilities of the REs and some of the challenges they face.

2. Implementation: Scope Development on a Megaproject

The Scope Statement is the primary document that is used to develop the work break down structure (WBS) from which everything else follows including the schedule, milestones, the cost estimation, the budget, and the risk assessment.

In this section, we contrast the work breakdown structure template for London's Crossrail Project and Boston's Big Dig to see different considerations in the construction of these documents but also some of the similarities. As explained in the prior chapter, both Crossrail and the Big Dig had integrated project organizations (IPOs) with the Big Dig focused on the intraorganizational structure among the project owner, the construction management consultant and the construction team overseen by a combination of owner and consultant staff. On Crossrail, one team brought together the talent from Crossrail Limited, the Program Delivery Partner and the Project Partner. The integrated team ensured that Crossrail, as an organization, was able to source the talent it required for project delivery. Both Crossrail and the Big Dig had a public corporate governance structure with a high-level oversight board and a program governance board representing the delivery team and the program management team.

On Crossrail, the WBS is owned and maintained by the delivery team, Crossrail Ltd. in conjunction with the program delivery partner. In planning terms, the first priority for Crossrail Ltd. (CRL), before any contracts could be procured, and alongside the design of the organization itself, was to build and implement a coherent WBS that would form the backbone of how the program would be packaged and delivered (Crossrail 2021). The WBS was created by CRL and applied across the entire program, forming a fundamental component of the contractual CRL works information document. It was envisaged that all parties responsible for direct works delivery would use the WBS consistently across their respective work scopes and data tool sets.

On the Big Dig, to properly execute the planning activities, in addition to the work breakdown structure for construction, the project developed a hierarchy of project reporting levels so that the appropriate information was available based on need and level of responsibility, as shown in Table 7.1. Major project objectives were established by the project owner (MHD) and represent the uppermost level in the control hierarchy. More detailed control points are in turn established to support the overall control objectives and further delineate scope, budget, or activities necessary to achieve the desired results. Information is received and monitored at these more detailed levels. The top or executive level of the WBS reporting structure provides overall or summary control information. This information is directed toward federal and state officials as well as senior project management. The next level of information is targeted for core managers who have oversight roles on the project. Further levels provide more detailed information depending on the needs of the various department managers, supervisors, and staff. Comparing the 10 levels of the WBS in Crossrail to the 10 levels on the WBS of the Big Dig it provides a broader understanding of how different approaches can be used to structure the work packages based on priorities of the project, organizational structure and governance structure and that one size does not fit all megaprojects. Though some of the classifications of work packages were similar, for example, by geographic area or by specific phase of the project, the order of prioritization showed different objectives and governance priorities.

Table 7.1 Crossrail and the Big Dig WBS comparison.

Crossrail work breakdown structure	Big Dig work breakdown structure
Level 0 – Crossrail Program Level	Level 0 – project cost centers – construction, engineering/design, program management, right-of-way, force accounts, insurance, and contingency
Level 1 – geographical area (Switched to Sector Disciplines at 60% complete)	Level 1 – construction division of the project – construction contracts
Level 2 – industry/delivery partner	Level 2 – geographic areas of the project – East Boston, Harbor Tunnel, South Boston, South Bay Interchange, Central Artery
Level 3 – asset type	Level 3 – contract packages for each construction area, sometimes consisting of as many as 20–30 packages
Level 4 – location	Level 4 – physical class of each bid item for the package, such as tunnels, viaducts, or utilities
Level 5 – function/phase	Level 5 – specific phase of the construction, which may include excavation, sheet piling, reinforcing, or concrete roads
Level 6 – contract	Level 6 – type of material used for the construction, such as epoxy coating or steel reinforcing
Level 7 – activity group	Level 7 – category of the work – labor, materials, equipment, or subcontract
Level 8 – activity	Level 8 – function classification – ironworker, operator, or truck flatbed
Level 9 – title	Level 9 – labor costs – wages, overtime, benefits, pension, insurance

Sources: Palczynski (2016) and Greiman (2013).

3. Implementation: Megaproject Schedules: The Long and Winding Road

Many large organizations have articulated that schedule and cost are very closely correlated. Observations, studies, and research into cost and schedule related topics have been performed by organizations such as NASA, the United States Department of Defense, Booz Allen Hamilton, and The Aerospace Corporation. These studies have repeatedly found close correlations to schedule delays and cost overruns, although correlation does not necessarily mean causality. Therefore, while we can observe that both cost and schedule may move in the same direction (i.e. overruns, late delivery), other factors may be responsible for driving the schedule delay which, in turn, contributes to the cost growth. Similarly, an earlier Aerospace Corporation study of 40 NASA science missions revealed that a 10% increase in schedule correlates to a 12% increase in cost (Bitten 2008).

Table 7.2 shows that inadequate definition and overoptimistic estimates have been recognized for most of NASA's history. In the 1990s, as the Federal budget became tighter, funding stability was introduced as another cause of cost growth. Likewise, in the same timeframe, as in-house technical opportunities disappeared, the lack of adequate project management training opportunities also became an issue for NASA. These various studies and observations support the idea that schedule delays and cost overruns are closely related and that while some factors and causes impact both cost and schedule, often problems will first manifest themselves in a time delay which, in turn, leads to cost growth. Additionally, there is one characteristic, unique to logic-based project schedules, which may actually create the conditions that directly lead to schedule delays on complex projects: merge bias. Merge bias is the impact of having two or more parallel paths of activities, each with its own variability or uncertainty, merge into one milestone or other activity. This somewhat arcane concept is not well understood by project teams, yet its effect can have serious consequences for the project schedule. Essentially, the more paths that merge into another task or milestone, the more there is to go wrong which could delay the schedule.

Table 7.2 NASA Advisory Council Meeting: reasons for cost growth in NASA projects.

Cost growth reasons	1970s	1980s	1990s	2000s
Inadequate definitions prior to agency budget decision and to external commitments	X	X	X	X
Optimistic cost estimates/estimating errors	X	X	X	X
Inability to execute initial schedule baseline	X	X	X	X
Inadequate risk assessments	X	X	X	X
Higher technical complexity of projects than anticipated	X	X	X	X
Changes in scope (design/content)	X	X	X	X
Inadequate assessment of impacts of schedule changes on cost		X	X	X
Annual funding instability			X	X
Eroding in-house technical expertise			X	X
Poor tracking of contractor requirements against plans			X	X

Source: Adapted from NASA Advisory Council Meeting (2009).

Schedule Management Philosophy

On the Big Dig, schedule milestones were initially developed based on aggressive progress without contingency. The project always put a high priority on meeting its schedule milestones, though it was not always successful in reaching these goals. Achieving schedule commitments minimizes public disruption, avoids even greater delay and costs, and will deliver the benefits of the completed project to the public and businesses as soon as practicable. Throughout the life of the project, management continued to enforce an aggressive schedule utilizing a schedule philosophy that requires establishment of schedule offsets when new issues impact the schedule (CA/T 2000). Schedule offsets include mechanisms to keep the schedule on track such as additional resources, changes in scope, or acceleration of the schedule. The early success of the philosophy was evidenced by the opening of one of the major project milestones, the Initial Leverett Circle, slightly ahead of schedule, in October 1999, and the later opening of another major milestone, the full I-93 Southbound tunnel on 5 March 2005. The project recognized this as a significant achievement because it was opened on the first day of the three-month schedule window identified in the Cost and Schedule Update (CA/T 2005). Although the project schedule achieved milestone success early on, the Big Dig did not always run-on-time. In 2003, the National Academies issued a report on the project delays citing a focus on short-term details of the project activities rather than a more holistic view of project risk (BICE 2003).

Milestone Management

The Big Dig project managers developed a dynamic unit, the Milestone Manager Group, to help manage and overcome project delays that resulted from the unpredictable nature of several aspects of the work (Box 7.4). To manage the schedule

---

Box 7.4   Recommended Practices for Preventing or Reducing Schedule Delay
• Because megaprojects have a long life and many unanticipated events and disruptions that impact scheduling, the following practices are recommended:
• Breakdown the megaproject into sub-projects that represent the major work packages.
• Set up milestones for each package that are constantly monitored, evaluated, and adjusted as appropriate.
• Hire experts in the scheduling field to form a milestone managers group, to help manage and overcome project delays that result from the unpredictable nature of megaproject work.

- Include mechanisms to keep the schedule on track such as additional resources, changes in scope, value engineering, or acceleration of the schedule.
- Focus on the project risk holistically rather than as short-term activities.
- Reduce uncertainty wherever possible.
- If fast tracking is used, be aware that this methodology increases risk and also the schedule due to increased claims requiring rework or moving ahead before essential design is completed.
- Develop innovative ways to reduce the likelihood of lost time and include contractual provisions to protect against unexpected events such as force majeure clauses that will excuse performance under certain conditions including political interference, changes in the law, or uncontrollable events including changes in weather conditions or other acts of God.

process, the project was subdivided into four major milestones. The milestone process provided real-time project performance data and was used to forecast project delays and develop new work sequences. Before the development of this organization, the I-90 connection to Logan Airport was thought to be 12 months behind schedule. The Milestone Manager Group initiated schedule accelerations that resulted in an estimated eight months of schedule recovery, resulting in the connection finally opening on 18 January 2003 (CA/T 2003).

The Big Dig Zakim Bridge. *Source:* U.S. Environmental Protection Agency/Wikimedia Commons/Pubic Domain.

Boston's Central Artery/Tunnel Project Timeline

Table 7.3   Boston's central artery/tunnel project timeline.

Date	Event
1975	The Federal Highway Administration and the House and Senate Public Works Committee approved the inclusion of the Central Artery/Tunnel Project in the 1975 interstate cost estimate (ICE)
1975	Interstate cost estimate (ICE)
1983	Work begins on Final Environmental Impact Statement/Report (FEIS/R)
1985	Final Environmental Impact Statement/Report approved
1986	Bechtel/Parson Brinckerhoff begins work as management consultant
1987	Relocation begins (no private homes taken)
1988	Final design process under way. Exploratory archaeology digs begin
1989	Preliminary/final design and environmental review ongoing
1990	Congress allocates $775 million to project
1991	Williams Tunnel and South Boston Haul Road
1992	More than $1 billion in design and construction contracts under way. Dredging and blasting for the Ted Williams Tunnel ongoing. Downtown utility relocation takes place. Archaeologists find seventeenth- and eighteenth-century artifacts
1993	South Boston Haul Road opens. All 12-tube sections for the Ted Williams Tunnel are placed and connected on the harbor floor
1994	Charles River Crossing design approved. New set of loop ramps open in Charlestown
1995	The first major milestone, the Ted Williams Tunnel, officially opens
1998	Peak construction years begin. Construction begins on the Charles River Crossing
1999	Overall construction is 50% complete. New Broadway Bridge opens. Leverett Circle Connector Bridge opens
2000	Nearly 5000 workers are employed on the Big Dig
2001	Overall construction is 70% complete
2002	Leonard P. Zakim Bunker Hill Bridge is completed, with phased opening of lanes from 2003 to 2005
2003	Three major milestones are reached: I-90 Connector from South Boston to Rt. 1A in East Boston opens in January. I-93 Northbound opens in March. I-93 Southbound opens in December
2004	Elevated Central Artery (I-93) is dismantled. The tunnel from Storrow Drive to Leverett Circle Connector opens, which provides access to I-93 North and Tobin Bridge
2005	Surface streets to I-93 North, are opened
2006	Substantial completion of the CA/T Project is reached in January
2007	Boston city streets are restored. Construction of the Rose Kennedy Greenway and other parks continues

Source: Adapted from Central Artery Tunnel/Project 2007 (2007).

Comparison of Project Timelines

The long timelines of the Central Artery/Tunnel Project and Crossrail is shown in Tables 7.3 and 7.4, respectively. The CA/T Project timeline stretched from 1975, when the funding for the I-93 tunnel was first approved by FHWA, to 1991 when the first Final Environmental Impact Statement/Report (FEIS/R) was approved (CA/T 1990), to the restoration of Boston city streets and the construction of the Greenway in 2007. Some might contend that rather than being a 23-year project from environmental assessment to completion, it was really a 30-year project that commenced upon initial approval of interstate funding by the FHWA and Congress in 1975.

The Crossrail Project had a shorter timeline from the approval of the program by the Secretary of State in 2001 to an initial opening in 2022. However, from the time of construction start in 2009 till completion of full end to end services anticipated in 2023, Crossrail will have completed a 14-year construction period which is almost identical to Boston's Big Dig. Both projects were also similarly more than US$4B over budget from early estimates and more than four years delayed from the original schedule (BICE 2003; NAO 2021). Although it is hard to generalize between two very different projects with different political structures, governance, levels of technical difficulty, and topography, it is not surprising that megaprojects of this size and complexity are difficult to estimate despite all efforts made by the very experienced delivery teams and governments overseeing these projects when projections are first made 15–20 years before project completion.

Crossrail Ltd. Project Timeline

Table 7.4 Crossrail Ltd. project timeline.

Date	Event
2001	Ministers announce Strategic Rail Authority and Transport for London to create a project organization
2002	Cross London Rail Links Ltd. Established as a joint venture between the Strategic Rail Authority and Transport for London
2002	Stakeholder Consultation Document published requesting views about routes that should be adopted
2003	Crossrail Business case submitted to Secretary of State
2004	Secretary of State announces decision to proceed with Crossrail
2009	Start of Crossrail Construction
2010	Crossrail announces award of major tunneling contracts
2011	Crossrail awards remaining tunneling contracts (Thames Tunnel and Connaught Tunnel refurbishment)
2011	First central London station contract awarded – Paddington
2012	Tunneling comes to complete stop on western section after spoil conveyor collapses

(Continued)

Table 7.4 (Continued)

Date	Event
2014	First Crossrail TBM completes journey
2014	Crossrail programme reaches half-way point
2014	National Audit Office declares Crossrail program on schedule
2014	Dft and TfL announce that Crossrail route will be extended to Reading
2015	End of Crossrail Tunneling
2016	HM Queen visits Bond Street and announces that the new railway will be named the Elizabeth Line
2016	Crossrail awards the last of its major contracts, signaling the end of one of the UK's largest ever procurement programs
2017	Stage 1 commencement of services
2017	Crossrail reveals a 1B price hike in tunneling contracts
2018	Crossrail reveals work running 20% over budget but remains on track for December opening date
2018	Government confirms that Crossrail budget to increase from $14.8 to 15.4B due to engineering challenges
2021	The opening of Crossrail has been delayed until the first half of 2022 and construction will need as much as another $1.1B to be completed
2021	COVID-19 is one of the reasons behind the new delay, due to less than 50% of the staffing levels before the pandemic
2021	A full Crossrail timetable will not be in operation until May 2023

Source: Adapted from National Audit Office.

4. Implementation: Megaproject Cost, Budgets, Baselines, and Contingencies

Durdyev (2021) investigated the causes of project cost overruns that have been reported in project management articles since 1985. The first assessment of the studies in terms of the countries they reported from reveals that this problem mainly occurred in developing countries, in which the project cost overruns are attributed to resource-related, economic/financial as well as political problems. Durdyev (2021) listed the top 10 causes of cost overruns with the highest number of citations to be: (i) design problems and incomplete design, (ii) inaccurate estimation, (iii) poor planning, (iv) weather, (v) poor communication, (vi) stakeholder's skill, experience and competence, (vii) financial problems/poor financial management, (viii) price fluctuations, (ix) contract management issues, and (x) ground/soil conditions.

Another research study by Herrera et al. (2020) pointed out the five most important and frequent causes of cost overruns, (i) failures in design, (ii) price variation of materials, (iii) inadequate project planning, (vi) project scope changes, and (v) design changes.

Boston's Big Dig was impacted by cost overruns throughout its long life; however, the reasons were not just due to delayed design development. The Big Dig was a schedule-driven project with little or no contingency budget for delays due to federal regulations that limited contingencies to 7% way less than what was needed for a complex inner-city project built on the Colonial shoreline. This problem became apparent in 2000, when $300 million was needed for schedule maintenance, requiring the infusion of additional funds. The term *schedule maintenance* deserves definition. It refers to money additional to the value of the contract that the owner pays the construction contractor to maintain the schedule required by the contract (Greiman 2013). Every megaproject requires a contingency not just for cost overruns but also schedule delay (Box 7.5). Project schedule changes, particularly extensions, caused by budget constraints or design changes can result in unanticipated increases in inflation costs even when the rate of inflation has been accurately predicted. An astounding 55% of the reported increase in cost as of 2002 was estimated by the project team and an independent board of the National Academies to be due to inflation (BICE 2003; Edwards 2002).

Box 7.5   The National Audit Organizations Review of Crossrail Delays

The Crossrail program was further from completion and more complicated than Crossrail Ltd. or the sponsors understood. This, and the COVID-19 pandemic, resulted in a further forecast cost increase of £1.9 billion and 10–20 months of delay since we last reported. There are encouraging signs that the program is now in a more stable position with a better understanding of the total amount of work required. However, there is still a significant volume of work to complete alongside testing trains, signaling and other assets. Completing the program relies now on Crossrail Ltd, RfLi, MTREL, TfL, Network Rail, and the Department working closely (NAO 2021).

Neither Crossrail Ltd., the sponsors nor the contractors appreciated how complex it would be to bring together all of the separate systems and assets required and assure them as safe and working, or how long it would take. The Elizabeth line will be the first fully digital railway to be built and operated in the United Kingdom. Bringing it into service requires Crossrail Ltd. and its contractors to complete and integrate around 500 000 physical and digital assets, such as fire safety systems or platform screen doors. They must be assured as safe and operational both individually and in combination. There must also be digital operating manuals, guidance, and processes for individual elements and the railway as a whole. The work to bring the railway into service was made more complex by the high number of contracts, bespoke designs and a lack of standardization throughout the program, as well as needing to integrate three different signaling systems with trains (paragraphs 2.13–2.17) (NAO 2021).

Contingency Reserves

A contingency is an additional amount/percentage set aside against cash flow for liabilities and exposures that may arise during the course of the project. On the Big Dig, there were three contingency accounts: one for construction, one for costs that arose in the project's non-construction cost centers, and a third for management reserves. A contingency reserve is a designated amount of time and/or budget to account for parts of the project that cannot be fully predicted. For example, it is relatively certain that there will be some rework, but the amount of rework and where it will occur in the project (or phase) are not known. These are sometimes called "known unknowns." The purpose of the contingency reserve is to provide a more accurate sense of the expected completion date and cost of the project (or phase). Some project managers separate contingency reserves from management reserves, while others combine the two into a single reserve. The better practice is to separate the contingency and management reserves, as they serve two different purposes. Reserves for changes and issues may be part of the contingency reserve or separate reserves.

Management Reserve

A management reserve is a designated amount of time and/or budget to account for parts of the project that cannot be predicted. These are sometimes called "unknown unknowns." Use of the management reserve generally requires a change to the total budget. Though management reserves are typically in the 2–5% range, on megaprojects management reserves can be as high as 25% based on the complexity of the project and the uncertainty in the political, technical, and operational aspects of the project. Eurotunnel, for example had a 25% cushion in its budget for unknowns which was seriously insufficient due to the numerous unexpected events and financial difficulties (Grant 1997).

Cost and Schedule Tracking and Control: Earned Value

One of the critical components in tracking cost and schedule progress is the application of the earned value methodology (EVM). EVM integrates project scope, cost, and schedule measures to measure project performance and progress. The calculation of earned value must be consistently based on the physical progress of the work. The consistency by which work is measured ensures earned value accuracy. Earned value is used to determine the current status of the project as well as long-term performance trends. On megaprojects, earned value is used to (i) forecast future performance, (ii) determine variances from budget, and (iii) project completion date and costs. PMI's *PMBOK* provides an excellent overview of the principles of EVM and its application to any industry.

Repairing Flawed Budgets and Baselines

Projects must establish clear and measurable performance goals and then monitor and enforce those goals to protect the public interest. When a project is over budget and behind schedule, we often look for reasons within the project organization but rarely point to project governance as the cause of the project's problems. Governance is often the root cause of all kinds of problems, including escalating budgets and failed baselines. The National Audit Office in the United Kingdom (2021) highlighted the following as reasons for Crossrail's large overrun during the final stages of the project:

1) Delay in the project schedule has been the most significant reason for the cost increase, at £934 million, 62% of the total cost increase of £1510 million.
2) Not having a sufficiently detailed delivery plan against which to track progress
3) Crossrail Ltd. and its contractors are still identifying new tasks that need to be done, which, alongside tasks taking longer than expected to complete, are increasing forecast costs.
4) The physical asset was different from that documented.
5) The work no longer met current regulations – for example, wiring in some stations, and fire systems.

Phase 4: Project Controls

> The Truth is incontrovertible. Panic may resent it, ignorance
> may deride it, malice may distort it, but there it is.
> — *Sir Winston Churchill (1916)*

Regulatory Process and Oversight

Oversight is one of the key characteristics of successful projects. When no one is monitoring a project, it will lead to schedule delays, cost overruns, quality control issues, and sometimes catastrophic loss of life. The government/project owner may be assisted in its oversight and monitoring function by third parties such as independent specialists or regulators including the Occupational Health and Safety Administration (OSHA) or the Nuclear Regulatory Commission in the United States or transport, energy, or water regulators in many countries. There are several models for oversight that have been highly effective. One such model was adopted for Boston's Big Dig and has since been used by other public agencies in the United States (see Figure 7.2).

All megaprojects have oversight authorities but they can vary widely in terms of structure and composition of its members. Sometimes the oversight authority is a government agency either connected to or independent of the governance body of

Audit and Oversight of the Big Dig

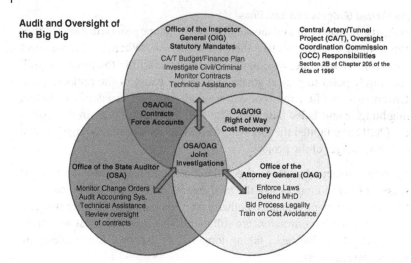

Figure 7.2 Central Artery/Tunnel Project (CA/T) Oversight Coordination Commission (OCC). *Source:* CA/T OCC (1996).

Box 7.6   The Central Artery/Third Harbor Tunnel (CA/T) Project Oversight Coordination Commission

The Massachusetts legislature established the Central Artery/Third Harbor Tunnel (CA/T) Project Oversight Coordination Commission (CA/T OCC 1996), to coordinate the scrutiny of potential cost savings and recommend opportunities for systemic improvement in the operations of the Big Dig. This unified, integrated, independent oversight commission was the first of its kind to oversee a megaproject in the United States. A strong project-independent audit program creates transparency, provides for accountability, and identifies areas for contract review and enforcement. Some lessons learned from the Big Dig concerning the use of independent state oversight agencies include the following:

- Assurance that the outside resources are truly independent and have the requisite expertise to know the questions to ask and the documentation necessary for an effective audit.
- Linking the auditors with insiders that are cooperative and willing to devote the time and resources necessary to provide all critical information so that the auditor's report does not provide a false sense of security.
- Recognizing that audits are not a substitute for good design, decision making, and project management; and
- Providing incentives for project personnel to support audits, recognizing that they may be the best security against a massive project failure.
- Focused and proactive oversight of a project is critical for safeguarding the public interest.

the megaproject. Figure 7.2 and Box 7.6 shows the structure of the legislatively created oversight authority on Boston' Big Dig.

Planning for Quality: Philosophy

> For you to sleep well at night, the aesthetic, the quality, has to be carried all the way through.
>
> *— Steve Jobs*

Success in the delivery of projects requires a commitment to the highest standards of quality. Quality can never be compromised or negotiated. It is an essential requirement of every project implementation, yet often it is assumed and there is amazement when the delivery of the project does not meet stakeholder's expectations of quality. Understanding a project's philosophy on quality is key to preventing and reducing requirements errors and must be understood from conception. Quality requires time, up-front expenditures, resources, support from project management, independent oversight, and an integrated project structure. The governance structure of a megaproject must be continually monitored, since a defective governance structure will surely doom the project, as evidenced by the large number of projects that fail due to huge cost overruns and failure of quality assurance. The *PMBOK Guide 7th edition (2021)* recognizes quality as one of its 12 project delivery principles – the other principles are: stewardship, team, shareholder's, value, holistic thinking, leadership, tailoring, complexity, opportunity and threats, adaptability and resilience, and change management. The new edition prescribes that quality should be built into a project's processes and results and does not distinguish between quality assurance and quality control (Box 7.7).

Quality may have several different dimensions, including but not limited to the following: (PMI PMBOK 2021):

Performance. Does the deliverable function as the project team and other stakeholders intended?

Conformity. Is the deliverable fit for use, and does it meet the specifications?

Reliability. Does the deliverable produce consistent metrics each time it is performed or produced?

Resilience. Is the deliverable able to cope with unforeseen failures and quickly recover?

Satisfaction. Does the deliverable elicit positive feedback from end users? This includes usability and user experience.

Uniformity. Does the deliverable show parity with other deliverables produced in the same manner?

Efficiency. Does the deliverable produce the greatest output with the least number of inputs and effort?

Sustainability. Does the deliverable produce a positive impact on economic, social, and environmental parameters?

Box 7.7   Key Questions That Must Be Answered in Designing a Quality Program Include the Following:

---

1) What is the project's commitment to quality? Is it a top priority? A constraint on cost and schedule? Does it provide for enforceable penalties for noncompliance? Will the project commit to quality for the entire life cycle?
2) How will quality assurance be structured?
3) Is quality assurance integrated in the project? Does quality assurance have a relationship to design, construction, risk management, geotechnical, safety and health, legal, claims and changes, and procurement?
4) How will quality be measured?
5) Are the quality standards and requirements broad or limited? Are there contractual provisions? Laws and regulations? Project policy and directives?
6) What are the quality control tools? Is a full range of tools going to be employed, including deficiency reporting, inspections, training, control checks, review of project changes, cause-and-effect analysis, corrective action plans, and insurance?

Safety and Health

All projects require controls and oversight of safety and health programs. The overarching risk mission at the Central Artery Tunnel Project was clear: to instill a project culture where safety was considered the most important value. It was a well-understood mandate that pervaded every aspect of project decision making from the early planning stages through project completion. Initial research from the U.S. Department of Labor (DOL) reflected that in 1992, when project construction began, the fatality rate for a construction project of this size was very high based on anticipated labor hours worked (DOL 2021). Considerable efforts were required to maintain a first-class loss prevention, risk mitigation, loss control, and safety and health integrated team with support from the U.S. Occupation Safety and Health Administration (OSHA). The program was based on principles that were regularly reviewed and enforced. One of the most significant impacts on worker safety was the Safety and Health Awards for Recognized Excellence (SHARE) Program. The program was an integral part of ensuring a safe workplace that not only prevented and reduced accidents but improved productivity, schedule, morale, behavioral changes and individual job satisfaction, and ultimately provided substantial cost savings to the owner and all project participants and became a model for other programs worldwide (BICE 2003). Box 7.8 shows examples of safety and health programs on the Project.

Box 7.8   Selected Safety and Health Programs

- Safety and Health Awards for Recognized Excellence (SHARE) – incentive program for workers to prevent and reduce accidents.
- The world's largest owner-controlled insurance program.
- Lost time and OSHA recordables safety performance evaluations
- Substance Abuse Prevention – drug testing program.
- Emergency response – prepare for and respond 24/7 to construction-phase project incidents.
- Continual root cause and risk analysis.
- Weekly health and safety inspections and report cards.
- Onsite medical and safety oversight teams.
- More than 1300 Environmental commitments and sustainability requirements.
- Electronic identification for all workers.
- Community and Business Artery Public Awareness Program and Utility Disruption Plan.

Claims and Changes

The challenges of managing change were described by Ted Weigle, of Bechtel/ Parsons Brinckerhoff, the Big Dig project's former program manager, this way:

> Managing [the Big Dig] project is like trying to crowd liquid mercury into a corner . . . Every time you think you have it contained; you realize it's leaking out the other side. Whereas most people in business like to think that 70 percent of their day is predictable, on a job like this only 30 percent of the day is predictable. It's a process of nonstop, hands-on management and of constantly trying to anticipate where and when the next challenge will occur.
>
> — *Weigle as cited in Rigoglioso et al. (1993)*

Change is inevitable in projects because that is why we create projects – to innovate, advance a new idea, and explore new ground. Megaprojects by their very nature are creatures of change. The environment of megaprojects – the pace, the pressure, the tension and uncertainty, the widespread fears and skepticism – combined with the sheer volume of difficult decisions create a herculean task that can quickly overwhelm even the most capable project managers and the most sophisticated organization. As if the challenges of the long up-front formation and planning phase are not difficult enough, projects are then faced with the difficult work of the execution phase. Either phase, if poorly led and managed, can prevent the progression of an otherwise successful project. For a project organization to successfully integrate multiple polices, processes, and procedures, leadership in change management is a key requirement.

On megaprojects, the need for change is driven by internal factors such as technological challenges, uncertainty, and risk, complexity, and external challenges such as the regulatory environment, stakeholder expectations, and unpredictable markets.

Project management was developed over many centuries as a means for organizations to manage change. Change management has been written about extensively in the literature as a result of technological advances and the focus on project sustainability and complexity over the last several decades. How and when we see the need for change is a critical component of project success.

Resistance to change is a common reaction in most organizations, and projects are no exception (Paton and McCalman 2008). Political change is particularly prevalent in large projects, as are ethical and legal change. Because change comes in different forms, and from different sources, a variety of systems, structures, and programs are needed to manage it.

Litigation, Arbitration, Partnering, and Problem Resolution

How we solve problems on megaprojects is one of the important unknowns about megaproject management, yet essential to better management and the future of these large-scale endeavors. The more we know about problem resolution the better prepared we will be to tackle the megaproject phenomena that will arise in the future. As described in this chapter, project change requests and how they are managed is one of the critical factors in cost increases in megaprojects (NAO 2019). If responsibility for risk is not clearly allocated in the project documents the project owner can pay a massive price for increased claims particularly during the peak years of construction in a complex megaproject with multiple engineers and contractors with complex interdependencies and multiple critical paths. Responsibility for costs as well as for delays should be allocated in the project contingency budgets.

Key to the success of the Big Dig's claims and changes program was the requirement of reassessing and revising it as the needs and priorities of the project evolved (Dettman et al. 2010). Hiring an independent engineer to review change order requests and making sure they align with contractual provisions and are typical for the construction industry is essential to cost control. For the CA/T project, change order requests included items such as unidentified utilities, design changes due to the desires of third parties, interface coordination with other contracts, minor alignment changes, revisions to entrance and exit areas such as ramp location and enhancements to original design, and some incomplete bid documents due to federal funding rules. During the peak years of construction, this program managed more than 13 500 claims, many of them involving nonroutine, complex contractual obligations, and unforeseeable events. As a result of this growing backlog of claims, project management had to take a new look at the dispute avoidance/resolution program. This involved establishing a claims resolution plan that called for a collaborative process between the contractors and the owner and a sophisticated dispute avoidance/resolution program that elevated issues to successively higher levels of management through a dispute review board (DRBF 2021).

Litigation

Global projects can raise major disputes between and among the numerous stakeholders including government sponsors, equity holders, lenders, contractors, designers, and local communities to name a few. Traditionally international commercial disputes have been resolved by litigation in national courts or arbitration in one of the many national or international arbitral tribunals. Bilateral investment treaties often include arbitration clauses as an alternative to litigation. Project agreements generally contain a forum selection clause, or an arbitration agreement that includes the selection of an arbitral tribunal. If a Bilateral Investment Treaty is involved the Treaty itself may have a default provision that provides for arbitration in a particular forum. For example, Article 25(1) of the ICSID Convention extends the jurisdiction of the ICSID Tribunal located in Washington, D.C. to any legal dispute arising directly out of an investment between a Member State or a constituent subdivision or agency designated by that State, and a national of another Member State. If a country is not a member of the ICSID Convention then the ICC International Chamber of Commerce or an UNCITRAL arbitration provide reasonable alternatives.

Choosing between Arbitration and Litigation is not always an easy decision and some of the advantages and disadvantages of each are highlighted in Table 7.5.

Alternative Dispute Resolution

In this section, we explore the advantages of using alternative dispute resolution (ADR) which are ways of resolving disputes that are usually voluntary, can occur as a predecessor to litigation or arbitration, can be far quicker (no waiting to schedule court time) and less costly (no legal fees and court costs) and do not result in a binding decision so it leaves the parties some options prior to adjudication by a binding authority.

Because intraorganizational relationships are so important on a megaproject, it is critical that the method of dispute resolution be consensual, open, transparent and fair so that it does not impede the ongoing working relationships so necessary to keep the project moving forward. As shown in Figure 7.3 litigation, arbitration, and ADR which includes all the methods in the inner circle and more, operate in different spheres. It is possible to resolve a dispute first by using

Table 7.5 Litigation v. Arbitration comparison.

Litigation	Arbitration
Case will be heard and decided by a court appointed judge who will have the power to decide the case based on the law chosen by the parties	Case will be heard and decided by arbitrators. Typically, one arbitrator is chosen by each party and the third is chosen by the two party appointed arbitrators. Arbitrators are chosen for their expertise on the matter in dispute
Courts can enter temporary orders including injunctive relief, liens, or specific performance	Arbitrators do not have the power to enter temporary orders so the parties must go to a court of law for temporary orders
Permits broad discovery to allow for testimony of witnesses and documentary evidence which can add considerable time to the process	The parties can better expedite and determine the procedures for the arbitration in accordance with the arbitral law and the arbitration rules
A court decision is generally subject to an appeal under the law and courts must follow precedent	It is much harder to overturn an arbitrator's decision as there are limited grounds for vacating a decision
Court decisions are almost always published and become part of the public domain	Arbitrator's decisions are not always published and can remain confidential as they are not required to follow the precedent of prior decisions of the courts or of arbitrators
Court decisions tend to be easier to enforce	Arbitral decisions can be enforced under the New York Convention of which 168 state parties are members, however, recognition of the decision can be held up by the courts for a long period of time in some countries

negotiation or a form of mediation for a limited period of time (30–90 days) before resorting to a binding method such as arbitration or litigation. The dispute review board is a formalized structure used by projects worldwide in an effort to resolve disputes consensually to reduce tension between the key stakeholders in the project.

Forms of Mediation

Mediation has been growing in popularity worldwide in recent years in international business and commercial practice. Facilitative mediation (FM) has progressively become the dominant form of ADR process in the United Kingdom and more widely in the United States and Europe. FM is now probably the dominant

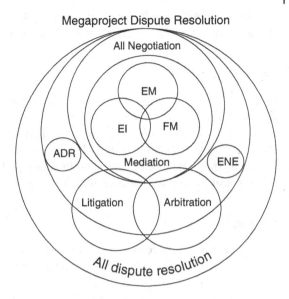

Figure 7.3 Megaproject dispute resolution. *Source:* Adapted from Clyft (2021).

form in major international disputes and the principal form of ADR chosen by the International Chamber of Commerce in Paris. The main reason is that it works: cases settle; problems are solved (Clyft 2021). Evaluative mediation (EM) An evaluative mediator assists the parties in reaching resolution by pointing out the weaknesses of their cases and predicting what a judge or jury would be likely to do. Because the evaluative mediator actually reaches a decision it can have an impact if the parties eventually take their case to binding arbitration or litigation.

Early intervention (EI), sometimes referred to as early intervention mediation, is a new form of mediation. It has evolved from the original concepts that have made facilitative mediation so successful, but with differences that can prove useful, in particular cases. It offers a wide range of methods to reach consensus and settlement.

Early neutral evaluation (ENE) aims to position the case for early resolution by settlement, dispositive motion or trial. It provides an assessment of the merits of the case by a neutral expert, and it may serve as a cost-effective substitute for formal discovery and pretrial motions. Although settlement is not the major goal of ENE, the process can lead to settlement.

Online dispute resolution (ODR) has existed from about 1999 onward. Software systems now offer the opportunity to conduct both traditional mediation and early intervention remotely, in a manner that broadly replicates the original concept in each case, but in a radically new way, as a new and enormously enhanced form of visual ODR. This change has occurred with staggering rapidity. The COVID-19 pandemic and technology have made online mediation, and online EI, both a necessity and a credible, workable, and effective new normal (Clyft 2021). Figure 7.3 places in context the three primary spheres of dispute resolution: arbitration, litigation, and ADR.

Expert Determination

Project disputes may involve highly technical matters and may require a quick determination. A form of dispute resolution that can be binding or non-binding is an expert determination. This type of dispute resolution clause would be included in a project agreement and would require the issues to be determined by this method, the number and qualifications of the experts, the appointment criteria, and the conditions upon which the decision will be binding.

Partnering

> I cannot monetize how much we saved as a result of partnering, but after awarding over $2 billion in construction, we have no lawsuits and no major disputes.
> — *Peter Zuk, former Central Artery/Tunnel project director (ENR 1996)*

The concept of *partnering* was first utilized by DuPont Engineering on a large-scale construction project in the mid-1980s, and the U.S. Army Corps of Engineers was the first public agency to use partnering in its construction projects. Partnering is now widely used by numerous government and construction entities around the world. It involves an agreement in principle to share project risks and to establish and promote partnership relationships to achieve business objectives. Partnering has been utilized to achieve the aspirations of early advocates of greater project integration, and it has become a widespread feature of global construction management practice (Wood and Ellis 2005).

The partnering process is an alternative team problem-solving approach, with the help of a neutral facilitator that is intended to eliminate the adversarial relationship problems between the owner and contractor by developing a single integrated team. More simply stated, it is a way of conducting business in which two or more organizations make long-term commitments to achieve mutual goals focused on interests, not position. Though often used in construction, it is valuable for almost any type of business relation. Partnering promotes open communication among participants, trust, understanding, and teamwork.

Dispute Review Boards

In conjunction with partnering, a good way to resolve disagreements between designers and contractors without getting entangled in time-consuming, expensive litigation is by providing for a dispute review board (DRB) in the project contract. From the data gathered by the DRB Foundation, dating back to 1975, 60% of projects with a DRB had no disputes, and 98% of disputes referred to a DRB did not result in subsequent litigation or arbitration (DRBF 2021). The review board, a panel of experienced, respected, and impartial reviewers, takes in all the facts of a dispute and makes recommendations based on those facts and the board's own expertise.

Dispute review boards have been successfully used in many megaprojects. Major multilateral organizations that have recommended the DRB include the World Bank, the Fédération Internationale des Ingénieurs Conseils, or later, the International Federation of Consulting Engineers (FIDIC), and the International Chamber of Commerce (ICC).

Phase 5: Transitioning a Megaproject to Operations

One of the most difficult phases in a megaproject is transitioning the project to operations where it will deliver the benefits promised and will provide a sustainable structure for a long period of time, sometimes 50 years or longer in the case of a bridge, a tunnel, or even a roadway. As discussed in an earlier chapter, Heathrow Terminal 5 when transitioned to operations failed to properly deliver the Terminal on day one due to a failure of connection with the Airport's baggage system which had not been properly integrated into operations resulting in cancellation of many flights and disruptions for an extensive period of time. To ensure a successful transition the following steps must be undertaken:

1) Identify well in advance usually in the early days of the project's life who on the project team will transition to operations when the project is substantially complete. It is important that expertise is available to assist in the difficult issues that will arise during operations in all disciplines including engineering, construction, quality control, risk, administration, commissioning, and governance.

2) Begin the project with the end in mind. Review frequently throughout the project life the expectations of stakeholders and then continuously test for realization of those expectations.

3) Verify final delivery is complete. After submitting the final deliverables, make sure that your client and stakeholders have received what they expected in the final product and confirm a formal receipt of the official acceptance or approval. As part of this step, review the project charter or concession agreement, which should include the original goals and the business case for the project. Did the final deliverable meet these goals or solve the intended problems? Taking this step can provide helpful insights into how the project achieved its original goals or evolved over time.

4) Provide training for the operators across disciplines to make sure all expertise has been properly assimilated into the new organization.

5) Prepare a guidebook for the operators, but also for the sponsors of the project so they will have valuable knowledge and information to pass onto future projects.

6) Ensure all project documents are made available to the operators including outstanding contracts and policies prior to the transition date. Some contracts may include requirements for storing or destroying files at the end of a project. Make sure all confidentiality requirements are observed.

7) Prepare for operations by testing all technical aspects of the project and ensuring that inspections will continue after the project is completed and throughout the operational life of the project. Too many accidents have occurred due to failure to inspect or a lack of knowledge of the operators as to what should be inspected and verified and in what time frame.

8) Transfer all policies, performance bonds, contracts, and insurance that may be continued when operations commence.

9) Conduct an extensive evaluation of the project with the client including lessons learned, areas for improvement in the future, and knowledge transfer to project operations. These lessons should be shared with all projects at the local, national, and regional levels.

Lessons Learned on Cost and Schedule

We conclude this chapter with some important lessons from two of the world's largest megaprojects – London's Crossrail and Boston's Central Artery Tunnel Project – based on project audits by external oversight authorities and organizations responsible for evaluating the status of the project at various phases. In both megaprojects, "change" and the management of "change" were major reasons put forth for cost increases and schedule delays and ultimately escalating budgets. In this section, we look at the problems and reasons for these increases as set forth in the oversight reports, and what future megaprojects need to consider before creating unrealistic expectations.

Reasons for Cost Increases on London's Crossrail Project

> Our vision is to help the nation spend wisely. Our public audit perspective helps Parliament hold government to account and improve public services
> — *Department for Transport, National Audit Office*

On 3 May 2019, UK's Department for Transport issued the following Report: *Completing Crossrail: Department for Transport, Report of the National Audit Office.*

Below are some excerpts from that Report that highlight some of the reasons for cost escalation as determined by the National Audit Office. "The Report is based on review and analysis of documents produced by Crossrail Ltd and sponsors, including reports on the program's progress, and interviews with key senior figures involved in the delivery and oversight of the program" (NAO 2019, p. 6).

Importantly as noted in page 6 of the report, "this report is not intended to apportion blame for what has happened to the Crossrail programme, but instead to learn why the program ran into difficulty and what needs to be done to deliver the promised benefits to passengers and the economy."

Problem: "In August 2018, Crossrail Ltd announced that the program could not be delivered on time and that they would not be in a position to open the central section through London in December 2018 as planned. In December 2018, the Department announced that cost increases on the program had resulted in an increase in funding to £17.6 billion (some £2.8 billion more than the level of funding announced in 2010), including more than £2 billion of loans from the government to TfL and the Greater London Authority" (NAO 2019, p. 5).

Reasons for Cost Increases: "Crossrail Ltds. reporting of the causes of the increase between 2013 (the time of the previous study) and 2018 are *change and delay*; dividing the reasons into four main categories: design change, the settlement of contractual compensation events, acceleration, or changing the sequencing of works to improve confidence in delivering on schedule, and additional scope" (NAO 2019, p. 18).

Management of Interfaces: "Crossrail Ltd did not require individual contractors to manage interfaces with other contractors, and so protected contractors from changes that were outside their control. Therefore, Crossrail Ltd had to compensate individual contractors for delays that occurred on other contracts, on which their work depended, and had to engage in costly change control negotiations" (NAO 2019, p. 7).

Multiple Activities in Parallel: Managing activities in parallel, "meant that some work to install systems required to operate the railway, and complete stations, would take place at the same time during the latter stages of the program. This created vulnerability on the critical path." The delivery approach, "led to increased compression in the program and increased risks. Crossrail Ltd. Executive Team executive team recognized the challenges but believed this was an exceptional team capable of delivering exceptional results and overcoming these challenges" (NAO 2019, p. 7).

Incomplete Delivery Plan: "Crossrail Ltd did not have a sufficiently detailed delivery plan against which to track progress. Crossrail Ltd started to produce a detailed, realistic, bottom-up plan in late 2018. Prior to this, from 2015, it had based its management of the program on an aspirational plan designed to improve progress by suppliers, rather than to provide a reality check on overall progress. Crossrail Ltd presented the plan as the critical path for completing the overall program. However, the plan did not adequately reflect interdependencies across the program. Consequently, Crossrail Ltd had a gap in its understanding of delivery risks and the likelihood of meeting the December 2018 opening date" (NAO 2019, p. 7).

Early Downsizing of Critical Staff: "Crossrail Ltd also reduced its central program and risk management capability during 2018, on the basis that they anticipated the program reaching completion in December 2018. It is currently rehiring staff now that it is clear that significant work remains, although it has faced challenges recruiting the skills it needs" (NAO 2019, p. 8).

NAO Report Conclusions: "Until the new services are open to passengers and the final costs of the program are known, it is not possible to conclude on overall value for money." The compressed schedule, the contractual model, the loss of downward pressure on costs, and the absence of a realistic plan were set against an atmosphere where "can do" became unrealistic. "All these factors and many more set out in this report have contributed to underachievement in terms of cost and progress so far" (NAO 2019, p. 9).

NAO Report Recommendations: Crossrail Ltd should continue to (i) refine its plan; (ii) work with sponsors, to establish a range of scenarios that set out the potential future impacts on the taxpayer, passengers, and businesses and develop plans for how further cost increases or delays in collecting revenue will be financed; and (iii) rebuild its capacity and capability to complete the program in a timely and cost-effective way (NAO 2019, p. 9).

Reasons for Cost Increases on Boston's Central Artery/Tunnel Project

Problem: In 1985, The initial cost estimate of $2.56 billion for the Central Artery Tunnel/Project, (the Big Dig) was based on preliminary concepts before detailed technical studies had been completed. By 2007 at substantial completion, the project had escalated to $14.8B (CA/T 2007). Explanations for the massive project cost increase abound and are documented in the many reports, audits and reviews conducted on the project range from excessive spending on improvements for nearby neighborhoods and private businesses to unanticipated subsurface conditions, to accounting assumptions related to inflation, to federal rules on change allowances, to environmental impacts and design development, and the unanticipated long duration of the project (Greiman 2013, p. 216).

Reasons for Cost Increases: In the years that followed, the project design was changed in response to environmental concerns and community demands. As noted by the project's management consultant, Bechtel Parsons Brinckerhoff (BPB) the process of developing cost estimates was therefore evolutionary, as it was not possible to anticipate all final design and program decisions that would be made by the state, the federal government and other interested parties, as well as the extent and nature of unanticipated conditions that impact cost and schedule (B/PB 2003). The final design concept was not approved until 1994 almost 11 years after the issuance of the first Environmental Impact Report (EIS), and at an additional cost of a billion dollars. Significantly, the McCormack Institute estimated in 1997 that about 30–35% of the cost of the project could be attributed to the required tunneling and interchange construction alone, about 25–30% to the measures needed to mitigate the impacts of the construction and meet required environmental standards, and up to 40% to account for inflation and cost escalation resulting from a 25-year design and construction period (McCormack 1997).

The Claims and Changes Program on the Big Dig was one of the largest operations on the project, with more than 100 employees. A *claim* was defined very broadly in the project's specifications to include "any written demand that seeks relief in any form arising out of or relating to the contract, the contract documents, or the work, including without limitation, all contract claims, statutory claims, equitable claims, and claims for extension of time . . ." (CA/T OCC 2003). As of September 2004, the project reported that differing subsurface conditions represented about 19% of the almost $2 billion in requested claims and changes CA/T (Central Artery/Tunnel Project) (2004) Finance Plan.

In 2003, the National Academies Board on Infrastructure and the Constructed Environment (BICE) issued a report on the project delays citing "a focus on short-term details of the project activities rather than a more holistic view of project risk" (BICE 2003) with the following critical analysis and recommendations:

> "[M]eeting the CA/T Project's schedule targets continues to be a problem. Despite an emphasis on reaching the milestones on time, slippage continues, thereby reducing public confidence in CA/T management" (BICE 2003). The Report further emphasized that the slippage was due to the focus on short-term details of the project's activities, rather than evaluating the project risk as a whole in advance of potential occurrence (BICE 2003). The problem was particularly evident due to the continuous postponement of the opening dates of two major highway connections, the I-90 turnpike and I-93. The Academies Report recommended, "that project managers should strategically evaluate future schedules by determining what critical tasks needed to be done without fail and how long these activities will likely take." Moreover, "published completion dates should be developed around realistic workflows and schedule risks, with modest allowances for unknown issues" (BICE 2003).

Also, in 2003, Deloitte and Touche (D&T), the company that conducted the Commonwealth of Massachusetts's independent assessment of the project's cost and schedule, found that the project did not adequately document the value of outstanding claims. Although D&T did not believe the project's cost estimate should be increased, it recommended that the project closely manage and monitor the settlement of claims so that project costs would not increase in the future (D&T 2003). The assessment further determined that the number of claims continued to grow despite project efforts to resolve claims. For example, from August through December 2003, an average of 418 new claims were received and an average of 406 claims were resolved. As a result, the total number of unresolved claims as of March 2004 had increased to 4805 (CA/T 2004, appendix E, 11).

Conclusion on Lessons Learned

Contrasting the Crossrail Project with the Central Artery/Tunnel Project is a research project unto itself. Megaprojects are constantly evolving and changing at a rapid pace creating future uncertainty. A focus on project evolution and the ever-changing nature of the project impacts the need for a holistic scope, cost, and schedule evaluation for both the short- and long-term.

As described in this chapter, to maintain the confidence of the project owners, sponsors, and the public generally, megaprojects require major change in the structure, planning and management of these massive undertakings. Engineers and contractors cannot solve these problems alone. Project sponsors must be willing to treat the megaproject as a massive research and development laboratory that will benefit not only the present project, but also the megaprojects of the future.

Summary

This chapter set forth many of the requirements of implementing a megaproject. As expressed in the introduction many of the aspects of managing programs will become more efficient through the use of advancements in technologies like AI and machine learning. Though it would take several lengthy volumes to include all of the processes and steps in planning, executing, controlling and transitioning a megaproject to operations, the intent of this chapter is to raise issues for further discussion and research in the context of a particular project. There are a vast number of resources available for understanding the success factors and pitfalls in the implementation of a megaproject and each project requires different strategies, policies, approaches, and solutions. Guidance can be obtained from government reports and standards and project evaluations as well as the extensive literature available from private industry and research journals.

Lessons and Best Practices

1) The megaprojects of the future will require implementation through advanced technology and the relationship between artificial intelligence and humans will be the major focus of all projects.

2) Global megaprojects require critical upfront planning that may take several years as compared with smaller projects.

3) Understand the various delivery models and conduct a cost/benefit analysis and feasibility study to determine which models will work best in the context of the project.

4) Document recommendations given to the project's owner and the basis for the decisions to avoid conflicts or disputes down the road.

5) Understand the impact of schedule driven projects through the experience of other megaprojects and build in flexibility and contingency to address unexpected events.

6) Measure the impact of change on your project including design development, expanded scope and political change.

7) Establish a partnering environment and dispute resolution mechanisms early in the project and require compliance with the methods chosen.

8) Review project schedules and timelines to determine potential hazards and disruptions that may arise and determine methods for mitigating those exposures.

9) Make sure potential adverse impacts to the environment are mitigated or prevented throughout the project life.

10) Provide for a contingency for both cost and delay.

11) Look for opportunity in every negative risk and mitigate uncertainty.

12) Use earned value management to control and forecast costs and schedule.

13) Critically evaluate the project cost and schedule estimate including allowances for known elements and bring on board qualified experts to review and validate those estimates.

14) Review the legal issues involved in dispute resolution to determine whether a binding decision is preferable or whether a nonbinding negotiation or mediation process might help you get to a faster and a more mutually satisfying decision for all parties.

15) Evaluate the drivers of uncertainty and address the impact of uncertainty in the project financial reports.

16) Seek approval for a management reserve that is sufficient to cover potential unknowns until the uncertainty has been identified and can be managed as a risk under the project's contingency budget.

Discussion Questions

1 How will you manage the interdependencies among the project management areas, i.e., cost, scope, schedule, budget, quality, and risk?

2 Why must projects have both incentives and disincentives to ensure early contract completion? What is the difference between these two strategies? Do you think incentives or disincentives would be more effective, and why?

3 How will you track, review, and regulate progress to meet performance objectives?

4 What processes and procedures will you need to perform integrated change control?

5 To maintain project schedule, what types of data should be collected during reporting periods?

6 If the project needs to be accelerated, what kinds of activities would be the primary focus? Why? If the project needs to be delayed, what are the major financial risks and impacts that the project faces and how can they be mitigated?

7 How will you account for uncertainty in the project management reports?

8 What are some cost monitoring tools that you could utilize to provide further advance indications as to the accuracy of cost projections through Project completion that are not discussed in this chapter?

9 When can earned value be a useful tool in tracking project performance, and when will it be of no use in tracking cost and schedule?

10 Describe three vital signs that would indicate that your cost and schedule control processes are out of control.

11 Research has shown that large-scale projects are beset by an overly optimistic bias at their inception. How can projects prevent this systemic underestimation of cost so that the public is aware of the true cost of the project before commitment of any funding?

12 How can we better manage unanticipated events in future projects?

References

American Institute of Architects (2009). Standard form multi-party agreement for integrated project delivery, AIA C191, and "General conditions of the contract for integrated project delivery." AIA A295. https://www.aiacontracts.org/contract-doc-pages/27166-integrated-project-delivery-ipd-family.

Bechtel Parsons Brinckerhoff (B/PB) (2003). The Boston Globe's Big Dig: A Disservice to the Truth. Reply from B/PB. February 20.

Belay, A.M. and Torp, O. (2021). Construction cost performance under quality-gated framework: The cases of Norwegian road constructions. *International Journal of Construction Management* 21: 1–14. https://doi.org/10.1080/15623599.202 1.1949670.

Bitten, R.E. (2008). *Perspective on NASA Mission Cost and Schedule Performance Trends*. Tempe, AZ: Future In-Space Operations (FISO) Colloquium, Arizona State University, July 2.

Board on Infrastructure and the Constructed Environment (BICE) (2003). *Completing the Big Dig: Managing the Final Stages of Boston's Central Artery/Tunnel Project*. National Academy of Engineering, National Research Council, Transportation Research Board of the National Academies, Committee for Review of the Project Management Practices Employed on the Boston Central Artery/Tunnel ("Big Dig"). Washington, D.C: National Academies Press.

CA/T (Central Artery Tunnel Project) (1990). *(FEIS/R) Final Supplement Environmental Impact Report, Central Artery (I-93)/Tunnel (I-90) Project*. Boston: Commonwealth of Massachusetts: Department of Public Works. November.

CA/T (Central Artery/Tunnel Project) (2000). *Finance Plan*. Boston: Massachusetts Turnpike Authority.

CA/T (Central Artery/Tunnel Project) (2003). *Finance Plan*, 21–25, October. Boston: Massachusetts Turnpike Authority.

CA/T (Central Artery/Tunnel Project) (2004). *Finance Plan*. (October). Boston: Massachusetts Turnpike Authority.

CA/T (Central Artery/Tunnel Project) (2005). *Finance Plan*, 6. Boston: Massachusetts Turnpike Authority.

CA/T (Central Artery/Tunnel Project) (2007). *Cost and Schedule Update*, May. Boston: Massachusetts Turnpike Authority.

CA/T OCC (Central Artery/Tunnel Project) (1996) (Central Artery/Third Harbor Tunnel Project Oversight Coordination Commission). 1996. Established by Massachusetts General Laws, Section 2B of Chapter 205 of the Acts of 1996.

Clyft, R. (2021). COVID-19, Facilitative mediation, early intervention, and online dispute resolution – part 2. *The Journal of International Maritime Law* 27 (4): 189–203.

Crossrail – a progress update Department for Transport, Transport for London and Crossrail Limited (2021). *Session 2021-22 9 JULY 2021 HC 299 R*. Report by the Comptroller and Auditor General. https://www.nao.org.uk/wp-content/uploads/2021/07/Crossrail-a-progress-update-2.pdf. London: National Audit Office.

Deloitte & Touche (D&T) (2003) Central Artery/Tunnel Project, CSU10, Project Assessment. Prepared for Massachusetts Executive Office for Administration and Finance. October 17.

Dettman, K., Harty, M.J., and Lewin, J. (2010). Resolving megaproject claims: Lessons from Boston's 'Big Dig'. *The Construction Lawyer* 30 (2): 1–13. Washington, DC: American Bar Association.

DOL (U.S. Department of Labor) (2021). *Injuries, Illnesses and Fatalities (IIF)*. Bureau of Labor Statistics www.bls.gov/iif/.

DRBF (Dispute Review Board Foundation) (2021). Estimated Use of DRBs and DBs, Success Rate of the DRB Process). *Dispute Review Board Practices and Procedures*. Seattle, WA: Dispute Review Board Foundation.

Durdyev, S. (2021). Review of construction journals on causes of project cost overruns. *Engineering, Construction and Architectural Management (ECAM)* 28 (4): 1241–1260.

Edwards, W. (2002). *Project History: Reasons for Cost Growth on the Big Dig*. Presentation to the National Research Council and the National Council of Engineering Committee for Review of the Project Management Practices Employed on the Central Artery/Tunnel Project, 21 October 2002.

Engineering News Record (ENR) (1996). One project, many parts, innovative answers. *ENR* 237 (8): The McGraw-Hill Companies, Inc., New York, NY.

Grant, M. (1997). Financing Eurotunnel. *Japan Railway Transport*.

Greiman, V.A. (2013). *Megaproject Management: Lessons from the Big Dig on Risk Management and Project Management*. Hoboken: John Wiley & Sons.

Herrera, R.F., Sánchez, O., Castañeda, K., and Porras, H. (2020). Cost overrun causative factors in road infrastructure projects: A frequency and importance analysis. *Applied Science* 10 (16): 5506.

McCormack Report (1997). *Managing the Central Artery/Tunnel Project: An Exploration of Potential Cost Savings*. Boston: The John W. McCormack Institute of Public Affairs. University of Massachusetts.

Miller, R. and Hobbs, B. (2005). Governance regimes for large complex projects. *Project Management Journal* 36: 42–50. https://doi.org/10.1177/875697280503600305.

Miller, R. and Lessard, D. (2001). *The Strategic Management of Large Engineering Projects: Shaping Risks, Institutions and Governance*. Cambridge, MA: MIT Press.

National Aeronautics and Space Administration (NASA) (2009) Advisory Council Meeting: Report of Audit and Finance Committee, Kennedy Space Center, February 5, 2009. https://www.nasa.gov/offices/nac/meetings/09-02_presentations.html.

National Audit Office (NAO) (2019, 3 May). *Completing Crossrail.* Report by the Comptroller and Auditor General, Department of Transport. HC 2106 SESSION 2017–2019.

National Audit Office (NAO) (2021, 9 July). *Crossrail – A Progress Update.* Department for Transport, Transport for London and Crossrail Limited.

Norway (2016). The Act on Public Procurement (the Procurement Act) of 17 June 2016 and the Regulation on Public Procurement (the Procurement Regulation) of 12 August 2016.

Norway (2021). Standard terms and conditions on European Economic Area and Norway Grants (Art. 1.1.) Innovation Norway. Updated: 2 June 2021.

Palczynski, R. (2016, September 27). *Work Breakdown Structure Document.* London: Crossrail Learning Legacy, Crossrail Ltd.

Paton, A.R. and McCalman, J. (2008). *Change Management: A Guide to Effective Implementation*, 3rde. London: Sage Publications.

Pinto, J.K. and Winch, G. (2016). The unsettling of 'settled science': The past and future of the management of projects. *International Journal of Project Management* 34 (2): 237–245.

Price Waterhouse Cooper (2017). Sizing the prize. What's the real value of AI for your business and how can you capitalize? PwC North America.

Project Management Institute (2021). *Project Management Guide to the Body of Knowledge.* Newtown Square, PA: (PMBOK) PMI.

Rigoglioso, M., Emmons, G., and Hogg, C. (1993). *By Land and by Sea, Boston Rebuilds for Tomorrow.* Harvard Business School Bulletin (June).

Schlack, E. and Varas, P. (2019). Peripheral urbanization and the UNCTAD III building in Santiago, Chile: Continuity and disruption in grassroots engagement. – DIE ERDE. *Journal of the Geographical Society of Berlin* 150 (2): 86–100.

Wood, G.D. and Ellis, R.C.T. (2005). Main contractor experiences of partnering relationships on UK construction projects. *Construction Management and Economics* 23 (3): 317–325.

Wong, J. (2011). Edificio UNCTAD III. In: *275 días: Sitio, Tiempo, Contexto y Afecciones Específicas* (ed. P. Varas and J. Llano). Santiago: 275 días.

Xie, H. and Liu, H. (2017). Studying contract provisions of shared responsibilities for integrated project delivery under national and international standard norms. *Journal of Legal Affairs and Dispute Resolution in Engineering and Construction* 9 (3): 04517009.

Ying, F.J., Zhao, N., and Tookey, J. (2022). Achieving construction innovation in best value procurement projects: New Zealand mega projects study. *Construction Innovation* 22 (2): 388–403. https://doi.org/10.1108/CI-11-2020-0182.

8

Megaprojects and Mega Risk: Opportunity, Risk, and Resilience Management

All human actions have one or more of these seven causes: chance, nature, compulsion, habit, reason, passion, desire

— Aristotle (384–322 BCE)

Introduction

Natural and man-made disasters around the globe, including earthquakes, pandemics, cyclones, oil spills, tsunamis, devastating floods, and political unrest, remind us that catastrophic risk is all around us and, as project managers, we must be vigilant and constantly prepare for the unexpected. However, it is not just the catastrophic risk that is of concern on megaprojects, but the everyday routine performance of tasks potentially resulting in serious harm or damage claims that can exceed potential catastrophic loss. For instance, the U.S. Department of Labor statistics indicate that on an annual basis slip and fall cases constitute the biggest single cause of loss on construction projects (DOL 2021).

Megaprojects bring tremendous value to the countries, regions, and municipalities where they are built yet also bear great risk to the many people involved in building these projects. To illustrate,

> The Panama Canal brings in almost $3B annually for Panama but between 1904 and 1913 some 5,600 workers died due to disease or accidents. (McCullough 1977)
>
> The South North Water Transfer Project in China which began in 2002 has a 48-year construction schedule and, when completed, it will supply 44.8B cubic meters of water each year to the local population. However, like China's Three Gorges Dam project, the diversion scheme has provoked many major

Global Megaprojects: Lessons, Case Studies, and Expert Advice on International Megaproject Management, First Edition. Virginia A. Greiman.

risks including environmental concerns, the loss of antiquities, the displacement of people and the destruction of pasture-land. (Wilson et al. 2017)

Globally, the International Space Station (ISS) research is contributing answers to challenging problems such as medical treatments, water resourcing, and disaster relief. But the risk from meteoroids and debris is unacceptably high, primarily because of the inadequate shielding of the Russian modules. (NASA 2021; NASEM 1997)

If one were to look closely at the potential risks that are present in a megaproject one would wonder how they ever get built? Here are a few examples:

- *The English Chunnel* – this tunneling presented a major engineering challenge, not least because underwater tunnels face the ever-present risk of major water inflow due to pressure from the sea above.
- *The Hoover Dam* – before construction began on the dam, the Colorado River had to be redirected through four diversion tunnels that were driven through the canyon's walls. They measured 56 feet in diameter and their combined length was more than three miles. During construction, more than 100 lives were lost due to heat exhaustion, drowning, and pneumonia. But most people died from being hit by falling objects before thousands of hard-boiled hats (now known as hard hats) were available.

The effective management of anticipated risks in large, complex megaprojects involving multiple organizations, stakeholders and technological challenges goes far beyond the steps in the risk management methodology of traditional project management. There are many tools and techniques available for managing risks in large construction projects (Goh et al. 2013), and there is no single successful approach to risk management in infrastructure projects (Brady and Davies 2014). Instead, there is a range of approaches to managing risks that we will explore in this chapter.

Emerging Risks on a Global Scale Impacting Megaprojects

> There are costs and risks to a program of action, but they are far less than the long-range risks and costs of comfortable inaction.
> — *John F. Kennedy, 35th President of the United States*

Risk management is increasingly important for projects and business generally. It has even become mandatory in data protection in Europe under the Council of Europe's General Data Protection Regulations (GDPR 2016). According to the Global Risks Report 2022, published by the World Economic Forum, the global economy is facing increased risks in many areas and therefore the societal need for protection from harm is also increasing (WEF 2022). Over a 10-year horizon, the health of the planet dominates concerns. The environmental risks are perceived to be the five most critical long-term threats to the world as well as the most potentially damaging to people and the planet, with "climate action failure," "extreme weather," and "biodiversity loss" ranking as the top three most severe risks. The WEF survey also highlighted "debt crises" and "geoeconomic confrontations" as among the most severe risks over the next 10 years. Understanding these risks are critical to the management of megaprojects because these projects depend upon environmental and ecosystem protection, financing and economic viability, and trade and geoeconomic relations. Each one of these risks should be at the forefront of megaproject risk assessment and analysis.

At a regional level, "cybersecurity failure" ranks as a top five risk in East Asia and the Pacific as well as in Europe, while four countries – Australia, Great Britain, Ireland, and New Zealand – ranked it as the number one risk. Many small, highly digitalized economies – such as Denmark, Israel, Japan, Taiwan (China), Singapore, and the United Arab Emirates – also ranked the risk as a top-five concern (WEF 2022).

Geopolitical Risk

Widening geopolitical fractures risk is another force for global divergence. Competition between the United States and China is increasing. China's growing military prowess is changing the balance of power in the Western Pacific (Cordesman and Hwang 2021). The United States is strengthening alliances focused on the Pacific in response, most recently with the Australia–UK–US security pact (AUKUS). Other states, such as Russia and Turkey, are also showing greater capability and willingness to project power abroad as evidenced by Russia's recent invasion of Ukraine. Meanwhile, other key global and regional powers are testing boundaries of international law and cooperation by invasion of sovereign borders. Competition is intensifying in newer dimensions and geographies, as shown in the militarization and weaponization of space and in developments in cyberspace, where already-sharp tensions between governments impacted by cybercrime and governments complicit in their commission will continue to rise (WEF 2022, p. 17).

Risk Perception Across Countries

Country risks typically referred to as political risk is an important consideration in project finance. Perceptions of risk by the investors and the host country may look very different, but it is important to understand both perspectives before embarking on a multi-billion-dollar international investment.

Table 8.1 presents the top five prioritized risks from the perspective of leaders for 10 of the 124 economies surveyed by the World Economic Forum's Executive Opinion Survey (EOS) between May and September 2021. Over 12 000 leaders answered the following question: "What five risks will pose a critical threat to your country in the next two years.?" They were asked to select these from a list of

Table 8.1 Perception of risk across 10 countries.

Economy	Risk 1	Risk 2	Risk 3	Risk 4	Risk 5
Argentina	Prolonged economic stagnation	Employment and livelihood crises	State collapse	Price stability	Digital inequality
Bangladesh	Employment and livelihood	Digital Inequality	Human made environmental damage	Climate action	Geopolitization of strategic resources
China	Extreme weather events	Asset Bubble Burst	Infectious disease	Collapse of social security	Geopolitization of strategic resources
France	Erosion of social cohesion	Debt crises in large economies	Climate action failure	Failure of cyber security	Geopolitization of strategic resources
Philippines	Prolonged economic stagnation	Digital inequality	Extreme weather events	Employment and livelihood	Failure of public infrastructure
Poland	Human-made environmental damage	Infectious diseases	Interstate conflict	Fracture of interstate relations	Prolonged economic stagnation
Qatar	Climate action failure	Digital inequality	Collapse of a systematically important industry	Fracture of interstate relations	Infectious diseases
Ukraine	Prolonged economic stagnation	State collapse	Climate action failure	Failure to stabilize price trajectories	Natural resource crises
Russian Federation	Interstate conflict	Price stabilization	Infectious diseases	Employment and livelihood crises	Interstate conflict
United Kingdom	Failure of cybersecurity measures	Debt crises in large economies	Prolonged economic stagnation	Infectious diseases	Severe commodity shocks
United States	Asset bubble bursts in large economies	Climate action failure	Extreme weather events	Debt crises in large economies	Extreme weather event
					Employment and livelihood crises

Source: Adapted from World Economic Forum Global Risks Report (2022).

35 risks, with no particular order. The responses were collected from 8 September to 12 October 2021 from the World Economic Forum's multistakeholder communities (including the Global Shapers Community), the professional networks of its Advisory Board, and members of the Institute of Risk Management. The perception of risk across countries representing various regions of the world shows divergence in the top five priorities. The annual survey is important because it demonstrates that perception of risk matters as it causes countries and stakeholders to focus on those areas of risk that they perceive as most critical. For example, if climate action failure is a higher priority there will be a tendency to focus on climate action mitigation through investment and alliances with others who support a similar mission. It also impacts megaprojects in terms of the priorities that a host country may demand before agreeing to accept investment in their city or region. In this chapter, we discuss these and other global risks to learn how these risks are being managed within the context of large-scale project endeavors.

Defining Risk and Risk Management

A Risk is an identifiable state or condition which lies in the future and may or may not occur. This risk is invisible and therefore not capable of being solved. A risk can also be an opportunity. Risk management is commonly defined as a process to identify, assess, allocate, control, and monitor exposures arising from operations. It assists organizations in making informed decisions to mitigate or avoid risks to achieve better outcomes, and to contrast cost with benefits of action or inaction.

Shanghai Highway Interchange. *Source:* Stef Hoffer/Adobe Stock.

The Shanghai Interchange has attracted attention due to its magnificence and complexity. It has a structure of 5 levels and 15 lanes, extending to a total of 8 directions creating the potential for major risk if traffic exceeds the planned capacity.

All megaprojects need a clearly defined risk strategy. Project management is naturally success oriented; thus, risk management must focus on opportunities that will eliminate project failure modes, as well as develop opportunities to enhance the project's sustainable development. This means a continuous, proactive process of identifying and assessing program risk strategies and plans, monitoring processes that are deficient, and developing opportunities for success.

A few examples of risk strategies used on the world's largest megaprojects to reduce and mitigate risk include (Greiman 2013):

- Use of partnering approaches to improve understanding of stakeholder needs and concerns.
- Creation of an innovation program within the megaproject to search for and implement opportunities.
- Development of a risk intelligence program to ensure that data is collected, analyzed and made available to track risk exposures and to mitigate or eliminate risk wherever possible.
- Focus on resilience as essential to risk management.
- Initiation of scenario planning to stimulate innovation and test current processes.
- Development of an emergency response and critical infrastructure protection program to monitor risk, and implement an owner-controlled insurance program to better control risk.
- Integration of the project risk, safety, health, and loss control teams to build understanding and relationships.
- Mandatory reporting of near misses, failure of critical safety systems, unsafe conditions and at-risk behavior and penalties for failing to report when knowledge is available.
- Training of project safety staff on the benefits of incentive programs in motivating workers.
- Employment of root-cause analysis to prevent the same and similar accidents from occurring in the future.

Structure of Megaproject Risks

Figure 8.1 is critical to assessing risks across the project enterprise from the project level to the strategic level of the organization of which the project is a part. *Strategic* risks involve project selection and project finance, external stakeholders demand, market and economic risk, and political risk. Although risk analysis

Figure 8.1 Megaproject risk structure.

tends to focus on the project and program level risks, the strategic level risks often can cause the most damage including the termination or abandonment of the project. In the next chapter we will explore the risks at the strategic level more closely as we examine troubled projects, particularly those that are not able to recover from unanticipated or unplanned for events. Below the strategic level are the *programing* risks which must address the complexity and uncertainty of large-scale megaprojects that involve multiple organizations, employees, stakeholders, and governing structures. The *sector risks* such as oil and gas, nuclear, transportation, dams and mines, and critical infrastructure are concerned about technological challenges, supply chain, and catastrophic potential and also face many challenges in risk management due to the number of high impact risks. *Project risks* often focus on cost and schedule containment while *contract risks* involve dealing with conflicts both internal and external to the program and projects, and enforcement of contracts which often can be a challenge particularly in the developing world where the legal structure may still be in its early stages of development. Examples of these risks will be discussed throughout this chapter.

Resilience and Risk Management

In the context of accelerated global change, the concept of resilience, with its roots in ecological theory and complex adaptive systems, has emerged as the favored framework for understanding and responding to the dynamics of change and risk. Its transfer from ecological to social contexts, however, has led to the concept being interpreted in multiple ways across numerous disciplines causing significant challenges for its practical application (Davidson et al. 2016). Today's world complexity leads researchers and practitioners to focus on the ability of the organizations to recover from setbacks, adapt well to change, and keep going in the face of adversity. For instance, resilient infrastructures – whether they be nuclear power plants, tunnels, or hospitals – must be able to prepare for, and adapt to, changing conditions. This ability enables resilient infrastructures to withstand and recover rapidly from disruptions. "It is not just the physical structure which must withstand a blow and come back, we need resilient staff, resilient management, resilient plans, and planning" (Manto and Lockmer 2015, p. 199).

Resilience is a term that has been referred to in the context of a cyber-attack. It has been defined as the ability to adapt, recover, and reimagine a new normal (Greiman and Bernardin 2021). Resilience has also been referred to in other disciplines including medical care, agriculture responses to climate change, and responding to a disaster and saving lives before, during, and after the disaster. As an example, natural hazard-induced disasters (NHID), such as floods, droughts, severe storms, and animal pests and diseases have significant, widespread and long-lasting impacts on agricultural sectors around the world. With climate change set to amplify many of these impacts the focus has moved from a risk management to a resilience management approach (OECD 2021).

These trends mean that a traditional approach to disaster risk management in agriculture cannot continue if the focus is to increase the sustainable agricultural productivity growth needed to meet the triple challenge of (i) feeding a growing global population, (ii) providing livelihoods along the agri-food chain, and (iii) improving the sustainability of the agricultural sector, and support progress towards sustainable development. Governments and agricultural sector stakeholders need to shift from an approach that emphasizes coping with the impacts of disasters, to preventing and mitigating the adverse impacts of disasters *ex ante*, and being better prepared to recover from disasters, and to adapt and transform in order to be better placed to manage future disasters. That is, to move from a risk coping to a resilience approach.

Academic research and practice have been increasingly centered around the broad umbrella term of resilience, reflecting the desire to understand the ability of ecological, technological, and social systems to perform despite predictable and unknown changes. Resilience analysis and management is today an integrated part of the risk field and science, and risk analysis in a broad sense is needed to increase relevant knowledge, develop adequate policies, and make the right decisions, balancing different concerns and using our limited resources in an effective way (Aven 2019, p. 1196). Resilience analysis needs to be further highlighted, but a risk analysis framework is required to ensure that the right questions, concerning threats, hazards, and opportunities, are asked and the resources are used in the best possible way. "Risk analysis and resilience analysis should join forces to improve the research basis and increase impact. It is urgent that the separation trend is stopped" (p. 1202).

Resilient Responses to Risk

The steps in risk management to preventing and mitigating harmful unforeseen events that can impact the very core of the project's sustainability require going beyond the traditional seven step process reviewed in this chapter. It requires adhering to the following additional steps:

1) Taking a proactive stance to predicting a risk then preventing it from ever happening.
2) Improving operational performance. This requires modernizing operations and adjusting to the latest trends in risk mitigation.
3) Detecting and then assessing where the vulnerabilities exist within the project and the organization. This means recognizing and improving upon your weaknesses. For example, a weakness may lie in managerial experience or knowledge of technology, or behavioral theory. Wherever the weakness lies it must not be ignored but instead recognized and removed.
4) Respond and recover – assess risks when and where they occur and minimize their impact.
5) Reimagine what the new normal may look like after a risk event recovery.

International Standards for Global Risk Management

In the absence of mandatory laws and regulations on risk management, international standards play an important role in the development of a compliance regime for projects in the management of global risk. Standardization of risk management through compliance with industrial standards allows organizations to demonstrate their efforts in this area. The most widespread standards organization for risk management is the International Organization for Standardization (ISO). Importantly, a recent study shows that the ISO standards on risk management are not based on risk science and not aligned with scientific literature. For effective risk management guidance, the ISO standards need updating and alignment with the latest scientific literature on risk management (Björnsdóttir et al. 2021).

It is a challenge to find one (golden) standard approach to model complex systems and identify their potential risk. Economic and trade uncertainty, changing societal expectations, the impacts of climate change, urgency for sustainability and digital transformation were the main disruptive forces highlighted at ISO's

General Assembly in 2019 that will affect the organization moving forward (Bird 2019). Keeping up with scientific theory on risk management will be a major challenge in identifying, assessing and responding to risk in the decades ahead if countries are to realize their goals of providing a safer, healthier and more sustainable world both economically and socially.

What Is Project Risk Intelligence?

As governments around the world establish sophisticated intelligence agencies for the purpose of gathering intelligence to protect a nation's security, so must projects think about how they gather information and whether that information is reliable in protecting projects from all kinds of emergent events. These unplanned events, whether it be pandemics, terrorism, climate change, hurricanes, tsunamis, health and safety, gas explosions, or a myriad of other dangers require a project structure that produces intelligence across disciplines in the project. There is no greater need for intelligence than in managing project complexity, uncertainty and risk. Figure 8.2 highlights the seven kinds of intelligence commonly found in governments and businesses throughout the world.

Although megaprojects, similar to all projects, requires the development of a risk organizational structure and tools and techniques for managing risk it rarely refers to the concept of risk intelligence. Financial executive and Columbia University professor Leo Tilman defined risk intelligence in his 2012 book, *Risk Intelligence: A Bedrock of Dynamism and Lasting Value Creation.*

According to Tilman (2012),

> Risk intelligence is *"the organizational ability to think holistically about risk and uncertainty, speak a common risk language and effectively use forward-looking risk concepts and tools in making better decisions, alleviating threats, capitalizing on opportunities and creating lasting value."*

As explained in earlier chapters, in complex projects discerning between ambiguity, conflicts, uncertainty and risk on a project can be essential to not only ensure projects are meeting cost and schedule requirements, but it also requires an understanding of risk at a much deeper level than standard one-off projects. Risk intelligence comes not just from understanding risks internal to the project but also the intrusions and uncertainties that come from a complex external environment. Figure 8.2 shows the various types of risk intelligence that have been used by national governments to protect important interests including the defense of a country, national intelligence and the protection of innocent members of society from potential harm.

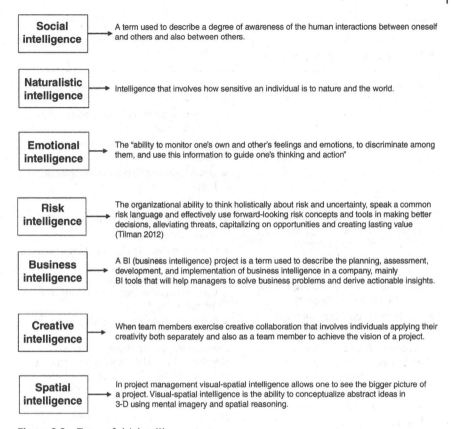

Figure 8.2 Types of risk intelligence.

Project risk intelligence requires the following inputs: (i) knowledge of the internal and external environment; (ii) continuous root cause analysis; (iii) extensive data on the underlying causes and impacts of risk mitigation and transfer; and (iv) emerging trends in risk events.

Sources of Project Risk Intelligence

Though the sources of project risk intelligence are many, a few of the most important are listed in Table 8.2. Data collection is particularly significant in terms of being able to measure the risks you are managing. Understanding the data you will need, the collection methods, the analysis methodology, and the communication of that data will be a key aspect of the assimilation and dissemination of risk intelligence.

Table 8.2 Selected sources of risk intelligence.

Selected risk intelligence organizations

The Bond Rating Agencies – Standard and Poor's Global Ratings, Moody's, Fitch Ratings

Credit Agencies and Micro Finance Organizations
- International Finance Corporation (IFC) Microfinance
- Consultative Group to Assist the Poor (CGAP)
- United Nations Capital Development Fund

Government Agencies and multilateral organizations – involved in national security, international security and critical infrastructure protection
- Ministries of Defense, National Security, NATO, Organization for Cooperation and Security in Europe (OSCE), Homeland Security
- U.S. International Development Finance Corp (Risk Intelligence)

The Global Supply Chain
- Association for Supply Chain Management
- Council of Supply Chain Management Professionals
- The International Association of Marine Consultants and Surveyors (IAMCS) was formed in 2015.
- Multinational Corporate Alliances

Intergovernmental Development Organizations
- The United Nations Development Program
- Transparency International
- Interpol
- The World Food Program
- The International Monetary Fund

Professional Project Management Organizations – The Project Management Institute, Association of Project Management, International Project Management Association

Non-profit and Academic Research Institutions

Multilateral Banks and Financial Institutions

The World Bank, The Asian Development Bank, The European Bank for Reconstruction and Development, The Asian Infrastructure and Investment Bank, The Islamic Development Bank, the Inter-American Development Bank, the African Development Bank, the European Investment Bank and each country's Central Banking System

Insurers, Reinsurers and Underwriters
- Group Underwriters Association of America (GUAA)
- Lloyd's of London, Munich Re, Swiss Re,
- American International Group

Developing a Risk Management Framework: A Shared Vision of Risk

As illustrated in Figure 8.3, risk management depends upon the establishment of a risk framework that provides clarity to the program's mission, principles, and strategy. To enable robust and proactive oversight of risk at all levels, the risk management framework comprises (i) a mission that is based on mutual cooperation and shared values, (ii) a formal risk management organization that involves the creation of multiple interdependent teams with shared principles, and (iii) a defined strategy that focuses on continuous improvement, shared lessons learned, and the implementation of best practices.

The overall risk profile must be strategically managed to ensure the project achieves its vision and business strategy. It is critical that the risk program be established with support from the top of the organization's management structure and recognized as central to the project's success. The most important aspects of developing a state-of-the-art risk management program are an integrated approach and an organizational structure that ensures sound decision making and the free

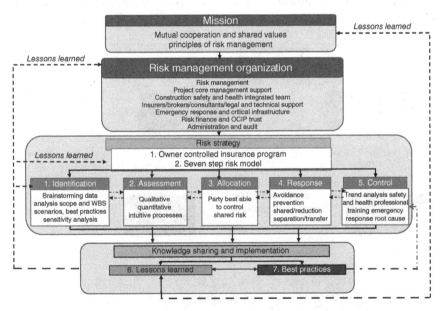

Figure 8.3 Risk management framework. *Source:* Greiman (2013)/with permission of John Wiley & Sons.

flow of information among all project participants and project stakeholders, both internal and external, throughout the life of the project.

In complex projects, involving public/private collaboration, it is important to recognize that "risks are manageable and that public officials can mitigate these risks if they take prudent and reasonable steps to ensure that they are . . . performing necessary due diligence before committing to projects . . ." (USDOT 2008).

Rigid oversight must be applied not only in the conceptual stage but throughout the project's development and execution stages. Effective risk management requires that the project organization have a clear overarching vision of risk, including the right team of participants, and that it structures the organization to ensure that risk is the first and most important priority.

Risk Management Organization

As shown in Figure 8.4, an integrated risk management structure was developed at the Big Dig (CA/T) to provide oversight of the risks the project faced by assembling a team of worldwide risk experts, brokers, and insurers to identify, assess, mitigate, and control risk. Risk management should be at the core of every project regardless of size, industry, or complexity. The CA/T's risk mission was the operation of a world-class risk management program for engineering and

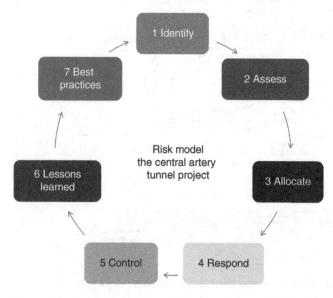

Figure 8.4 Megaproject risk model. *Source:* Greiman (2013)/with permission of John Wiley & Sons.

construction, loss control, and safety and health that focused on both opportunities and threats. Regular meetings were held with brokers, insurers, and contractors to discuss current and future risks in order to make sure appropriate resources and processes were in place to respond to and control these risks. For every contract on the Big Dig, contractor meetings were held with all interested stakeholders before construction commenced in order to discuss risk identification, assessment, response, and control with the contractor's risk manager and health and safety managers as well as with the contract resident engineer and area manager. Detailed plans were developed for dealing with identified risks, and expert advice was sought concerning the people, processes, and tools and techniques for controlling these risks. Risk Management also worked closely with the project's chief financial and budget officer, and the Owner Controlled Insurance Program (OCIP).

The Development of a Risk Model

Figure 8.4 shows a risk management model describing the shared risk concept based on approaches used by various organizations including the Project Management Institute, the National Institute of Science and Technology, and the U.S. Department of Energy. Risk Management is an iterative progression that involves a sequence of analysis utilizing various methodologies, tools and techniques, and processes and procedures. Thus, each time a new risk is identified, the sequence of analysis starts again and continues throughout the life cycle of the project and beyond, until all responsibilities of the risk manager are transitioned to operations (Greiman 2013).

Step 1: Identify: Risk Identification

Effective risk management relies on the identification of risks, particularly in the early phases, before the project concept has been finalized. In the early phases, it is more important to identify all the potential types and sources of risk than to actually identify individual risk events. Having a rich coalition of project participants and a large network external to the project that the owner can draw on in the search for information and solutions is essential. Risk identification involves continuous analysis and evaluation of potential risks based on the nature, severity, and possible impact on the project. "It is, for example, impossible to predict for the individual project exactly which geological, environmental, or safety problems will appear and make costs soar . . . [b]ut . . . it is possible to predict the risk, based on experience from other projects, that some such problems will haunt a project and how this will affect costs" (Flyvbjerg et al. 2002).

Big Dig Image of Fort Point Channel. *Source:* Virginia A. Greiman (Author).

The Fort Point Channel in Boston Harbor involved one of the more complex feats of engineering on the Big Dig Project. Project planners turned to a European technique and decided to build a concrete immersed tube tunnel. The six tubes were manufactured in a casting basin – a hole 305 m (1000 feet) long, 91 m (300 feet) wide, and 18 m (60 feet) deep. The basin was sealed off from the water by a series of steel cofferdams filled with crushed stone. When the sections were completed, the basin was flooded and the tunnel boxes – the largest weighing 45 350 metric tons – were floated out of the basin and put in position to be lowered into a trench dug in the channel bottom. Positioning the tubes was done precisely (13-mm [1/2-inch] tolerance) because they cannot be moved once they're in place.

But exact positioning wasn't the only issue. The Red Line Subway sits 60 feet below the floor of the channel. It is the oldest subway system in America and the fourth-oldest in the world. To mitigate any loss, 110 concrete shafts were drilled as much as 145 feet into bedrock on both sides of the subway tunnel. These tunnel tubes and caissons fit together like Legos and match up within 1/16 inch of perfection. Two of the tunnel tubes permanently rest exactly 4 feet above the Red Line. To add to this engineering feat, the westernmost portion of the concrete immersed tube tunnel serves as the foundation for a ventilation building. These were both first-of-their-kind engineering solutions (Greiman 2013).

Step 2: Assess: Risk Assessment

Once risks are identified, the second step in the process is the assessment or analysis of the risks. Risk analysis can be a lengthy process, usually resulting in a formal report known as a *risk assessment*. A simple explanation of the process is that risk assessment deals with answering four questions: What can go wrong? How frequently does it occur? (probability) What are the consequences? (impact) How can the risk be mitigated and what are the opportunities?

Though slip and falls are certainly a preventable risk, they are often ignored. There were 1008 fatal injuries in the construction industry in 2020 in private industry and government, many of which could have been prevented. A worker died every 111 minutes from a work-related injury in 2020. Total workplace fatalities in the United States in 2020 were (4764). Falls, slips, and trips are the most frequent type of fatal event in the construction industry (DOL 2021). In the next section the characteristics of a qualitative risk assessment and a quantitative risk assessment are compared.

Qualitative Risk Assessment Risk includes not only the exposures that evolve from the design and construction activities but also the risks inherent in the overall financial structure of the project and the market forces that bear on that risk. Close attention should be paid first to those risks that fall in the probability and impact ranges of high/high, high/medium, and medium/medium. A more detailed matrix can be used to identify the specific type of liability that may occur, such as damage to property, bodily injury, general liability, builders' risk, and environmental and professional liability.

Figure 8.5 shows the qualitative risk assessment for a typical developing world country project where poverty eradication is the primary goal. Based on assessments produced by the United Nations Development Program and the multilateral banks poverty remains a high priority but also a high risk in terms of probability and impact (CRS 2020; DESA 2021). Closely aligned to the risk of poverty eradication is the weakness in public institution administrative capacity and social inclusion. Start up and activation of projects remains a low risk while high-risk elements such as climate change, project finance needs, and the accountability of institutions continues to be a major focus of financial institutions and equity investment.

Quantitative Risk Assessment Quantitative risk assessment is about estimating specifics for risks – focused primarily on the highest-priority exposures. Quantification of risk consists of evaluating risks and interactions to assess the range of possible project outcomes. The measurement and evaluation phase involve the application of probability theory, statistical analysis, and loss forecasting methodologies, and predictions. A single risk factor can result in many outcomes. For example, collapse of a major portion of project work may result in

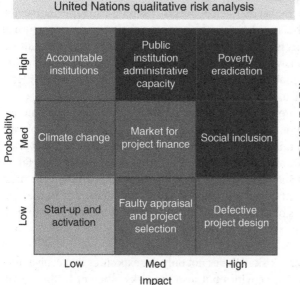

Figure 8.5 United Nations qualitative risk assessment.

flooding, explosion, property damage, personal injury cleanup, and replacement costs. Common tools for quantitative risk assessment include decision trees, influence diagrams, and Monte Carlo simulation and sensitivity methods (Meredith et al. 2022). A decision tree is a diagram that depicts key interactions among decisions and related chance events as they are understood by the decision maker. Like decision trees, simulations can also estimate both threats and opportunities, as well as sequential events.

A variety of methodologies are used to quantify risks on megaprojects, including trend analysis based on loss histories for similar projects within the same area, insurance industry informal indications for expected losses for workers' compensation insurance, expert judgment, and Monte Carlo simulations and sensitivity analysis. The benefits of quantitative risk analysis for both public and private projects include (i) evaluating the uncertainty in the requirements and the overall risk that this places on stakeholders, (ii) establishing contingency levels, (iii) improving the accuracy of project cost estimates, (iv) determining the impact of retained and transferable risks, and (v) choosing between alternative technologies or approaches with different risk profiles (Cooper et al. 2005).

In every project, contingency reserves represent the quantitative analysis of risk. It represents an allocation of the budget to known unknowns – Potential for Hurricanes, delays due to internal and external environmental factors, changes in scope and other factors, while management reserve represents a negotiated allocation of the budget for unknown unknowns.

Step 3: Allocate: Risk Allocation

Traditionally, virtually all risk associated with the design, construction, financing, operation, and maintenance of a transportation project is borne by the public sector. However, in recent years public projects have increasingly used public–private partnerships (PPPs) to share financial, technological, and operational risks, and to encourage the private sector to come forward with creative ideas for improving the quality of public transportation infrastructures. Proper allocation of project risk to the parties (public or private) best able to manage the risks has lowered overall risk, reduced project costs, and accelerated project delivery. Proper risk allocation can also increase the public sector's ability to manage a large number of projects simultaneously. When risks are understood and their consequences are evaluated, decisions can be made to allocate risks in a manner that minimizes costs and delays, promotes project goals, and aligns the construction team with the interests of the local citizens.

Step 4: Respond: Develop Risk Response Strategies

Risk response is the process by which projects reduce the likelihood that risks and resulting damages will occur from known events. It is critical that risk response begin in the early stages of project planning. Risk response strategies are the approaches that risk management takes to deal with the risks that have been assessed and quantified. In the section on risk quantification, we discuss evaluating the risk in terms of its impact and probability so that identified risks can be prioritized in their order of importance. This is called *severity*, the combination of impact and probability.

Risk response strategy is really based on risk tolerance. Risk tolerance in terms of severity is the point above which a risk is not acceptable and below which the risk is acceptable. Risk tolerance also plays a major role in what risks will be accepted and what risks will be transferred or even avoided. On public projects, risk tolerance tends to be lower because of the duty to protect the public interest. Because risk is about probability, response must be determined using processes such as root-cause analysis to determine source, and a combination of likelihood and consequence to assess options such as risk avoidance, mitigation, and transfer.

Step 5: Control: Risk Monitoring and Control

Risk monitoring and control is commonly defined as (i) the process of identifying, analyzing, and planning for newly arising risks; (ii) keeping track of the identified risks and those on the watch list; (iii) reanalyzing existing risks; (iv) monitoring trigger conditions for contingency plans; (v) monitoring residual risks; and (vi) reviewing the execution of risk responses while evaluating their effectiveness. As risk exposures are identified and quantified, appropriate means for managing each exposure must be selected in order to minimize the cost of risk. Effective control requires technical knowledge of the exposure, processes, and procedures that are regularly reviewed and enforced. Based on the risk appetite of the project's stakeholders and the prevalent regulatory requirements, risk control measures need to be monitored and reviewed throughout the project life cycle.

Some risks may be avoided entirely through decisions not to engage in certain activities, thus giving up the potential benefits of the risky activity in order to avoid the potential loss (terminate risk). The risk monitoring and control process applies techniques, such as variance and trend analysis, which require the use of performance data generated during project execution. Important strategies and tools for risk control on the project are as follows:

1) Establishment of a world-class health and safety program
2) Use of root-cause analysis and evaluation to prevent recurrence of similar events
3) Protection of critical infrastructure through security and emergency preparedness programs

Step 6: Lessons Learned

Lessons learned are among the most important process assets generated by projects and are recognized by the Project Management Institute as an essential influence on a project's success (PMI 2021). Unfortunately, these valuable assets are often collected either too late in the project's life to be of value or are forgotten before the organization's next project begins. All projects should have a comprehensive, coordinated, systemized, integrated lessons-learned documented program that cuts across all project areas and links with the umbrella organization. For instance, early in a project, a representative group of core managers who meet weekly to discuss and document all lessons learned from the previous week should be formed and supported with a best practices' implementation plan (Greiman 2013).

Step 7: Best Practices: Identify Opportunities and Implement Best Practices

Every project strives to identify best practices by comparing actual or planned project practices to those of comparable projects (PMI 2021). To avoid dooming a project to failure, you must consistently look at the best practices available (Kendrick 2009). Most effectively used as a benchmark, a "best" practice can evolve to become better as improvements are discovered on the project. Best practices are used to maintain quality in addition to mandatory legislated standards. Implementation of these practices can occur in various forms including policy memorandums, codes of conduct, or directives from senior project management.

Just as important as identifying and implementing best practices is the process of changing "worst practices." Ceasing to do things wrong can be more beneficial than doing more things right. Risk management should continuously identify both best- and worst-case practices.

Projects succeed generally because their leaders do two things well. First, leaders recognize that much of the work on any project – even a high-tech project – is not new. Lessons learned on earlier projects can be a road map for identifying and avoiding potential problems. Second, they plan project work thoroughly, especially the portions that require innovation to understand the challenges ahead and to anticipate many of the risks (Kendrick 2009).

Turning Risks into Opportunities   Project risk management has been extensively explored in the literature on project management due to the significant influence of risk management on the outcome of projects (Aladağ and Isik 2018; Galli 2018; Williams 2017; Yazdani et al. 2019). However, there has been very little research on the exploitation of opportunity as a response to project risk and threats. Opportunities arise frequently but can be overlooked because of a focus on the negative side of risk and risk reduction rather than a focus on the positive. Hillson (2002) advocates that:

> ". . ."Uncertainty" is the overarching term, with two varieties: "risk" referring exclusively to a threat, i.e., an uncertainty with negative effects; [and] "opportunity" which is an uncertainty with positive effects." (p. 235) [Hence,] "a single extended risk management process can effectively handle both opportunities and threats, and that there is therefore no need for a separate process focused exclusively on opportunities" (p. 239). Risk opportunity comes in many forms but a few examples are:

- Supply chain risks create opportunities to cut costs of materials . . ..
- Safety risks create opportunities to develop a world class health and safety program . . ..
- Health risks create opportunities to provide a safer and healthier work environment . . ..
- Compliance risks create opportunities to improve testing and audit processes.

Qazi et al. (2021) on their study of risk attitude in construction projects establishes the extent to which risk managers underestimate or overestimate overall project opportunity (risk) and expected project performance owing to their risk attitude. The findings necessitate developing a comprehensive uncertainty assessment framework that can capture the risk attitude of experts involved in assessing project uncertainty. Further, risk management tools such as risk registers and risk matrices must be tailored to explicitly account for such deviations from the actual risk and opportunity exposure of projects and to adjust the assessment of experts using appropriate factors. To improve project performance, the risks which occur across the entire project lifecycle must be recognized and managed (Dandage et al. 2019). The term uncertainty had been proposed by Ward and Chapman

Table 8.3 Turning risks into opportunities.

Risk	Opportunity
The project has a serious accident	Develop a new more rigorous health and safety program
The project has ethical problems	Establish better ethical standards in the project contract and create enforcement mechanisms
The project is behind schedule	Reach out to the project's partners and experts to develop better scheduling techniques
The project cannot implement the planned technical solution	Create an innovation program to enhance options for technical problems
The project has an environmental risk on its work site	Conduct an extensive environmental assessment to identify all emerging risk

(2003) instead of risk to indicate that it can be an opportunity or a threat. Table 8.3 highlights how risks can be turned into opportunities recognizing that it will require a commitment on the part of the project sponsors to reduce risk by making an investment in opportunity that will provide a greater return in the end.

The responses below show how a negative risk can be developed as an opportunity that may not only reduce risk but provide financial, technical, political or social benefits to the projects. Figure 8.6 summarizes the difference in response to a negative risk v. the types of responses that can elicit opportunity.

Responses to Threats and Opportunities

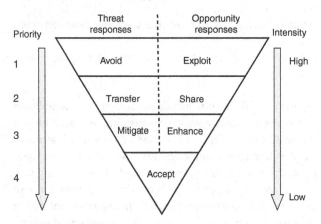

Figure 8.6 Threat responses and opportunity responses. *Source:* Adapted from Hillson (2002).

Threat/Opportunity Responses

Avoid/Exploit (Priority 1 High Intensity)

The aim of this risk response strategy is to eliminate the uncertainty associated with a particular upside risk. An opportunity-risk is defined as an uncertainty that if it occurs would have a positive effect on achievement of project objectives. The exploit response seeks to eliminate the uncertainty by making the opportunity definitely happen. Whereas the threat-risk equivalent strategy of avoid aims to reduce probability of occurrence to zero, the goal of the exploit strategy for opportunities is to raise the probability to 100% – in both cases the uncertainty is removed. This is the most aggressive of the response strategies and should usually be reserved for those "golden opportunities" with high probability and potentially high positive impact, which the project or organization cannot afford to miss.

In the same way that risk avoidance for threats can be achieved either directly or indirectly (see Hillson 2002), there are also direct and indirect approaches for exploiting opportunities. Direct responses include making positive decisions to include an opportunity in the project scope or baseline, removing the uncertainty over whether or not it might be achieved by ensuring that the potential opportunity is definitely locked into the project, rather than leaving it to chance.

Transfer/Share (Priority 2 Medium Intensity)

One common objective of the Risk Response Planning phase is to ensure that ownership of the risk response is allocated to the person or party best able to manage the risk effectively. For a threat, transferring it passes to a third party both liabilities should the threat occur and responsibility for its management. Similarly, sharing an opportunity involves allocating ownership to a third party who is best able to handle it, both in terms of maximizing the probability of occurrence, and in increasing potential benefits should the opportunity occur. In the same way that those to whom threats are transferred are liable for the negative impact should the threat occur, those who are asked to manage an opportunity should share in its potential benefits.

A number of contractual mechanisms can be used to transfer threats between different parties, and similar approaches can be used for sharing opportunities. Risk-sharing partnerships, teams, special-purpose companies or joint ventures can be established with the express purpose of managing opportunities. The risk-reward arrangements in such situations must ensure equitable division of the benefits arising from any opportunities that may be realized.

Mitigate/Enhance (Priority 3 Low Intensity)

For risks that cannot be avoided/exploited or transferred/shared, the third type of response strategy aims to modify the "size" of the risk to make it more acceptable. In the case of threats, the aim is to mitigate the risk to reduce probability of occurrence and/or severity of impact on project objectives. In the same way, opportunities can be enhanced by increasing probability and/or impact, by identifying and maximizing key risk drivers. The probability of an opportunity occurring might be increased by seeking to facilitate or strengthen the cause of the risk, proactively targeting, and reinforcing any trigger conditions that may have been identified.

Risk enhancement responses are likely to be specific to the individual opportunity-risk identified since they address the particular causes of the risk and its unique effects on project objectives. It is therefore not possible to provide a comprehensive list of actions under this strategy, and a considerable variety of actions are to be expected.

Accept (Priority 4 Low Intensity)

Residual risks are those that remain after avoid/exploit, transfer/share, and mitigate/enhance responses have been exhausted. They also include those minor risks where any response is not likely to be cost-effective, as well as uncontrollable risks where positive action is not possible. The common terminology adopted for threats in these categories is to accept the risk, with application of contingency where appropriate, and ongoing reviews to monitor and control risk exposure. One way in which these opportunities can be included in the project baseline without taking special action to address them is by appropriate contingency planning. It is also important for the project team to remain risk-aware, monitoring the status of identified opportunities alongside threats to ensure that no unexpected changes arise, and the use of an integrated risk process to manage both threats and opportunities together will assist in achieving this goal (Hillson 2001).

Challenges to Risk Analysis

Infrastructure UK, a unit with the UK Treasury that works on long-term infrastructure priorities undertook a review of the cost of infrastructure projects in the United Kingdom in 2013 to learn about best practices for cost reduction and risk mitigation. The review culminated in a report from the Infrastructure Risk Group 2013 on leading practices and improvements (IRG Report 2013). One of the challenges uncovered in their analysis was the tendency of project sponsors to adopt assumptions to favor a project's acceptance. This means either developing an optimistic bias or pricing the project too low with the hope of gaining acceptance of the project referred to in the literature as strategic misrepresentation (Flyvbjerg 2008).

Although it is not clear this is a concern in all projects, the uncertainty and complexity generated in large scale infrastructure projects creates an environment where project sponsors may adopt assumptions which will favor a project, based on their interests, while financial managers will tend to exert pressures to reduce contingencies. This paradox can cause an overstatement of risk to enhance contingencies, while understating the cost of the project. Referred to in the Infrastructure Risk Group Report as "Gaming" Behaviors." All megaproject participants should be aware of these behaviors which requires a constant vigilance of best risk management practices to ensure that contingencies are realistic, while at the same time investigating the sponsors position on project costs particularly during the early stages when uncertainty abounds on highly complex megaprojects.

Global Risk Factors in International Projects

Political Risk

According to the World Bank's 2019 Survey on *Retention and Expansion of Foreign Direct Investment: Political Risk and Policy Responses*, political risk was the most important constraint for Foreign Direct Investment (FDI) in developing economies. According to the survey data analyzed in the responses, the rate of investors divesting from developing countries because of irregular government conduct is approximately 25% (WBG 2019). The types of political risk include factors such as adverse regulatory changes, breach of contract, and transfer and convertibility restrictions. The study finds that regulatory risk, as measured in this framework, matters for investment.

Similarly, the 2018, World Bank's Global Investment Competitiveness Survey found that 45% of investors rated investment protection guarantees as critically important or deal breakers when investing abroad, notably, this was the highest among all investment climate factors. Lack of transparency and predictability in dealing with public agencies and delays in obtaining the necessary government permits to start or operate a business were also identified as factors that significantly impact investment retention and expansion (WBG 2018).

A recent study presents a new framework to measure regulatory risk that is linked directly to specific legal and regulatory provisions (Hebous et al. 2019). It captures features of countries' regulatory frameworks that can limit the potential for unexpected losses due to arbitrary government conduct. It is organized around three pillars – (i) transparency and predictability, (ii) investment protection, and

(iii) recourse mechanisms – drawing on existing indicators and newly constructed data on the content of selected legal instruments. Specifically, the measure covers the following questions:

- Is there transparency in the content as well as the process of making laws and regulations that apply to investors?
- What is the extent of legal protection provided to investors against arbitrary and non-transparent government interference?
- Do investors have access to effective mechanisms for recourse?

Disaster Risk Management

Disasters hurt the poor and vulnerable the most. Of all of deaths from weather, climate, and water hazards, 91% occurred in developing economies, according to the United Nations country classification from 1970 to 2019. The proportion remains similar for the World Bank country classification, according to which 82% of deaths occurred in low- and lower-middle-income countries.

Since 1980, more than 2.4 million people and over $3.7 trillion has been lost to disasters caused by natural hazards globally, with total damages increasing by more than 800%, from $18B a year in the 1980s to $167B a year in the last decade. The Bank's *Shock Waves* (2016) report states that almost 75% of the losses are attributable to extreme weather events. Climate change threatens to push an additional 100 million people into extreme poverty by 2030 (Hallegatte et al. 2016). The Bank's Unbreakable (2017) report shows that natural hazards have had large and long-lasting impacts on poverty (Hallegatte et al. 2017).

Population growth and rapid urbanization are driving the increase in disaster risks. The UN estimates that more than two-thirds of the world's population will live in cities by 2050, and that this trend could put 1.3B people and $158 trillion in assets at risk from river and coastal floods alone.

International Risk Response Strategies

International projects very commonly experience failure due to various factors at the global level. Especially, large projects at the international level virtually have no chance of meeting scope, time, cost, and quality. This fact has been underlined by most of the international surveys and published literature. Effective risk management plays a vital role in preventing projects from failure by implementing appropriate risk response strategies. The success of risk management will be based on the understanding of various risk categories which specifically affect international projects. This requires analysis of their interdependence, prioritization based on importance, and development of strategies for risk management based on the prioritization (Dandage et al. 2019). The literature and analysis from international projects provides examples of risks that typically pervade these projects. Among the most common risks reflected in the literature and case studies are: political instability and corruption both at the government and individual levels, cultural risk based on the multitude of international organizations and companies involved in the project, contractual risk due to a mixture of legal systems and problems of interpretation, financial risk due to investor and lender concerns, and economic risk due to currency fluctuations, cash flow shortages, and rising costs that impact return on investment.

Business Continuity Planning in the Management of Risks

To try to be safe everywhere is to be strong nowhere.
— *Winston Churchill*

The treatment of enterprise risk management would not be complete without discussing business continuity management (BCM) in the event that disasters or complex and unique risks materialize. This is similar to the need to view risk management along the value chain (internal and external customers and suppliers) in an integrated way to ensure mutual collaboration from multiple interdependent teams and their understanding and execution of common goals.

Business continuity management seeks to identify potential risks or threats to an organization and allows it to plan and develop ways to react and recover from major risk events (Hiles 2010). Today's business continuity management is tied closely to crisis management that systematically deals with a disaster or a risk event as it arises. PMI's PMBOK project management methodology for post disaster reconstruction, the Disaster Recovery Institute International (DRII) and the Business Continuity Institute (BCI) formulate the common body of knowledge that provides a structured and systematic approach to business continuity management (Howe 2007).

Root Cause Analysis

Root cause analysis has become of increasing importance to all projects no matter the location or the type of project involved. Root Cause is the cause that, if corrected, would prevent recurrence of this and similar occurrences. The root cause does not apply to risk occurrence only but has generic implications to a broad group of possible occurrences, and it is the most fundamental aspect of the cause that can logically be identified and corrected.

There may be a series of causes associated to a risk that can be identified, one leading to another. This series should be pursued until the fundamental, correctable cause has been identified (DoE 2008). For example, in the case of a leak, the root cause could be management not ensuring that maintenance is effectively managed and controlled. This cause could have led to the use of improper seal material or missed preventive maintenance on a component by the maintenance team, which ultimately led to the leak. In the case of a system misalignment, the root cause could be a problem in the training program, leading to a situation in which operators are not fully familiar with control room procedures and are willing to accept excessive distractions.

A root cause analysis (RCA) is a structured, facilitated process used by the use Department of Energy, Office of Environmental Management (EM) to identify root causes of an event(s) that resulted in an undesired cost and schedule performance (DoE 2020, p. 23). The RCA process provides EM with a way to identify and address the underlying causes of cost overruns, schedule delays, missed or postponed milestones, and performance shortcomings and it describes how to prevent future events from occurring. EM uses RCAs to find out what happened, why it happened, and determine what changes need to be made. An RCA is supported by a corrective action plan (CAP) and is an early step in a performance improvement plan to help identify what needs to be changed to improve EM cleanup performance. Figure 8.7 shows the progression of a risk event leading to

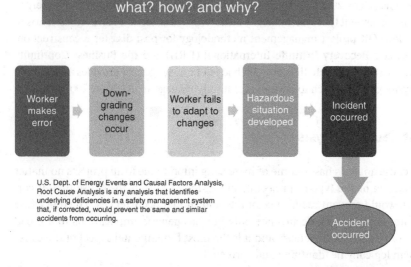

Figure 8.7 Steps in root cause risk analysis. *Source:* Adapted from Root Cause Analysis DOE Environmental Management (2008).

the root cause that explains what, how and why of the event. The (DoE) events and causal factors analysis identifies underlying deficiencies in a safety management system that, if corrected, would prevent the same and similar incidents from occurring.

Root-Cause Evaluation: Management Oversight and Risk Tree (MORT)

Table 8.4 highlights the Department of Energy's top 10 risk-related issues and root causes.

To improve safety in the nuclear industry, the MORT system was developed by Bill Johnson in the 1970s for the U.S. Atomic Energy Commission, now the U.S. Department of Energy (DoE). The system contains approximately 1500 items arranged into a large/complex fault tree used primarily for accident investigation, but also for inspection, audit, or appraisal purposes.

Events and causal factors analysis (ECFA) are an integral and important part of the MORT-based accident investigation process. It is often used in conjunction with other key MORT tools, such as MORT analysis, change analysis, and energy trace and barrier analysis, to achieve optimum results in accident investigation.

Table 8.4 Department of Energy DoE root cause analysis top 10 risk-related issues and root causes.

Issue	Root causes
1. DoE often does not complete front-end planning (project requirements definition) to an appropriate level before establishing project baselines	Insufficient resources, lack of skills, limited time, reliance on contractor, no benchmarks, ineffective integration, limited budget
2. DoE does not have an adequate number of federal contracting and project personnel with the appropriate skills (e.g. cost estimating, scheduling, risk management, and technical expertise) to plan, direct, and oversee project execution	Budget, prioritization, low government compensation, ambiguity in roles, training
3. Risks associated with projects are not objectively identified, assessed, communicated, and managed through all phases of planning and execution	Staff numbers, training, no direction, skills and resources

(Continued)

Table 8.4 (Continued)

Issue	Root causes
4. Failure to request and obtain full funding or planned incremental funding results in increased risk of project failure	Suboptimum portfolio management, prioritization and resource allocation
5. Contracts for projects are too often awarded prior to the development of an adequate independent government estimate	Lack of policy or standards, qualified personnel, database historical info
6. DoE's acquisition strategies and plans are often ineffective and are not developed and driven by federal personnel. DoE does not begin acquisition planning early enough in the process or devote the time and resources to do it well	Qualified staff, priorities, conflicts, budget, integration, lessons learned, roles
7. DoE's organizational structure is not optimized for managing projects	Priorities, alignment in authority and accountability, optimized org structure not understood
8. DoE has not ensured that its project management requirements are consistently followed. In some instances, projects are initiated or carried out without fully complying with the processes and controls contained in DoE policy and guidance	Conflicting priorities, resources, training, project reviews
9. Ineffective DoE project oversight has sometimes resulted in failure to identify project performance issues in a timely manner	Budget, resources, portfolio management, field oversight
10. DoE is not effectively executing its ownership role on some large projects with respect to the oversight and management of contracts and contractors	Expectations, definition of federal ownership role, lack of experience, limited authority of FPDs, lack of accountability

Source: Adapted from U.S. Department of Energy (2008) Root Cause Analysis.

Characterizing Risk

There are many things of which we are completely unaware—in fact, there are things of which we are so unaware, we don't even know we are unaware of them.

— *Donald Rumsfeld, 13th and 21st U.S. Secretary of Defense*

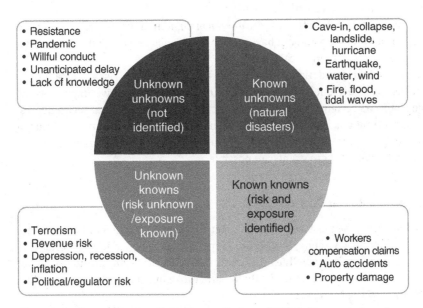

Figure 8.8 Four categories of risk. *Source:* Greiman (2013)/with permission of John Wiley & Sons.

Although risk can be characterized in many different ways through classification systems, by type of risk, or amount of risk, or cause of risk, Figure 8.8 highlights classification by extent of knowledge of risk, the probability it will occur, and the damage that will be done if the event occurs. The project management literature differentiates between events that occur totally by surprise, and "outcomes or events that actors have identified as possibly existing, but do not know whether they will take place or not" (Geraldi et al. 2010, p. 553). This spectrum of growing uncertainty is frequently known by the labels of known knowns, known unknowns, unknown knowns, and unknown unknowns (Cléden 2009; De Meyer et al. 2002; Sanderson, 2012; Winch 2010; Winch and Maytorena 2011).

Former U.S. Secretary of Defense, Donald Rumsfeld, distinguishes between four categories of risk (Rumsfeld 2012). These categories are also recognized in the project management literature though not always described in the same way.

The first category is identified as *known knowns* describing the things we know we know. Examples could be a project's location, the type of project, the stakeholders that are impacted by the project and other realities.

The second category is defined, as *known unknowns* describing the things we know are uncertain. Examples could be how many workers are needed to complete a particular task or unpredictable weather conditions. Known unknowns can be estimated in a *probabilistic* sense.

The third category is defined as the *unknown unknowns*. This category describes uncertainties that we could not have known in advance and let alone foresee their consequences, e.g., natural cataclysms.

The last category is defined as *unknown knowns* describing risks that cannot be identified precisely due to multiplicity, but whose total negative impact on the project appears certain. These include an act of terrorism, criminal acts, a depression or revenue risk. We may not know when or how these events will occur or how much damage or destruction will result from the event (unknown), but we do know that all of these events are possible.

The Ariane 5 Rocket Disaster

On 4 June 1996, the massive *Ariane 5* rocket, launched on its maiden voyage from the European Space Agency's center in French Guiana, exploded 39 seconds into flight, destroying four satellites on board (ESA 1996). At an altitude of about 3700 m, the launcher veered off its flight path, broke up, and exploded. According to the Inquiry Board Report:

> The failure of the Ariane 501 was caused by the complete loss of guidance and altitude information 37 seconds after start of the main engine ignition sequence (30 seconds after lift-off). This loss of information was due to specification and design errors in the software of the inertial reference system.

How could such an unexpected outcome befall a project that took 10 years, $7B, and tens of millions of hours of human labor and expertise?

Despite the series of tests and reviews carried out under the program in the course of which thousands of corrections were made, shortcomings in the system approach concerning the software resulted in failure to detect the fault. It is stressed that alignment function of the inertial reference system, which served a purpose only before lift-off (but remained operative afterwards), was not considered in the simulations and that the equipment and system tests were not sufficiently representative (Mark et al. 2016).

In a postmortem analysis, the European Space Agency (ESA) noted that the problem was due to a malfunction in the rocket's guidance system software, and it acknowledged that the system was not fully analyzed or understood, perhaps because the software had worked successfully with *Ariane 4*. More generally, the disastrous outcome was due to an uncertainty of which the Agency was unaware prior to the launch – an unknown unknowns (or "unk" in common industry parlance). Yet was this uncertainty truly unknowable, or was it potentially knowable but just escaped recognition by the project management team?

The Space Shuttle Challenger

On 28 January 1986, the Space Shuttle Challenger exploded about 73 seconds after lift-off and cost the lives of seven astronauts. After this incident President Ronald Reagan launched an official investigation (the Rogers Commission) to examine the cause. The commission determined that exposure of the launch vehicle to three abnormally cold nights caused an O-ring on one of the solid rocket boosters to became brittle. When the launch occurred, hot gas had exited the SRB and pierced the primary fuel tank, which caused the explosion. This determination derived from physicist Richard Feynman dunking a piece of O-ring into a cup of ice water. This exercise demonstrated lost resiliency at such a low temperature, which proved that abnormal cold affected the O-ring in the Shuttle (Mark et al. 2016). The engineering risks were well publicized but the governance failure has barely received attention in the scholarship as described in Diane Vaughan's (1990) article describing the concept of "normalization of deviance."

Normalization of Deviance Related to Risk Discovery

Diane Vaughan (1990) has developed the concept of "normalization of deviance" to explain how technical flaws (unknown/unknown risks) can escape the scrutiny of the various safety bodies within NASA over time. In many cases, unanticipated problems continue to occur even though nothing particularly catastrophic happens during a given Shuttle mission. This leads to the very pragmatic notion of "acceptable" deviance. That is, "it was often very expensive and time-consuming to root out the cause of a given anomaly with some problems being incorporated into the regular maintenance cycle of the Shuttle without detailed examination."

The NASA shuttle challenger disaster highlights the concept that Vaughan has developed that design flaws are usually recognized well in advance of an accident.

- NASA – knew five years prior about the rocket booster O-ring failures in the challenger.
- NASA – knew about Debris shedding from the external fuel tank 20 years prior.

Engineer's Intuition

One of the concerns in the shuttle challenger and other large-scale engineering projects is the need for a mechanism for engineers to be able to bypass the bureaucracy and hierarchy, especially in the pre-launch process. Hall (2003) explored this issue in his research on the shuttle challenger launch decision process and the NASA Columbia disaster and raised the following questions: What would have been the alternative if the engineers had succeeded in getting their point across in the case of challenger? Probably challenger would have had to been taken off the launch pad and the SRBs disassembled to replace the damaged O-rings (p. 244). This would have been expensive but not nearly as costly as the loss of crew and vehicle. If an engineer has a special request for a certain type of data, there should be a way to request exceptions to formal bureaucratic procedures to focus on getting the data. Engineers have many intuitions and hunches that take time and resources to translate into analysis and data. "These intuitions need to be respected, given credence, explored and welcomed by upper management" (p. 244).

The following three theories – institutionalization, socialization, and rationalization have been found to provide some reasons for the occurrence of deviant behavior which increase the possibility of unobserved risks. These theories have been applied to deviant behavior in the criminal system to enable corrupt practices to evolve in white collar organizations (Ashforth and Anand 2003) and to deviant behavior in the health care system (Banja 2010); though the health professional's "deviance" is virtually never performed with criminal or malicious intent (p. 4). Although they certainly do not represent all possible forms of deviance, understanding why deviance occurs and how to proactively find ways to prevent it is an important first step.

Institutionalization – exposes newcomers to deviant behaviors, often performed by authority figures (modeling).

Socialization – mediated by a system of rewards and punishments aimed at determining whether the newcomer will or will not join the group's deviant behaviors (incentives).

Rationalization – enables system operators to convince themselves that their deviances are not only legitimate, but acceptable and perhaps even necessary. The justification for deviating from a project standard, process or procedure includes the belief that:

1) The rule is irrational
2) The workers are afraid to speak up and confront violators
3) There are too many rules – no one in their right mind would follow them all
4) The rule is inefficient – too bureaucratic
5) The work environment is complex – dynamic, unstable, and unpredictable

Key Questions for Unknown Unknowns

From the perspective of a PM, key questions are:

1) *What are the driving factors in a project that increase the likelihood of its encountering unknown unknowns?*
2) *How should an organization invest to uncover unknown unknowns?*
3) *Are there approaches that a project can take to reduce the likelihood of unknown unknowns?*

Lessons on Unknowns from High-reliability Organizations (HRO)

Answers to these questions can help a PM allocate appropriate resources toward the conversion of unknown unknowns to known unknowns (e.g. risk identification). A major challenge in project management is dealing with the uncertainties within and surrounding a project that give rise to outcomes that are unknown or known only imprecisely (Ramasesh and Browning 2014). Uncertainties can have a positive or negative effect on one or more project objectives.

Since unknown unknowns are unanticipated outcomes they bear some connection to accidents, safety and reliability, especially in hazardous industries, which have been the subject of prominent sociological theories such as the *normal accident theory* (NAT) and the theory of *high-reliability organizations* (HRO).

A HRO is an organization that has succeeded in avoiding catastrophes despite a high level of risk and complexity. Specific examples that have been studied, most famously by researchers Karl Weick and Kathleen Sutcliffe, include nuclear power plants, air traffic control systems, and naval aircraft carriers. Recently healthcare organizations have moved to adopt the HRO mindset as well. In each case, even a minor error could have catastrophic consequences. According to Weick and Sutcliffe (2015) the so-called HROs demonstrate particular characteristics in the way they operate: anticipating problems (being aware of what is happening in the work system; being alert to ways in which an incident could occur; looking beyond simplistic explanations for incidents); and containing problems (being prepared to deal with contingencies; using relevant expertise regardless of where it is situated within the organizational hierarchy). In their research on "high-reliability organizations" they analyzed how highly regulated and standardized organizations such as nuclear plants, aircraft carriers and firefighting units achieve resilience in a complex environment. They found that technical checklists were not the key to success but instead a list of cultural features they define as "mindful organizing." They identified five principles that include (i) preoccupation with failure, (ii) a reluctance to simplify, (iii) a sensitivity to operations, (iv) a commitment to resilience, and (v) a deference to expertise as a shared set of values that foster resilience through constant communication and recalibration in the face of unknowable risks (Weick and Sutcliffe 2015).

Vogus and Sutcliffe (2013) proposed "mindful organizing" as a collective mental orientation in which the organization continually engages with its environment, reorganizing its structures and activities as necessary, rather than mindlessly executing plans in ignorance of the prevailing circumstances. This is a dynamic, social process, consisting of specific actions and interactions between those engaged in frontline organizational work. It creates the context for thought and behavior across the organization but is relatively transient and so needs to be actively maintained. Hopkins (2014) argues that the characteristics of an HRO and the components of safety culture (reporting; flexibility; learning; and fairness) are broadly equivalent from the perspective of the field of safety science.

HRO, asserts that accidents – even in complex organizations that operate hazardous technologies (such as air traffic control and aircraft carriers) – are avoidable by creating appropriate behaviors and attitudes that would increase "heedful interrelations and mindful comprehension" and with emphasis on safety, attention to problems, and learning (Weick and Roberts 1993).

Safety Critical Design Questions for All Megaprojects:

- Are cost and schedule pressures detracting from safety critical design and/or design verification?
- Is there an effective pathway to express your concerns?
- Are whistleblowers protected from retaliation and other fears?
- Are safety critical maintenance activities being identified and conveyed to others by the proper authorities?
- Are inspections implemented in a timely manner?
- Have all stakeholders worked to understand root causes associated with any unexpected results or off-nominal behaviors in development, testing or integration?
- Are you assuming engineering accountability or are you delegating?

Catastrophic Loss Potential: Natural and Manmade Disasters

In this section, we look at a few natural and manmade disasters that have led to catastrophic loss and impacted and disrupted megaprojects globally. Major disasters can occur because of wildfires, nuclear and oil industry failures, attacks on control systems, widespread electrical power outages, or major system failures. To ensure the prevention of manmade disasters we need to be better prepared for these unknown unknown risks to ensure a faster project recovery. After Hurricane Katrina the federal government in the United States developed a long list of preparatory steps to mitigate the impact of hurricanes. These steps have been applied and results have shown the effectiveness in mitigating some of the losses that might otherwise have occurred (The White House 2006).

Since the early 1990s, total economic losses from natural catastrophes in the United States alone have averaged tens of billions of dollars per year. These disasters cause death and injury, damage property and the natural environment, interrupt business activities, and generally disrupts society. Damages from natural catastrophes in the United States are rising and are expected to continue to increase in the future. However, manmade disasters are also on the rise. Lloyd's City Risk Index 2015–2025 reports due to an increasingly interconnected and technologically dependent world, that nearly half of the total GDP at risk is linked to manmade threats, including market crash, human pandemic, cyber-attack, power outage and nuclear accident (Lloyd's 2015).

In the megaproject world, the spectrum for risk is broad. According to Munich Re (2021) Project risks are often inadequately insured, and many occur because of manmade rather than natural disasters. Although it is rare for infrastructure projects to collapse during the construction phase, when it does happen the financial loss can be immense. Insufficient planning may have a material impact on both costs and deadlines during the construction process. For example, investors fear the uncertainty that naturally comes with the numerous risks involved in construction projects.

According to global studies, the range of unplanned increases of construction costs is in the order of 20–30% of the project building cost. Collapsing bridges, leaky pipes, crumbling roads – infrastructure in the western industrialized countries is not getting any younger.

There are various reasons for this. Roads and bridges are being worn down by increasing, and heavier, vehicular traffic. Energy and water use is increasing, but water and sewage systems in major cities are relatively old. The same applies to public transport facilities, which often no longer meet today's standards. The problem is compounded by the growing number of natural disasters which place new strains on our infrastructure each day (Munich Re 2021).

In its *2021 Annual Report*, Munich Re reported that natural catastrophes caused US $280B in worldwide losses in 2020. At US $120B in insured losses, the 2021 figure is second only to the record year of 2017. In 2018, Lloyd's of London reported a second consecutive year of losses after a series of natural disasters including hurricanes Florence and Michael and the deadly California wildfires. The insurance market reported a £1B loss, halving its 2017 loss of £2.1B.

To understand the impact of catastrophic loss worldwide, the major impact of catastrophes over the last two decades is highlighted in Table 8.5, based on estimates

Table 8.5 Major catastrophic disasters and losses.

Event	Estimated loss
Fukushima Daiichi Nuclear Plant (2011)	$200B (not including decommissioning)
Hurricane Katrina (2005)	$82.39B
World Trade Center (2001)	$30–40B
BP Deep Water Horizon (2010)	$25–40B
California Earthquake (1994)	$12.5B
Indonesia Lusi Mudflow (2006)	$2.7B

Source: Adapted from Lloyd's of London, Munich Re, Swiss Re.

from the world's largest underwriters and reinsurers. What is significant is that three of the six disasters listed are manmade and include the Indonesia mudslide (Fallahi et al. 2017; Gibbens 2017; Judd 2005), the BP Horizon Deep Water Oil Spill (National Commission 2011), and the 2001 attacks on the World Trade Center (Tognini 2021).

> Hurricane Katrina was one of the worst natural disasters in our Nation's history and has caused unimaginable devastation and heartbreak throughout the Gulf Coast Region. A vast coastline of towns and communities has been decimated.
>
> — *President George W. Bush, 8 September 2005*

Hurricane Katrina

This hurricane was the most destructive natural disaster in U.S. history. The overall destruction wrought by Hurricane Katrina, which was both a large and powerful hurricane as well as a catastrophic flood, vastly exceeded that of any other major disaster, such as the Chicago Fire of 1871, the San Francisco Earthquake and Fire of 1906, and Hurricane Andrew in 1992. Though Hurricane Katrina in 2005 is one of the five deadliest hurricanes in the history of the United States some argue that this hurricane had identifiable risks including the failure of the region's flood-control system and the design of the levee structure (Knabb et al. 2005). This highlights that even natural disasters can be controlled to a certain extent, and that potential losses can be seriously reduced.

Catastrophic loss generally has a low probability of occurring because of the mitigation efforts that are undertaken, but high damages if the events do occur. Infrastructure development is often at risk, as it was during the 2004 tsunami in the Indian Ocean, which set back development in parts of Southeast Asia by more than 20 years (UNDP 2009). Catastrophic loss can be classified into four categories: human perils, natural perils, technological perils, and economic perils.

BP Deepwater Horizon Oil Spill

Oil spills are among the most visible man-made disasters of our times, and 2010s Deepwater Horizon incident is reputed to be the largest marine oil spill in history. An investigation of the risk found Deep Water Horizon to be a combination of technical and managerial risk (National Research Council 2014). As described in the *Report of the National Commission to the President on the BP Deepwater Horizon Oil Spill and Offshore Drilling* (2011) "the exploratory Macondo well deep under the waters of the Gulf of Mexico, began a human, economic, and environmental disaster" (p. xi). The Deepwater Horizon oil rig was a floating platform that was drilling an exploratory oil well around 18 300 feet (5600 m) below sea level in the Gulf of Mexico. On 20 April 2010, methane gas from the underwater well expanded and rose into the drilling rig, where it ignited and exploded. The explosion quickly engulfed the entire drilling platform, killing 11 workers, and 94 crew members were evacuated. Two days later, the rig had sunk. Eventually, the oil was contained and either dispersed or cleaned up using several different methods and thousands of volunteers. By then, though, the environmental impact was being felt on a global scale: the Deepwater Horizon spill took 11 lives, killed millions of animals, and the incident affected wildlife and ecosystems across several US states and beyond (National Commission 2011).

Indonesia Luci Mudslide

In 2006 in Sidoarjo, Indonesia, the world's biggest mud volcano occurred. Twenty lives were lost, 10 000 homes were destroyed, and nearly 40 000 people displaced, and metal from the mud flow has contaminated nearby rivers with damages topping $2.7B from the disaster, known as the Lusi mudflow (Fallahi et al. 2017). Typically, you think of a volcano as a natural disaster but this one was different. It was created by an explosion at a gas well drilled by an energy company, although company officials claim that an earthquake around 155 miles (250 km) away provoked the problem (Gibbens 2017). There are more than 1000 mud volcanoes around the world, but this Indonesian volcano is probably the only one caused by human activity (Judd 2005) as indicated by extensive geophysical research (Fallahi et al. 2017).

Fukushima Daiichi Nuclear Power Station

The 11 March 2011, Great East Japan Earthquake and tsunami sparked a humanitarian disaster in northeastern Japan causing more than 15 900 deaths and 2 600 missing persons as well as physical infrastructure damages exceeding $200B. Though Fukushima was initiated by a large earthquake, investigations on this case by the United States and other countries have revealed technical mishaps and regulatory oversight that might have mitigated the disaster sooner (National Research Council 2014). The earthquake and tsunami also initiated a severe nuclear accident at the Fukushima Daiichi Nuclear Power Station that caused an explosion of the released hydrogen that damaged three reactor buildings and impeded onsite emergency response efforts. The accident prompted widespread evacuations of local populations, large economic losses, and the eventual shutdown of all nuclear power plants in Japan and resulted in numerous investigations and stricter regulation of nuclear systems in the United States and other countries (National Research Council 2014). The Japan Center for Economic Research, a think tank, estimates that decommissioning alone would cost 70 trillion yen in 40 years ($587B) if the water is not released and tritium removal technology is pursued (Kobayashi et al. 2019).

The following table shows the losses for some of the largest catastrophic events globally.

Important Lessons Learned from Major Disasters:

1) Catastrophic potential on mega infrastructure projects calls for a shift from risk to resilience so that both frameworks are integrated to ensure continued survival of the project, adaptation to the catastrophic event and the reimagination of a new normal.

2) All organizations must prepare for the worst-case scenario and be prepared to implement disaster prevention and disaster relief plans at the earliest possible time frame.

3) Even natural disasters can be reduced and mitigated with the appropriate risk management procedures.

4) The national and local governments should review current laws, policies, plans, and strategies relevant to communications.

5) Most disasters are preventable and all megaprojects should assess ways of mitigating natural versus manmade disasters including the prioritization of saving human life.

6) All projects should obtain insurance for catastrophic loss and make sure that project stakeholders are protected from potential catastrophic loss.

7) All megaprojects should learn how "high reliability organizations" such as nuclear plants, aircraft carriers and firefighting units achieve resilience in a complex environment.

8) Plans should be laid out for responses to natural and manmade disasters along with extensive policies and procedures so that all participants have a uniform understanding of the processes that will be applied in the event of a major disaster.

Summary

Megaprojects require a nontraditional approach for managing risk that must be conceived and developed before project commitments are finalized in the conceptual phase. Risk intelligence should be identified and sought out to reduce the likelihood of serious loss based on inexperience or lack of knowledge. Risk management requires a shared vision and mutual cooperation among all project stakeholders. The project's organization must be structured to respond quickly to unplanned events and to establish open communication and a collaborative environment. Risk management must be linked to resilience as preparing for resilience is essential to adaptation and often long recovery periods after major unexpected events.

All participants must be educated and updated on the underlying assumptions and dynamics of the ever-evolving processes of risk identification, assessment, allocation, response, and control. Lessons learned and root-cause determinations must be communicated to all project participants, and best practices must be developed and integrated into all aspects of the project's organization. Finally, preventing catastrophic loss even caused by natural disasters should be considered for every known event, even if the probability of that event occurring is low. Human life should be protected at all cost.

Lessons and Best Practices
1) Risk management should be seen as a collaborative process, utilizing the skills and expertise of all parties to manage and mitigate risk.
2) Risks should be owned by the organization most able to manage the risk. This should be clearly documented in contracts.
3) Risk intelligence can assist in the identification, analysis, and control of risk and can be used to identify opportunities.
4) The approach to risk management should align to the delivery model, and be supported by clear policies, processes, procedures and systems to enable consistency.
5) Process maturity reviews should be used to drive performance improvement through collaboration and knowledge sharing.
6) A risk management framework with a shared vision, an integrated organization structure, and a strategic risk model must be developed from the conceptual phase of the project to ensure effective risk preparedness. Converting from a traditional contractor-controlled program to an owner-controlled program will be costly and will delay the implementation of project wide practices.

7) Owner-controlled insurance programs have proven to be effective mechanisms for large-scale projects with multiple contractors and complex governance systems if they are properly structured and integrated into the project mission and procedures.

8) Risk sharing is essential to mitigate losses, reduce cost, and maintain relationships both internal and external to the project.

9) A mission of safety and health must take priority and rebel against cost, schedule, and scope constraints, but never against quality.

10) To change behavior, workers must be incentivized through recognition and awards based on clearly understood criteria and expectations, as well as penalized for wrongful behaviors.

11) Political leaders, the public, the business sector, and local communities must be educated on the benefits of risk mitigation options and the impact if these options are not accepted.

12) Stakeholder participation is critical to understanding risk perception, and all project participants as well as the public must be educated to prevent faulty decision making.

13) Risk management strategies used in large-scale complex projects must expand upon traditional project management methodologies, tools, and techniques and introduce innovative practices not readily available in the project management literature.

14) Project organizational, governance, and contractual structure are key factors in mitigating risk.

15) Project culture is a by-product of organizational, governance, and contractual structure and must be continually assessed to ensure a safe and healthy project environment.

16) Creating a culture of fact finding rather than fault finding and rewarding behavioral change are important factors in risk mitigation and control.

17) Reporting of all incidents, unsafe conditions, and at-risk behavior is essential to foster learning and continuous improvement.

Discussion Questions

1 How do you avoid complacency when your project has repeatedly been successful at inherently hazardous or difficult tasks?

2 How would you conduct a quantitative and qualitative analysis of the consequences of a disaster, such as a fire in your plant? What type of data would you require? What would your standard form of measurement be?

3 What should the role of a risk manager be in a project versus in a program? Are the roles identical? What are the differences?

4 Should the risk manager in a megaproject be responsible for construction risks? Cost and schedule risks? Political risks? Environmental risks? Does it depend on the size of the project? The location of the project? The type of project (infrastructure versus research and development)? What factors would you consider in structuring your risk management services?

5 PMBOK states: project risk has its origins in the uncertainty present in all projects. It defines known risks as those that have been identified and analyzed, making it possible to plan responses for those risks. Specific unknown risks cannot be managed proactively, which suggests that the project team should create a contingency plan.

6 Risk managers often face ethical decisions concerning conflicts of interest, transparency, environmental obligations, and political risk. Provide an example of an ethical dilemma a risk manager might encounter and explain how you would resolve this dilemma.

7 How would you prioritize actions in a risk assessment? Would you look at: Impact of the risk? Probability of the risk? Public concerns? Perceptions of risk? Stakeholder expectations? Other concerns?

8 Distinguish between a hazard and a risk. As a risk manager, would you manage a hazard differently than a risk? Explain why or why not. Should certain risk issues have a higher priority than others?

9 How would you identify risk opportunities on a project?

10 How would you create a culture of fact-finding rather than fault finding on a project with serious safety concerns and near misses?

11 What are the most important factors to consider in changing behaviors?

12 Assume you were hired as director of risk management to develop a risk program for a new tunnel project in the United States. The project director has given you full authority to develop the high-level framework and structure for this project. Describe the top 10 questions you would ask the project director before beginning your responsibilities as director of risk management.

13 How can risk intelligence assist in managing and controlling risk?

14 How would you conduct a qualitative risk analysis in a developing country to ensure all risks were captured in your risk register and all possible risk management strategies were employed?

References

Aladağ, H. and Isik, Z. (2018). The effect of stakeholder-associated risks in mega-engineering projects: A case study of a PPP airport project. *IEEE Transactions on Engineering Management* 67 (1): 174–186. http://dx.doi.org/10.1109/TEM.2018.286626.

Ashforth, D.E. and Anand, V. (2003). The normalization of corruption in organizations. *Research in Organizational Behavior* 25: 1–52.

Aven, T. (2019). The call for a shift from risk to resilience: What does it mean? *Risk Analysis* 39 (6): 1196–1203. https://doi-org.ezproxy.bu.edu/10.1111/risa.13247.

Banja, J. (2010). The normalization of deviance in healthcare delivery. *Business Horizons* 53 (2): 139. http://dx.doi.org/10.1016/j.bushor.2009.10.006.

Bird, K. (2019). *Four Trends will Impact ISO's Future Strategy*. Geneva, Switzerland: International Standards Organization (IOS) https://www.iso.org/cms/render/live/en/sites/isoorg/contents/news/2019/09/Ref2436.html.

Björnsdóttir, S.H., Jensson, P., de Boer, R.J., and Thorsteinsson, S.E. (2021). The Importance of Risk Management: What is Missing in ISO Standards? *Risk Analysis*. Wiley Online. https://doi-org.ezproxy.bu.edu/10.1111/risa.13803.

Brady, T. and Davies, A. (2014). Managing structural and dynamic complexity: A tale of two projects. *Project Management Journal* 45 (4): 21–38. http://dx.doi.org/10.1002/pmj.21434.

Cléden, D. (2009). *Managing Project Uncertainty*. Gower, UK: Farnham.

Congressional Research Service (CRS) (2020). *Multilateral Development Banks: Overview and Issues for Congress updated February 11, 2020*.

Cooper, D., Grey, S., Raymond, G., and Walker, P. (2005). *Project Risk Management Guidelines: M,naging Risk in Large Projects and Complex Procurements*, 187–188. West Sussex, England: John Wiley & Sons, Ltd.

Cordesman, A.H. and Hwang, G. (2021). *Updated Report: Chinese Strategy and Military Forces in 2021*. Center for Strategic and International Studies https://www.csis.org/analysis/updatedreport-chinese-strategy-and-military-forces-2021.

Dandage, R.V., Mantha, S.S., and Rane, S.B. (2019). Strategy development using TOWS matrix. *International Journal of Managing Projects in Business* 12 (4): 1003–1029. http://dx.doi.org/10.1108/IJMPB-.

Davidson, J.L., Jacobson, C., Lyth, A. et al. (2016). Interrogating resilience: Toward a typology to improve its operationalization. *Ecology and Society* 21 (2): 27. http://dx.doi.org/10.5751/ES-08450-210227.

De Meyer, A., Loch, C.H., and Pich, M.T. (2002). Managing project uncertainty: From variation to chaos. *MIT Sloan Management Review* 43 (2): 225–257. http://dx.doi.org/10.1109/EMR.2002.1032403.

Department of Economic and Social Affairs (DESA) (2021). *Financing for Sustainable Development Report 2021*. United Nations: Interagency Task Force on Financing for Development.

Department of Labor (DOL) (2021). *Injuries, Illnesses and Fatalities (IIF)*. Bureau of Labor Statistics www.bls.gov/iif/.

European Space Agency (ESA) (1996). No 33–1996: Ariane 501 - Presentation of Inquiry Board Report.

European Union (EU) (2016). *The General Data Protection Regulation (EU) 2016/679 (GDPR)*.

Fallahi, M.J., Obermann, A., Lupi, M. et al. (2017). The plumbing system feeding the Lusieruption revealed by ambient noisetomography. *Journal of Geophysical Research: Solid Earth* 122: 8200–8213. https://doi.org/10.1002/2017JB014592.

Flyvbjerg, B. (2008). Curbing optimism bias and strategic misrepresentation in planning: Reference class forecasting in practice. *European Planning Studies* 16: 3–21.

Flyvbjerg, B., Holm, M.S., and Buhl, S. (2002). Underestimating costs in public works projects: Error or lie? *Journal of the American Planning Association* 68 (3): 279–295.

Galli, B.J. (2018). The future of economic decision making in project management. *IEEE Transactions on Engineering Management* 67 (2): 396–413. http://dx.doi.org/10.1109/TEM.2018.2875931.

Geraldi, J.G., Lee-Kelly, L., and Kutch, E. (2010). The Titanic sun, so what? *International Journal of Project Management* 28 (6): 547–558.

Gibbens, S. (2017). Why this Massive Mud Vocano Turned Deadly. *National Geographic*. https://www.nationalgeographic.com/science/article/mud-volcano-lusi-indonesia-video-spd.

Goh, C.S., Abdul-Rahman, H., and Samad, Z.B. (2013). Applying risk management workshop for a public construction project: Case study. *Journal of Construction Engineering and Management* 139: 572–580.

Greiman, V.A. (2013). *Megaproject Management: Lessons on Risk and Project Management from the Big Dig*. Hoboken, NJ: John Wiley & Sons.

Greiman, V.A. and Bernardin, E. (2021). *Cyber Resilience: A Global Challenge*. Reading, UK: Academic Conferences and Publishing International (ACPI).

Hall, J.L. (2003). Columbia and challenger: Organizational failure at NASA. *Space Policy* 19: 239–247.

Hallegatte, S., Bangalore, M., Bonzanigo, L. et al. (2016). *Shock Waves: Managing the Impacts of Climate Change on Poverty*. Washington, DC: *Climate Change and Development,* World Bank.

Hallegatte, S., Vogt-Schilb, A., Bangalore, M., and Rozenberg, J. (2017). *Unbreakable: Building the Resilience of the Poor in the Face of Natural Disasters. Climate Change and Development*. Washington, DC: World Bank.

Hebous, S., Kher, P., and Tran, T.T. (2019). *Regulatory Risk and FDI: Global Investment Competitiveness Report 2019*. Washington, DC: World Bank.

Hiles, A. (2010). *The Definitive Handbook of Business Continuity Management*, 3rde. West Sussex England, Hoboken, NJ: John Wiley & Sons.

Hillson, D. (2001). *Effective strategies for exploiting opportunities*. [Paper Presentation] Project Management Institute Annual Seminars & Symposium, Nashville, TN. Newtown Square, PA: Project Management Institute.

Hillson, D. (2002). Extending the risk process to manage opportunities. *International Journal of Project Management* 20 (3): 235–240. http://dx.doi.org/10.1016/S0263-7863(01)00074-6.

Hopkins, A. (2014). Issues in safety science. *Safety Science* 67: 6–14.

Howe, J. (2007). Project Initiation and Management. In: *The Definitive Handbook of Business Continuity Management* (ed. A. Hiles). Chichester, England, Hoboken, NJ: John Wiley & Sons.

Infrastructure Risk Group (IRG Report) (2013). *Managing Cost Risk and Uncertainty in Infrastructure Projects: Leading Practice and Improvement*. Report from the Infrastructure Risk Group 2013 and the Institute of Risk Management, London, UK.

Judd, A. (2005). Gas Emissions from Mud Volcanoes. In: *Mud Volcanoes, Geodynamics and Seismicity*, NATO Science Series (Series IV: Earth and Environmental Series), vol. 51 (ed. G. Martinelli and B. Panahi). Dordrecht: Springer https://doi.org/10.1007/1-4020-3204-8_13.

Kendrick, T. (2009). *Identifying and Managing Project Risk: Essential Tools for Failure-Proofing Your Project*, 2nde. New York: AMACOM.

Knabb, R.D., Rhome, J.R., and Brown, D.P. (2005). *Tropical Cyclone Report: Hurricane Katrina: 23–30 August 2005*. National Hurricane Center http://www.nhc.noaa.gov/pdf/TCR-AL122005_Katrina.pdf.

Kobayashi, T., Suzuki, T., and Iwata, K. (2019). *Accident Cleanup Costs Rising to 35–80 Trillion Yen in 40 Years*. Follow up Report of Public Financial Burden of the Fukushima Nuclear Accident. Japan Center for Economic Research.

Lloyd's (2015). *Lloyd's City Risk Index 2015–2025*. Cambridge, UK: Cambridge Centre for Risk Studies at the University of Cambridge.

Manto, C.L. and Lokmer, S. (ed.) (2015). *Planning Resilience for High-Impact Threats to Critical Infrastructure*, 199. Infraguard, Westphalia, Press.

Mark, A., Carpenter, C., Dipak, T., and Harkins, K. (2016). Causes of project failure: Case study of NASA space shuttle challenger. *Journal of Engineering and Economic Development* 3 (2): 23–31.

McCullough, D. (1977). *The Path Between the Seas: The Creation of the Panama Canal, 1870–1914*. New York: Simon & Schuster.

Meredith, J.R., Shafer, S.M., Mantel, S.J., and Sutton, M.M. (2022). *Project Management in Practice*, (7th ed.)e. Hoboken, NJ: John Wiley & Sons.

Munich Reinsurance Company (2021). *Annual Report*. Munich, Germany: Munich Re Group https://www.munichre.com/en.html.

National Academies of Sciences, Engineering, and Medicine (NASEM) (1997). *Protecting the Space Station from Meteoroids and Orbital Debris*. Washington, DC: The National Academies Press https://doi.org/10.17226/5532.

National Aeronautics Space Administration (2021, November 15). *NASA Administrator Statement on Russian ASAT Test*. https://www.nasa.gov/press-release/nasa-administrator-statement-on-russian-asat-test.

National Commission on the BP Deepwater Horizon Oil Spill and Offshore Drilling (2011). *Deep Water: The Gulf Oil Disaster and the Future of Offshore Drilling. Report to the President*. BP Oil Spill Commission Report.

National Research Council (2014). *Lessons Learned from the Fukushima Nuclear Accident for Improving Safety of U.S. Nuclear Plants*. Washington, DC: The National Academies Press https://doi.org/10.17226/18294.

Organization for Economic Cooperation and Development (OECD) (2021). *Development Framing Paper on Climate-resilient Finance and Investment*. Draft for review. 2021 Environment Directorate Environment Policy Committee. Working Party on Climate, Investment and Development.

Project Management Institute (2021). *Project Management Body of Knowledge*. Newtown Square, PA: PMI.

Qazi, A., Daghfous, A., and Khan, M.S. (2021). Impact of risk attitude on risk, opportunity, and performance assessment of construction Projects. *Project Management Journal* 52 (2): 192–209. https://doi.org/10.1177/8756972820985673.

Ramasesh, R.V. and Browning, T.R. (2014). A conceptual framework for tackling knowable unknown unknowns in project management. *Journal of Operations Management* 32 (4): 190–204. https://doi.org/10.1016/j.jom.2014.03.003.

Sanderson, J. (2012). Risk, uncertainty and governance in megaprojects: A critical discussion of alternative explanations. *International Journal of Project Management* 30 (4): 432–443.

The White House (2006). *The Federal Response to Hurricane Katrina: Lessons Learned*. https://tools.niehs.nih.gov/wetp/public/hasl_get_blob.cfm?ID=4628.

Tilman, L. (2012). Risk intelligence: A bedrock of dynamism and lasting value creation. *European Financial Review*.

Tognini, G. (2021). *20 Years and $20 Billion After 9/11, The World Trade Center Is Still A Work in Progress*. Forbes.

U.S. Department of Energy (DOE) (2008). *Root Cause Analysis: Contract and Project Management*. Washington, DC: DOE.

U.S. Department of Energy (DOE) (2020). *Issuance of the Environmental Management Program Management Protocol*. Washington, DC: Office of Environmental Management.

U.S. Department of Transportation (USDOT) (2008). *Innovation Wave: An Update on the Burgeoning Private Sector Role in U.S. Highway and Transit Infrastructure*, Section VI: Managing Risk in PPPs. Federal Highway Administration.

United Nations Development Program (UNDP) (2009). *The Tsunami Legacy: Innovation, Breakthroughs and Change Report*. Coordinated by Kuntoro Mangkusubroto, Chair of the Tsunami Global lessons Learned Steering Committee, April 24.

Vaughan, D. (1990). Autonomy, interdependence and social control: NASA and the Space Shuttle Challenger. *Administrative Science Quarterly* 35 (2): 225–257.

Vogus, T.J. and Sutcliffe, K.M. (2013). Organizational mindfulness and mindful organizing: A reconciliation and path forward. *The Academy of Management Learning and Education* 11 (4): 722–735.

Ward, S. and Chapman, C. (2003). Transforming project risk management into project uncertainty management. *International Journal of Project Management* 21 (2): 97–105.

Weick, K.E. and Roberts, K.H. (1993). Collective mind in organizations: Heedful interrelating on flight decks. *Administrative Science Quarterly* 38 (3): 357–381.

Weick, K.E. and Sutcliffe, K.M. (2015). *Managing the Unexpected*. Hoboken, NJ: John Wiley & Sons.

Williams, T. (2017). The nature of risk in complex projects. *Project Management Journal* 48 (4): 55–66. http://dx.doi.org/10.1177/875697281704800405.

Wilson, M.C., Li, X.-Y., Ma, Y.-J. et al. (2017). A review of the economic, social, and environmental impacts of China's south–north water transfer project: A sustainability perspective. *Sustainability 9*: 1489. https://doi.org/10.3390/su9081489.

Winch, G.M. (2010). *Managing Construction Projects: An Information Processing Approach*, 2nde. Oxford, England: Wiley-Blackwell.

Winch, G.M. and Maytorena-Sanchez, E. (2011). Managing Risk and Uncertainty on Projects: A Cognitive Approach. In: *The Oxford Handbook of Project Management* (ed. P.W.G. Morris, J. Pinto and J. Söderlund), 345–364. Oxford University Press https://doi.org/10.1093/oxfordhb/9780199563142.003.0015.

World Bank (WBG) (2019). *Retention and Expansion of Foreign Direct Investment: Political Risk and Policy Responses*. Washington, DC: World Bank Group https://openknowledge.worldbank.org/handle/10986/33082.

World Bank Group (WBG) (2018). *Global Investment Competitiveness Report 2017/2018 Foreign Investor Perspectives and Policy Implications*. Washington, DC: World Bank Group.

World Economic Forum (WEF) (2022). *Global Risks Report*, 17the. Geneva, Switzerland: WEF.

Yazdani, M., Abdi, M.R., Kumar, N. et al. (2019). Improved decision model for evaluating risks in construction projects. *Journal of Construction Engineering and Management* 145 (5): 4019–4024. http://dx.doi.org/10.1061/(ASCE)CO.1943-7862.0001640.

9

Megaprojects: Troubles and Triumphs

I haven't failed, I have just found 10,000 ways that won't work
— Thomas A. Edison

Introduction

It seems that failure travels faster than light. Failure is announced around the world, while success is rarely shared above a whisper. Why is that? We fear risk taking for fear of failure, but experience teaches us that failure serves as a foundation for success. It may arise from one mammoth failure or a series of failures – such as Edison's 10 000 attempts to create a light bulb, or the Wright brothers attempts to fly a plane before making history. There have been many great inventors who struggled to get their ideas, accepted in the world, people who had started in less-than-ideal situations, or who simply made countless mistakes only to eventually change the world with their new ideas and inventions. As shown throughout our history, failure usually precedes success. In this chapter we explore the world of success and failure in building amazing projects that have never been tried before. We can judge success or failure in many ways depending on our perspectives.

In this chapter we will explore project success and failure by responding to the following questions through a review of the literature, lessons from practice, and examination of the critical factors that enhance project success or ensure project failure:

1) What is the problem the project is trying to solve?
2) Are the solutions to the problem feasible?
3) What causes a project to escalate its commitment?

Global Megaprojects: Lessons, Case Studies, and Expert Advice on International Megaproject Management, First Edition. Virginia A. Greiman.
© 2023 John Wiley & Sons, Inc. Published 2023 by John Wiley & Sons, Inc.

4) How can a project destined to fail be terminated?
5) Why do some projects survive despite a weak economic analysis?
6) What benefits and impact will the project generate?
7) How will we know the project is successful?
8) Is failure always a bad thing?
9) Is success always a good thing?
10) Will the project focus on the most consequential or profitable things first?

What Is Meant by Success? Successful Megaproject Failures

As described by Samset (2013) in his analysis of successful failures and inefficient successes in megaprojects, measuring success in megaprojects is not a simple and straightforward undertaking, because the term "success," used as an indicator, is a highly complex and aggregated measure. Megaproject benefits tend to accrue over a long period of time and success is interpreted differently depending on the nature of their immediate benefits and long-term outcome. Also, different individuals tend to assess the success of the same megaproject differently due to their preferences, values and to the degree they are affected by the project (Samset 2013).

Defining success is complicated. It might seem straightforward to consider projects successful that come in on time and on budget, but what if the finished venture does not solve the problem it was meant to? If a high-speed rail system meets its time and budget commitments, for example, but cannot attract the ridership necessary to alleviate traffic congestion and improve air quality, it is still a failure (Greiman 2015).

On the other hand, comprehensive research has found that projects that are both late and costly can still be considered successful if they deliver what is promised and if associated socioeconomic benefits are accounted for. Usually, however, such benefits are not even identified. For example, Boston's Big Dig – a complex plan that rerouted an interstate and built a bridge, a tunnel, and a greenway – was a source of enormous frustration and controversy because it took many more years and many more billions of dollars than projected to finish. Now that it is done, residents and visitors are enjoying the benefits of dramatically reduced travel time, as well as improved waterfront access, wildlife conservation, new and expanded parks, and increased business development (EDRG 2006; Greiman 2015).

Successful Project Management v. Successful Projects

Cooke-Davies (2002) differentiated between project success and project management success. Essentially, project success is when the project achieves its business objectives, and project management success is when the project is finished to time, cost and quality or performance. Project-management success has traditionally been defined as meeting scope, schedule, and cost compliance requirements. These three factors constitute what is known as the "iron triangle" and are the traditional benchmarks used to evaluate most projects. Serrador and Turner (2015) identified that there is a 60% correlation between project success and project management success. Both of them judged by the perceived satisfaction of the stakeholders, not by making quantitative measures of the various parameters. But it confirms what is well known, not all projects that finish on time, cost, and quality go on to achieve business success, and many projects which finish late with serious cost overruns still achieve financial success. In their development of a new model for measuring the success of megaprojects, Turner and Yan (2018) defined four dimensions of the success of megaprojects:

1) Megaproject management success: the megaproject delivers output with desired functionality and performance, at a time and cost that makes the functionality and performance worthwhile.
   a) Megaproject success level 1a: the project delivers the desired outcome, and that is operated to deliver the desired benefit.
   b) Megaproject success level 1b: if the desired benefit is financial, the project delivers positive net present value, that is, the internal rate of return is higher than the cost of capital.
2) Megaproject success level 2: the project delivers the desired business or public need.

Many megaprojects also produce additional benefits of general public good which are often not measurable financially (Bornstein 2010). There are additional attributes, however, such as socioeconomic improvements, technological innovation, and improved environmental conditions that could and should be part of the equation in determining whether a project is a success (Greiman 2015).

There are different ways to measure these benefits. One method is to do qualitative and quantitative analysis of project and industry data, including stakeholder surveys, screening, and observation. Typically, there is a base-case cost-benefit analysis to which investment alternatives are compared. The analysis addresses these questions: What additional benefits will accrue if this alternative is chosen? And what additional costs will it incur? The objective is to translate the effects of an investment into monetary terms and to account for the fact that benefits play out over a long period while capital costs mostly arise up front (Greiman 2015). The World Bank Group, for example, seeks to link infrastructure-development projects to job creation, environmental improvements, and poverty reduction. Another empirical study has shown that government stability, law, and order, internal conflict, government effectiveness, regulatory quality, quality of bureaucracy, corruption, external conflict, investment profile, politics, religious tensions, and ethnicity are the key policy variables that are directly and strongly correlated with all strategies and measures to reduce poverty (Shahid et al. 2021).

The San Francisco Oakland Bay Bridge was damaged during the 1989 earthquake and reopened in 2013, was $5B over budget and took 10 years longer than originally projected. But the bridge was built to last for 150 years – much longer than the typical 50 years of service – and to withstand earthquakes and seismic activity of the highest magnitude. Both factors will support substantial savings down the line (TRB 2014).

The pressure to incorporate sustainability principles and objectives as shown in Figure 9.1 into policies and activities is growing, particularly in megaproject management. A successful project cannot disregard any of the three triple bottom line (TBL) sustainability pillars (economic, social, and environmental).

Stakeholders representing each of those pillars have to be satisfied to a certain degree in each successful project, even if the way of balancing the three pillars varies depending on project mission and goals.

Megaprojects are recognized as an important part of public life, but they are not always popular. The members of a local community may view the project as a failure based on the media reports which tend to focus on what went wrong rather than the successes of a project. In the project management literature, cost overruns are well reported, but the stories of success are rarely told. However, once the project is completed and serving a necessary public purpose the failures of the project during construction are long forgotten.

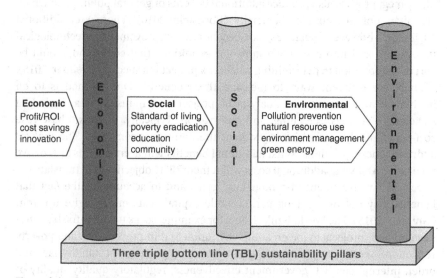

Figure 9.1 Sustainability pillars: the triple bottom line.

Comprehensive Benefits Assessment

In recent years, "comprehensive benefits assessment" has been used to attract financing for big public projects and to build community support. This term means that all benefits, tangible and intangible, are taken into consideration in assessing a project's justification. Intangibles include skill development, alleviation of poverty, knowledge sharing, and institution building. Undertaking a comprehensive benefits assessment is becoming the norm for determining the likely long-term success of projects and influencing decisions about priorities and resource allocation.

All projects should incorporate a comprehensive assessment from the start and develop practices to implement and measure these benefits. For example, the worth of technological improvements can be measured by the increased value of the intellectual-property portfolio. There are different ways to measure these benefits. One method is to do qualitative and quantitative analysis of project and industry data, including stakeholder surveys, screening, and observation.

Typically, there is a base-case cost-benefit analysis to which investment alternatives are compared. The analysis addresses these questions: What additional benefits will accrue if this alternative is chosen? And what additional costs will it incur? The objective is to translate the effects of an investment into monetary terms and to account for the fact that benefits play out over a long period while capital costs mostly arise up front (Greiman 2015).

Success of Project Management but Failure of Project Strategy

In applying success criteria to a megaproject, it is important to distinguish between the megaprojects' tactical and strategic performance as shown in Box 9.1. Tactical success typically means meeting performance targets, such as keeping within budget and schedule requirements. Tactical success is about project management while, strategic performance, includes the broader and longer-term considerations of whether the megaproject would have a sustainable impact and remain relevant and effective over its lifespan. This is essentially a question of getting the business case right or choosing the most viable megaproject concept (Samset 2013). If the wrong strategy is used in project selection no matter how successful the project is in attracting investment or meeting schedule and budget, the project will ultimately fail. An example of this is the Iridium Project.

What Is a Successful Megaproject?

In project evaluation, five analytical criteria are commonly used to provide a comprehensive yet simple picture of the status of a project. They constitute the key analytical elements in the definition of the term "evaluation" originally formulated by the U.S. Agency for International Development and later adopted by the

---

Box 9.1   The Iridium Case Study

In mid-1998, Iridium was one of Wall Street's favorites having more than tripled in stock price in less than a year. The Iridium project was designed to create a worldwide cell connection with the ability to communicate anywhere in the world at any time. Motorola executives regarded it at the time as the eighth wonder of the world. Among the reasons for failure were lack of a business case, technical inefficiency, wrong strategy, wrong technology, and bad timing (Esty 2004).

Poor Strategy
Due to Iridium's elaborate technology, the concept-to-development time was 11 years. During this period, the competition, cellular networks, grew to cover the overwhelming majority of Europe and even migrated to developing countries such as China and Brazil.

In November 1998, 11 years after engineers developed the concept for Iridium, the company launched its service. By April 1999, however, Iridium had only 10 000 customers and its CEO, Edward Staiano, resigned under pressure.

Poor Execution
Moreover, management launched the service before enough phones were available from one of its two main suppliers, Kyocera, which was experiencing software problems at the time. Ironically, this manufacturing bottleneck meant that Iridium could not even get phones to the few subscribers that actually wanted one.

Three forces combined to create Iridium's business failure. First, an "escalating commitment," particularly among Motorola executives who pushed the project forward in spite of known and potentially fatal technology and market problems. Second, for personal and professional reasons Iridium's CEO was unwilling to cut losses and abandon the project. And third, Iridium's board was structured in a way that prevented it from performing its role of corporate governance (Finkelstein and Sanford 2000).

Key Takeaway: If the Business Case fails so does the project!

Sources: Esty (2004) and Finkelstein and Sanford (2000).

United Nations (UN), the Organization for Economic Cooperation and Development (OECD) and the European Commission (EC). They comprise five requirements or success factors that have to be fulfilled, as shown in Box 9.2.

Megaprojects that score highly on all the five success criteria in Box 9.2 are those that perform successfully both tactically and strategically. Project efficiency and effectiveness are resolved at the tactical level through project management,

---

Box 9.2   Megaproject Success Factors

- Efficiency
- Effectiveness
- Relevance
- Impact
- Sustainability

---

while project relevance, benefits and impacts evolve throughout the life cycle and beyond (Miller and Hobbs 2009).

The Organization for Economic Cooperation and Development (OECD) in its study on the governance of large infrastructure projects found the following attributes to be significant for successful outcomes in public investments:

Institutional capacity, public procedures, institutions, and tools – and the development of a coherent and integrated national framework (OECD 2015).

An important study of success in Mega Transport Projects (MTPs) by the Omega Center at University College London, revealed a great deal about the power of context and how this colors judgements about "success." Decision-making for MTPs should transparently include a much wider set of complex considerations than those traditionally acknowledged by formal procedures that are 'Iron Triangle-led' (Dimitriou et al. 2012). (The "iron triangle" means meeting scope, schedule, and cost requirements). Acceptance of MTPs as "open systems" with powerful "agent of change" functions necessitates the need for them to be seen as "organic" phenomena requiring time and space to evolve and adapt in response to changing contextual influences that exert themselves over the (often lengthy) project lifecycle (Dimitriou et al. 2012). Megaprojects hardly ever end with complete success or failure, but somewhere in between. Therefore, the performance of megaprojects should be assessed relative to the promised benefits delivered.

A Holistic View of Megaproject Success

Because it is the whole network that has value, this larger value should be reflected in a broader understanding of cost/benefit analysis, one that is shared with the public just as cost overruns and schedule delays are widely discussed and become a focal point for project success or failure. There needs to be a framework to help governments – and the public – understand the larger benefits of a project and to include the impact of economic and social development. Megaprojects must be understood in the context of a national strategy or the larger ecosystem of which they are a part (Greiman and Sclar 2019).

Until governments and sponsors better articulate the benefits of long-term value creation through network development as part of the megaproject discussion, the focus on cost and schedule as a measurement of success or failure will remain. The risk in this is to badly undervalue and hence forgo the opportunities that these projects are capable of creating. Governments can win and maintain public support for the benefit of future generations if they are able to develop a broader understanding of what these projects can mean and integrate stakeholders into active participation in their ongoing creation. Projects like the Big Dig demonstrates that this is socially and politically possible (Greiman and Sclar 2019).

The subject of cost overruns and cost escalation is a topic extensively covered in the project management literature with various theories as to why megaprojects are consistently over budget (Merrow 2011; Morris 2013; Winch 2013; van Marrewijk and Smits 2016). Comparative studies on megaprojects tend to focus on quantitative analysis and the percentage that projects exceed their original budget (Flyvbjerg 2014; Merrow 2011). Furthermore, the ability to determine appropriate and accurate cost estimates is not simply a failure of technical expertise, but also attributable to fundamental psychological biases and political motives of project promoters (Flyvbjerg et al. 2002). Flyvbjerg additionally argues that planners and promoters deliberately underestimate cost to increase the likelihood that their projects, and not those of competitors, are chosen (Flyvbjerg 2014). Some scholars have argued that outcomes are less a product of deliberate deception than a result of ". . . professionals and civil servants who, while managing to the best of their abilities, are faced with complexities, uncertainties, paradoxes, and ambiguities" (van Marrewijk et al. 2008, p. 597).

Whether deception or complexity is the cause, no one disputes the fact that early cost estimates on megaprojects are often inaccurate, and the reasons behind these inaccuracies can be explained best through a more holistic view of project shaping that comes from actual 'lived experience' and investigation (Cicmil et al. 2006; Smith and Winter 2010). Case studies, provide a deeper, richer explanation for cost escalation that identifies emerging organizational and environmental factors that not only are a cause of cost escalation, but can contribute substantially to delays over the project life (Pinto and Winch 2016; van Marrewijk et al. 2008). Megaprojects are not only moving targets in their own right for such reasons but, more importantly, they go forward in institutional environments that are themselves dynamic (Greiman and Sclar 2019).

Characteristics of Failed Projects

Merrow assessed the success or failure of 318 industrial megaprojects (projects that make a product for sale with a total capital cost of more than US$1B), based on a detailed analysis of their performance. In his assessment of these megaprojects, Merrow used measurable criteria to assess project success or failure. He characterizes an industrial project as a failure if it exceeds one or more of the following cost, time, and quality thresholds:

- Actual final cost exceeds the cost estimate made at the project sanction stage by more than 25%.
- Project spends more than 25% greater than other projects with similar scopes.
- The execution schedule slips more than 25% from the schedule forecast at project approval.
- The project execution schedule is greater than 50% more than the length of execution relative to similar projects; or
- There is significantly reduced production relative to that planned into the second year after mechanical completion.

The reported outcome of this assessment is quite surprising. Only 35% of the megaprojects were successful, and these tended to be genuinely excellent projects on the time, cost, and quality criteria. However, the 65% of megaprojects that were considered to be failures:

- Averaged 40% cost overrun whilst being very expensive in absolute terms.
- Slipped their execution schedule by an average of 28% whilst being 15% slower than a competitive schedule; and
- Averaged only 65% of planned production in the first year.

Table 9.1 provides a sampling of critical factors in megaproject success or failure based on a review of several case studies by scholars and professionals in the field of project management. It highlights that no one criterion is suitable for measuring the success of every project, and that the most important criteria are those selected by the project owner and sponsors based on a review of the best parameters for measuring success in the context of the project's goal, mission and objectives.

Strategies for Learning from Failures

Megaprojects provide great case studies for understanding the reasons for project success and failure but analysis of the reasons for failure or success is often incomplete and, in some cases, never shared with projects that could benefit in the future.

The wisdom of learning from failure is incontrovertible. Yet, organizations that do it well are extraordinarily rare. High-reliability-organizations (HROs) are an example of organizations that have developed practices to help prevent catastrophic failures in complex systems like nuclear power plants through early detection. Electricité de France, which operates 58 nuclear power plants, has been an exemplar in this area: It goes beyond regulatory requirements and religiously tracks each plant for anything even slightly out of the ordinary, immediately investigates whatever turns up, and informs all its other plants of any anomalies (Edmondson 2011).

Table 9.1 Critical factors in megaproject success or failure.

Success factors	Literature review
Clear Strategic Vision, Total Alignment, and Adapting to Complexity	Shenhar, A. and Holzmann, V. (2017) The three secrets of megaproject success. *Project Management Journal, 48* (6) 29–46.
Success factors for public construction projects in India: Awareness of and compliance with rules and regulations; pre-project planning and clarity in scope; effective partnering among project participants; and external monitoring and control.	Tabish, S. and Jha, S.K. (2011) Identification and evaluation of success factors for public construction projects. *Construction Management and Economics, 29* (8) 809–823.
An exploratory factor analysis for World Bank Projects shows that five factors (monitoring, coordination, design, training, and institutional environment) are correlated to project success.	Ika, L.A., Diallo, A., and Thuillier, A.D. (2012) Critical success factors for World Bank projects: An empirical investigation. *International Journal of Project Management, 30* (1) 105–116.
A literature review of data obtained through a comprehensive search across research center databases, mainly in Europe determined that competency of project managers was a critical factor in project success.	Mišić, S. and Radujković, M. (2015) *Critical drivers of megaprojects success and failure.* Operational Research in Sustainable Development and Civil Engineering. EURO working group and 15th German Lithuanian-Polish colloquium (ORSDCE).
A study of 15 megaprojects in Europe (The NETLIPSE research) found that common success factors included: a clear vision and a strong political will; an independent and stable project delivery organization; a realistic business case and systematic stakeholder management. A 2016 update of this study reflected the following challenges for the future: Network focus and asset management; sustainable development; value driven projects; adaptive management; integrated approach; T-shaped professionals; structured learning.	Hertogh, M., Baker, S., Staal-Ong, P.L., and Westerveld, E. (2008) *Managing Large Infrastructure projects.* AT Osborne BV, The Netherlands. Staal-Ong, P.L., Kremers, T., Karlsson, P.-O., and Baker, S. (2016) *10 Years of Managing Large Infrastructure Projects in Europe: Lessons Learnt and Challenges Ahead.* NETLIPSE, The Netherlands.
Two factors play a critical role in determining megaproject success or failure: modularity in design and speed in iteration.	Flyvbjerg, B. (2021) Make megaprojects more modular. *Harvard Business Review,* Nov–Dec, 58–63.

Table 9.1 (Continued)

Success factors	Literature review
A study of experiences from 50 Norwegian Megaprojects to improve major public investment projects concludes that up-front investments in quality improvements tend to pay back to an exceptional degree in terms of cost reduction, increased utility, and economic benefits.	Magnussen, O.M., and Samset, K. (2005) *Successful Megaprojects: Ensuring Quality at Entry,* EURAM, 1–12. Responsible Management in an Uncertain World, May 4–7th 2005, TUM Business School, Munich, Germany.
Research on megaprojects in rural Kenya reflects that the analysis of megaprojects should consider not only material outcomes but also their symbolic dimension for desirable futures pointing to aspiration as a driver of investment and change.	Müller-Mahn, D., Mkutu, K., and Kioko, E. (2021) Megaprojects—Mega failures? The *European Journal of Development Research,* 33, 1069–1090.

By contrast the National Aeronautics Space Administration has been responsible for some of the most horrific disasters in U.S. history. These include the 1984 explosion of the Challenger space shuttle 73 seconds after takeoff which killed seven astronauts, the 2003 explosion of the Columbia Space Shuttle which also killed seven astronauts and the Apollo 1 27 January 1967, disaster when the crew of three were killed in a fire in the Apollo Command Module during a preflight test at Cape Canaveral. Apollo 13's aborted mission to the moon in 1970 is referred to as a "successful failure" by NASA because the crew members survived a catastrophic accident and experience was gained in rescuing the crew (NASA 2009). In each of the four NASA instances the accident was preventable, but post event analysis established that the NASA culture, and failure to follow systems and procedures were the primary cause. Despite NASA's many failures the International Space Station as described in previous chapters is considered one of the most successful projects in history for its contributions to mankind.

Research on megaprojects in developed countries tend to focus on their failures, problems, complexity, cost overruns, delays, stakeholder conflicts, alternatives, ambiguities, schedules and governance. However, there are great benefits that are rarely discussed. Many areas of megaprojects management in these countries remain largely uncharted. These include such challenges and solutions as engineering, human development, managerial, political, and sustainability. Studies have shown that projects in developing countries tend to be poorly justified in feasibility studies and under-resourced in operations, making them particularly prone to high failure rates (Rose et al. 2017).

Despite the best practices all around the globe, numerous challenges can move a megaproject from a programmed success to a failure scenario. A successful megaproject can spur economic growth in Lesser Developed Countries (LDCs), while a failure can set development back for years. The media's current interest in the issue of cost overrun is associated with the cost-efficiency issue: cost overruns imply wastage of public resources that could have been otherwise used for productive purposes elsewhere. Another issue is of course that if the initial cost estimation of major projects with large cost overrun had been accurate and realistic, they might not have been implemented and more viable projects could have been given priority (Magnussen and Samset 2005).

Reasons for Megaproject Failure

Though the media projects a large number of project failures, rarely does the public have the opportunity to understand why these projects failed. Figure 9.2 highlights some of the major failures of global megaprojects and the reasons for these failures. Once the reasons are disclosed, it follows that corrective actions should not only be implemented but shared with the general public so that the projects of the future can use this information to prevent a similar reoccurrence.

PROJECT: Gautrain High Speed Railway Service

Sponsor: Johannesburg, Gauteng Dept. of Finance and Economic Affairs, PPP Unit, South Africa

Description: The $30B Rand project is Africa's first rapid rail network, connecting Pretoria, Johannesburg, and the OR Tambo International airport

Reason for Failure: The project is criticized for consuming an unprecedented amount of public funding, while the train is aimed primarily at an elite class of people within the region (Thomas 2013). The environmental benefits of the project are also disputed, and the environmental impact assessment (EIA) revealed that Gautrain would at best be environmentally neutral. Despite qualifying as a failed transport megaproject (the project is complete, and it works but it does not deliver expected benefits). The Gautrain project is considered a success by the elite within the country and plans are currently underway to expand the system. Given that the capacity of the system is currently constrained, the project experienced major time and budget overruns, ridership numbers were not met, and the expected economic benefits were not realized, expansion of the current system should be questioned (Rose et al. 2017).

Figure 9.2 Case Studies in Megaproject Failures.

PROJECT: Gyeongbu Expressway

Sponsor: Seoul, South Korea

Description: The Gyeongbu Expressway is the second oldest and most heavily travelled expressway in South Korea, connecting Seoul to Suwon, Daejeon, Gumi, Daegu, and Busan.

Reason for Failure: Though the expressway has provided a physical infrastructure through which the capital's circulation could be reduced, it has also led to serious problems such as uneven regional development, the expansion of non-place or alienated place, ecological destruction, and pollution (Choi 2012).

PROJECT: Coega Industrial Development Zone

Sponsor: Eastern Cape Province, South Africa

Description: The project is a multibillion-rand initiative, comprising an Industrial Development Zone (IDZ) and a deep-water port at Port Elizabeth. The initiative aims to position South Africa as a platform for global manufacturing and export through foreign and local investments

Reason for Failure: Popularly presented as being indispensable in the context of South Africa and the Eastern Cape, the project has nonetheless evoked conflict and controversy from the start. This casts serious doubt on the public participation processes conducted on behalf of the project. Evidence from analyses of the processes and outcomes of the public participation exercise illustrates these have not met the requirements of best practice (Bradshaw and Burger 2005). Despite the investments into the region, the estimated 8% growth from the Coega IDZ development however has not materialized, with a mere 4.8% reported during the 2015 Eastern Cape economy. The economy largely dependent on the automotive and agricultural processing industries have failed to assist with widespread economic reform, that has been listed as a common issue with IDZ's. The Coega IDZ, at most designed to attract FDI, has struggled to live up to its potential and has fallen victim to the overestimation of its success (Rose et al. 2017). Similarly, the Coega IDZ, which should have achieved radical economic transformation in the province, has also not achieved its aims, primarily due to overestimation of the benefits in the planning phase and a failure to operate the project at appropriate levels, suggesting inappropriate skills specifications and perhaps a lack of political will (Rose et al. 2017).

PROJECT: King Shaka International Airport (KSIA)

Sponsor: Surban, South Africa

Description: King Shaka International Airport was developed as part of the Dube Trade Port in order to facilitate economic growth.

Figure 9.2 (Continued)

Reason for Failure: Since the opening of KSIA, it has become evident that the build was not warranted, as the previous airport had been upgraded at a cost of R158 million in 2005 and international passenger numbers have not materialized. The airport was also massively over budget and built without a feasibility study. On the positive side, the research finds that the King Shaka International Airport has generated direct employment for the local labor force, that passenger and flight numbers have increased, and that local and national (large) companies directly benefit from increased business. On the other hand, the social-economic costs associated with the project have negatively affected society, not only through the relocation of business and jobs, but also through land expropriation, noise pollution, and the subtraction of public resources from social needs programs.

PROJECT: Hyderabad Metro Rail

Sponsor: India

Description: This was the longest rapid transit metro line opened in one go in India. It is estimated to cost ₹18 800 crore (US$2.5B)

Reason for Failure: The metro rail operator has had to ask the state government to extend financial help and also arrange soft loans for them as they incurred huge losses due to the pandemic and low ridership especially on JBS and MGBS corridor.

PROJECT: Airbus 380

Sponsor: SOGEPA (France): 11.0% GZBV (Germany): 10.9% Capital Group Companies (10.1%) SEPI (Spain): 4.1% The Vanguard Group (2.19%) PRIMECAP Management Company (2.1%) Invesco (1.1%) Others: 74.0%

Description: The development of the A380 "superjumbo," aircraft resulted in a significant cost overrun, schedule delays, and a major firm crisis.

Reason for Failure: Severe problems to successfully integrate the aircraft system, i.e., to ensure that its different parts function together well and thus yield high overall performance. The root causes of these problems were two basic organizational flaws. One was that even though several national aerospace firms were merged into an integrated Airbus company in 2000, the former independent national firms essentially still prevailed, each sticking to various idiosyncratic structures, cultures, and processes. Nonetheless, prior (yet smaller) aircraft families (e.g., A320 or the A330) had been developed successfully despite this particular, decentralized organizational set-up (Dörfler and Baumann 2014). Hence, Airbus was confident that the A380 would extend its list of successful projects. In consequence, the firm underestimated the high risks that the A380 program entailed owing to its much higher degree of complexity.

Figure 9.2 (Continued)

PROJECT: UK Ministry of Defense

Sponsor: Ministry of Defence

Description: £200 million Joint Personnel Administration (JPA) IT platform for paying the armed forces.

Reason for Failure: Failure resulted in major part due to different rules in the Army, Navy and Air Force. Only about 70% of pay and allowances were harmonized (Kelman 2009).

PROJECT: British Petroleum Deepwater Horizon

Sponsor: U.S. Department of Energy

Description: An ultra-deep water, dynamically positioned, semi-submersible off-shore drilling rig owned by Transocean. On 20 April 2010, while drilling at the Macondo Prospect, a blowout caused an explosion on the rig that killed 11 crewmen and ignited a fireball visible from 40 miles (64 km) away.

Reason for Failure: Failure resulted from three main reasons: (i) the decision to use a well design with few barriers to gas flow; (ii) the failure to use a sufficient number of "centralizers" to prevent channeling during the cement process; (iii) the failure to run a cement bond log to evaluate the effectiveness of the cement job; (iv) the failure to circulate potentially gas-bearing drilling muds out of the well; and (v) the failure to secure the wellhead with a lockdown sleeve before allowing pressure on the seal from below. The common feature of these five decisions is that they posed a trade-off between cost and well safety (U.S. Congress 2011). The Findings of the Presidential Commission are: (i) The explosive loss of the Macondo well could have been prevented. (ii) The immediate causes of the Macondo well blowout can be traced to a series of identifiable mistakes made by BP, Halliburton, and Transocean that reveal such systematic failures in risk management that they place in doubt the safety culture of the entire industry, (iii) Deepwater energy exploration and production, particularly at the frontiers of experience, involve risks for which neither industry nor government has been adequately prepared, but for which they can and must be prepared in the future (U.S. Congress 2011).

PROJECT: Dabhol Power Plant

Sponsor: Enron, Ratnagiri Gas and Power Private Limited Maharashtra, India

Description: This plant was built through the combined effort of Enron as the majority shareholder, and GE, and Bechtel as minority shareholders. From 1992 to 2001, the construction and operation of the plant was mired in controversies related to corruption in Enron and at the highest political levels in India and the United States

Figure 9.2 (Continued)

Reason for Failure: Fast track entry into India without competitive bidding, lack of ethical corporate decision making and good stakeholder management practices (Rangan et al. 2004).

PROJECT: Hong Kong International Airport (Chek Lap Kok Airport)

Sponsors: Key stakeholders in the project were the Hong Kong Government, the Peoples Republic of China Government, PAA, AA, and Mass Transit Railway Corporation (MTRC)

Description: The airport project, one of the biggest public works undertaken in the world, completed and commenced operation on 2 July 1998. The project was extremely complex by some technical or logistic standards in different aspects including land reclamation, dual-runways and a huge passenger terminal building.

Reason for Failure: Reasons for project failure included failure to consider external environment (inflation), lack of a contingency plan, insufficient test runs, so many variations to the contract resulted in substantial claims from the contractor totaling $89 million, compared to the original project cost of $230 million (Hong Kong 1998).

PROJECT: Berlin Brandenburg Airport

Sponsor(s): The German states of Berlin and Brandenburg (37% each), with the remaining shares being held by the Federal Republic of Germany (represented by the ministries of Transport and Finance).

Description: Around 30 years after its initial conception, the $7.3B Euro Berlin Brandenburg International Airport finally began operations in October 2020 after a troubled project marred by delays and scandals.

Reason for Failure: None of the members of the political leadership had any actual experience with such large construction projects. Supervision ended up almost exclusively in the hands of politicians, civil servants, and union officials. The lack of expertise and technical knowledge of the supervisory board prevented proper and rigorous oversight of the planning and progress, resulting in delays, confusion, and major cost overruns (OECD 2016). Germany is a prosperous nation with a total GDP (2015) of US$3.363 trillion. Set against the cost of US$7.3B), while this is not an insignificant amount, it only represents 0.002% of the country's GDP, if the project had been located in a developing country, such cost over-runs would potentially be crippling for the country. In February 2022, Germany secured European Commission approval to provide Berlin Brandenburg airport with a €1.7B ($1.9B) bailout as it was struggling to survive after opening midway through the pandemic.

Figure 9.2 (Continued)

PROJECT: Shoreham Nuclear Power Plant

Sponsor: Long Island Lighting Company

Description: The idea was developed in 1966 by the Long Island Lighting Company (LILCo) and the U.S. Atomic Energy Commission authorized construction in 1970. In 1966 the original cost was estimated to be about $65 million with the final estimate at $5.5B. A negotiated agreement with New York State finally resulted in Long Island Lighting's abandoning Shoreham without its ever-having begun operation. Decommissioning of the plant cost around $182 million

Reason for Failure: In 1983 state officials determined that the plant could not be safely evacuated in the event of a nuclear crisis. The reasons for failure included failure of project selection, lack of governance, organizational escalation, and misrepresentation of the reality of the feasibility of the project (Ross and Staw 1993).

Figure 9.2 (Continued)

Reasons for Megaproject Failures:

1) Over Escalation of Commitment (EOC)
2) Optimism Bias v. Strategic Misrepresentation
3) The Impact of Corruption, and
4) The Failure to Assess Quality at Entry

Sometimes projects fail for reasons very different than cost overruns, schedule delays, technical problems, or insufficient project finance. These reasons are more common than one might think and include irrational behavior and misrepresentation. In this next section, we review a few of these behaviors and learn how one can prevent projects from taking the wrong path and if the wrong path is taken how to end the project before devastating results can occur.

During the investigation of the NASA Space Shuttle Challenger disaster where seven astronauts lost their lives, the *Presidential Commission* concluded that economic and political pressures "caused rational men to make irrational decisions." Economic pressure, "is the choice of efforts to improve project performance based on economic, and, in particular, cost/benefit considerations" (Andersen and Vaagaasar 2009). The project's management team spent more time identifying cost saving procedures than risk avoidance. The economic pressure of the late 1970s pressured NASA into streamlining the shuttle process. The US Naval Quality Assurance Program, which arose around critical shipboard systems due to the Challenger disaster, emphasizes the importance of transparency and responsibility of the entire chain of command to address all potential risks whether they are real or perceived. When performing maintenance on a critical system, any person involved in the process, from the technician performing the work to the executive in charge of the overall program, must stop and give appropriate attention to any potential critical risk, no matter how seemingly insignificant. It is proper for the workers to feel empowered to raise any questions they might have about risk.

1) Over Escalation of Commitment (EOC)

One of the most frustrating decisions that investors must make is whether to continue projects in the face of uncertain prospects (Keil et al. 2000). Sometimes, to maintain their own reputation and retrieve losses, investors choose to continue by escalating commitment in response to sunk costs (including considerable past efforts and major existing investments) in social-economic activities (Hafenbradl and Woike 2018; Montealegre and Keil 2000).

> *Such irrational behavior is a general phenomenon known as escalation of commitment* (EOC) (Staw 1981).
> Simply stated, escalation of commitment causes people to continue a failing course of action.

EOC by private investors is also a common behavioral trap in PPP projects. For example, in the Taraso Fukuoka project, estimates of the number of project users were overoptimistic, and when faced with the low actual income, investors revised the project's facilities, operating methods, and content, and also replaced the external operating companies to reverse the situation, which despite their efforts ended in bankruptcy (Song et al. 2018).

Similarly, when the Channel Fixed Link and Quanzhou Citong Bridge encountered operational difficulties, investors extended the schedule and budget, only to find themselves trapped, leading to an enormous waste of resources and substantial losses for project stakeholders and even the public (Song et al. 2018; Winch 2013). EOC by investors not only results in an increase in project risks and a decline in performance but also leads to project failure in severe cases (Feldman and Wong 2018).

Escalation of commitment (EOC) is a common behavior among investors who receive negative feedback (NF) in public-private partnership (PPP) projects, and this behavior typically leads to sizable losses (Gao and Liu 2020). However, Escalation of Commitment is not a robust phenomenon and typically occurs only in the early stages of PPP projects when investors are attempting to interpret negative feedback (McCain 1986). With the increased negativity, projects are likely to fail more obviously, and the losses borne by investors also increase. Once losses accumulate to a certain extent (the so-called threshold), investors will terminate their EOC, thereby abandoning projects and transferring them to governments to receive some compensation or residual asset value. Investors usually have the right, but no obligation, to require governments to take over projects in advance as initially agreed, which can be defined as an abandonment option and also a put option (Xiong et al. 2015).

Trope and Liberman (2010) in their studies of escalation of commitment found that psychological factors are the strongest driver of escalation of commitment behavior. A potentially important psychological factor that has remained unexplored thus far, is construal level. Construal level involves the degree of psychological distance that people experience at the time of deciding. Psychological distance increases as the object that is being construed mentally, is further away (e.g., temporally, geographically, socially, or in hypotheticality).

Escalation of Commitment raises the important question of how do we get organizations moving, and how do we get them stopped once they are moving in a particular direction? Straw and Ross (1987) in their early research proposed that organizational escalations may involve the interplay of four sets of forces over time (Table 9.2):

Table 9.2 Determinants of escalation.

Determinants of escalation	Description
Project	Under this rubric are objective aspects of a project, such as its closing costs, its salvage value, the causes of setbacks to its completion, and the economic merits of pursuing or dropping it (Staw and Fox 1977);
Psychological	This category includes "reinforcement traps" (Platt 1973), such as difficulties in withdrawing from a previously rewarded activity; individual motivations, such as the need for self-justification; decision-making errors, such as trying to recoup "sunk costs" (resources already invested in a project); and biases in information processing, such as tendencies to slant data in the direction of preexisting beliefs (Conlon and Parks 1987).
Social	Included here are interpersonal processes that may lead to excess commitment, such as desires to justify losing projects to potentially hostile audiences (Fox and Staw 1979), modeling of others' behavior in similar circumstances (Brockner 1992), and cultural norms favoring consistent, or strong, leadership (Staw and Ross 1980).
Organizational	Under this category come such variables as the level of political support for a project within an organization (Pfeffer 1981), the level of economic and technical "side-bets" incurred by the organization with respect to the project (hiring of staff, development of expertise) and the extent of the project's institutionalization within the organization-how tied it is to the firm's values and objectives (Goodman et al. 1980).

Source: Adapted from Staw and Ross (1987).

Helga Drummond (2017) in her research on escalation proposed four questions:

1) How does escalation start?
2) What drives decision makers to persist with economically poor megaprojects?
3) What can decision makers do to curb project runaways?
4) What are the most promising directions for research?

She concludes that most troubled megaprojects get finished eventually, though they usually end up costing a lot more than expected, and the promised benefits may be slow to materialize. Importantly, she urges organizations to be careful what they get into. "Regardless of what drives escalation, the simple truth is that once megaprojects gain traction, they are almost unstoppable" (p. 210). "The real cost of any project is what we could have had instead" (p. 210).

1) **What does it take to kill a megaproject?**

The reverse of escalation of commitment is the desire to end a project before it is completed. Cornelio et al. (2021) explored the question of how political polarization fosters escalation of commitment leading to unfinished infrastructure megaprojects? Despite their relevance, unfinished megaprojects are scarcely discussed in project studies as if they do not exist. Through a longitudinal case study of the $13B Mexico City New International Airport, terminated during the construction phase they show how political narratives can portray a misleading escalation of commitment to justify project termination, even when it is not rational. The research describes how political polarization can lead to unfinished megaprojects.

Project termination has been described as natural and unnatural. (i) Natural termination is when (the project completes as it achieved the original objectives), and (ii) unnatural (the project is stopped for a problem and does not achieve the original objectives). Unnatural termination is mostly studied for IT projects (Keil et al. 2000), and it is underexplored in construction megaprojects.

As examples, the Shoreham Nuclear Power Plant (was stopped for an inability to provide evacuation plans); Iridium (was terminated due to technical difficulties and the insolvency of the sponsor); and the Dabhol Power Project (was terminated due to lack of transparency, environmental exposure, and alleged fraud). However, despite serious financial losses, the Channel Tunnel between London and Paris survived because the Banks had loaned so much money rather than lose it all they would take the risk the debt would be repaid over time.

2) **Optimism Bias v. Strategic Misrepresentation in Cost Estimation**

Flyvbjerg (2011) analyses deeply the roots of failure causes, identifying two possible root causes: Optimism bias and strategic misrepresentation. Optimism Bias refers to biased estimations that managers form based on delusional optimism instead of on a rational cost-benefits analysis. In this situation, the decision-making process is biased because of the underestimation of costs and the overestimation of benefits. Strategic misrepresentation is a behavioral bias that consists in the tendency to deliberately and systematically distort or misstate information for strategic purposes. It is sometimes also called political bias, strategic bias, power bias, or the Machiavelli factor. In his extensive studies of megaproject costs, Flyvbjerg has found that strategic misrepresentation has proved especially important in explaining outcomes in large-scale policies and projects (Flyvbjerg 2011, 2021). Importantly, this research shows that projects fail not just because of poor estimation of costs or unforeseen events or project complexity, but because they strategically misrepresented the cost of a project from its inception.

3) The Impact of Corruption

The European Union megaproject cost research concluded that these are the characteristics of megaprojects: colossal, captivating, costly, controversial, and complex (Brooks 2015). There is, however, one word beginning with the letter "C" they failed to include. "Corruption" in construction projects and megaprojects is prevalent in every country but particularly developing countries, such as South Africa, Nigeria, Pakistan, and India (Transparency International 2021). Corruption can occur at any time in the life cycle of a project and is prevalent during project initiation, procurement, implementation, and even after transfer to operation. Developing countries face greater challenges in preventing corruption due to the lack of rule of law, court systems and enforcement agencies. Corruption can lead to project delays, cost overruns and sometimes termination of the project. It also can cause the building of substandard projects and impact the delivery of promised benefits. A lack of a strong legal system will lead to an increase in corruption impacting financing from the multilaterals and the private sector.

> Transparency International's (TIs) 2021 Corruption Perceptions Index (CPI) paints a grim picture of the state of corruption worldwide. While most countries have made little to no progress in tackling corruption in nearly a decade, more than two-thirds of countries score below 50. This analysis shows corruption not only undermines the global health response to COVID-19 but contributes to a continuing crisis of democracy. Countries that perform well on the index invest more in health care, are better able to provide universal health coverage and are less likely to violate democratic norms and institutions or the rule of law.

4) The Failure to Assess Quality at Entry

One huge paradox in the front-end management of megaprojects is that many of these, even the largest public investment projects, originate as a single idea without systematic scrutiny or consultation. In addition, in too many cases, the initial idea will remain largely unchallenged and therefore survive and end up as the preferred concept – even in cases where it subsequently proved to be a strategic underperformer or failure. Improved front- end management is therefore likely to pay off if seen in a wider life cycle perspective (Miller and Lessard 2001). There is much to be gained from improving quality-at-entry at the earliest stage of the process. This can be achieved by challenging the initial ideas and applying simple analyses, extracting, and making use of previous experience from similar undertakings, and consulting and involving stakeholders.

Whenever demand analyses, goal setting, and impact assessment at the strategic level are skipped, and instead the project level is initiated within the framework of a given solution, the initiators' needs can easily be confused with those of the society (Samset 2013). Hence, the wishes of special interests for financial gains, prestige, or ideologically preferable solutions may take precedence over top-down political goals and the needs of broader social groups. Such constraint of planning at a premature stage is a commonplace weakness in the planning of large, public investment projects (Magnusen and Samset 2005; Olsson and Samset 2006; Samset 2013).

Characteristics of the Most Successful Projects

Organizational Resilience and Collective Mindfulness

Organizational resilience is critical to megaprojects due to their unprecedented risks and uncertainties. These risks and uncertainties include internal dynamics such as incomplete contracts, varied governance structures, long-term contract periods with multiple stakeholders (Wang and Pitsis 2019); as well as external uncertainties, such as political, economic, legal, social, natural, and project-specific factors. The extant literature in business and organizational behavior has explored the relevance of collective mindfulness for organizational resilience (Oeij et al. 2018)

Defined as the "capacity of a social system to proactively adapt to and recover from disturbances that are perceived within the system to fall outside the range of normal and expected disturbances" (Comfort et al. 2010, p. 9), organizational resilience deals with adversities and crises (Linnenluecke 2015). Constant mindfulness is required to successfully deal with adversities and crises by detecting risks that threaten megaprojects, deciding quickly on an appropriate response, and following through until the threat is mitigated (Ogliastri and Zúñiga 2016).

Weick et al. (1999) first introduced the term mindfulness in the literature on organizations and crises management. Collective mindfulness refers to organizational processes or practices that help organizations detect, categorize, and respond to unexpected events and errors. Collective mindfulness is about both the quality of attention and conserving scarce attention (Weick and Sutcliffe 2006); it suppresses organizational tendencies toward inertia, maintains an organization's commitment to resilience, and improves organizational effectiveness when disruptions occur (Weick and Sutcliffe 2006). The extant literature in business and organizational behavior has explored the relevance of collective mindfulness for organizational resilience (Oeij et al. 2018; Ogliastri and Zúñiga 2016).

A crisis is defined as a process of weakening or degeneration that can culminate in a disruption event to the actor's normal functioning (Williams et al. 2017, p. 739). It is "a low probability, high impact event that threatens the viability of the organization and is characterized by ambiguity of cause, effect, and means of resolution, as well as by a belief that decisions must be made swiftly" (Pearson and Clair 1998, p. 60). Crises differ in both risks (i.e., identifiable events with negative consequences) and disasters (i.e., events that pose serious threats to society, including human loss and material damage that potentially led to the collapse of social structures and/or functions) (Iftikhar and Müller 2019).

Crises in megaprojects manifest themselves as an accumulation of defects and weaknesses perceived to threaten the objectives of both the megaproject and its stakeholders (Wang and Pitsis 2019). Crises can be internal or external and social or technical/economic (Mitroff et al. 1987) and may be differently perceived, understood, and responded to by different stakeholders (Cuppen et al. 2016). Embedded in political systems, government, and private business sectors megaprojects involve multiple stakeholders and complex communication channels in a time of crisis. Stakeholders can include any group or individual who can affect or is affected by the achievement of the megaprojects objectives and impacted by its outcomes. Adequate stakeholder engagement is critical to megaprojects (Loosemore 1998) when a crisis occurs due to the importance of the unified efforts of multiple stakeholders. However, stakeholders' distributed attention structures and divergent institutional demands (Matinheikki et al. 2019) have not been fully considered in earlier studies of megaprojects (Wang et al. 2021).

Three mechanisms through which collective mindfulness impacts organizational resilience have been identified: (i) awareness allocation, (ii) emotional detachment, and (iii) attention alignment (Wang et al. 2021). The awareness allocation mechanism of collective mindfulness is enacted during the pre-crisis period, where readiness and preparedness for organizational resilience are applied. The emotional detachment mechanism of collective mindfulness is enacted during the in-crisis period, when response and adaption of organizational resilience occur. The attention alignment mechanism appears during the post-crisis period, when recovery and adjustment of organizational resilience are performed.

Megaprojects present global challenges that include geopolitical, macroeconomic, cultural, and interorganizational elements that create enormous problems for governments, policy makers, and project managers. The complexity, internal and external risks, and the significant social impact of megaprojects make organizational resilience particularly important. To survive potential adversities, megaproject organizational resilience depends on collective mindfulness. Understanding the mechanisms of collective mindfulness for megaproject organizational resilience is a process that functions prior to, during, and after recovery from crises.

Summary

The focus in this chapter is on how project planners should conceptualize projects and their delivery in order to achieve success. The ever-present political dynamics that arise between competing interests and explaining why such a rational approach to project management is not always followed is essential to achieving project success. A broad conclusion in this chapter is that starting with a well formulated strategy may be an advantage but is no guarantee of the best choice when the final decision is made. Project success must be ensured through a long-term strategy, rather than a focus on short-term achievements. To determine a project's success the whole value of the project must be considered including social, economic, and institutional returns, and not just the short-term benefits from cost and schedule adherence. The characteristics of sustainability, resilience, and mindfulness must be ever present throughout the long life of a megaproject in order to ensure successful outcomes.

Lessons and Best Practices

The most important lessons learned in this chapter are the responses to the questions introduced in the introduction to the chapter:

1) Identify the problem the project is trying to solve and then make sure you have assessed the need for the project prior to execution of the project.
2) Make sure the solutions are technically, economically, and strategically feasible.
3) Beware of escalating commitments and understand how a project can be terminated before it is too late.
4) Focus on the project's economic analysis and business case and not just the project's budget and schedule.
5) Understand the benefits and impact the project will generate and then evaluate both throughout the project's life cycle.
6) Identify the project's success factors and then evaluate those factors to ensure they have been achieved.

> 7) Recognize that sometimes failure is not a bad thing; and that success is not always a good thing.
> 8) Will the project focus on the most consequential or profitable things first?
> 9) The characteristics of sustainability, resilience and mindfulness must be ever present throughout the long life of a megaproject.

Discussion Questions

1 How can you find success in a project that has failed dramatically to meet its budget and schedule?

2 How can escalation of commitment be controlled before it is too late? Provide an example of an actual case study.

3 Define the meaning of resilience and mindfulness in achieving project success.

4 Give an example of a project that was difficult to kill and the reasons why?

5 Assuming you are the Project Director of a large megaproject, what strategy would you use to make sure your project succeeds in delivering the expected benefits.

6 How can you mitigate the likelihood of corruption, criminal activity and fraud?

7 What are the best strategies to use in terminating a project that has no identifiable benefits to continue?

8 What are the critical drivers of megaproject success and failure?

9 What does the research teach us about the causes of success and failure?

10 When might success be a bad thing and failure be a good thing?

References

Andersen, E.S. and Vaagaasar, A.L. (2009). Project management improvement efforts-creating project management value by uniqueness or mainstream thinking? *Project Management Journal* 40 (1): 1–27. https://doi.org/10.1002/pmj.20096.

Bornstein, L. (2010). Mega-projects, city-building, and community benefits. *City, Culture and Society* 1 (4): 199–206. https://doi.org/10.1016/j.ccs.2011.01.006.

Bradshaw, G. and Burger, W. (2005). Public participation, sustainable development and public disputes: Public scoping processes in the Coega Deep Water Port and Industrial Development Zone. *Africanus* 35 (1): 44–58.

Brockner, J. (1992). The escalation of commitment to a failing course of action: Toward theoretical progress. *Academy of Management Review* 17 (1): 39–61. https://doi.org/10.5465/amr.1992.4279568.

Brooks, N. (2015). *Delivering European Megaprojects: A Guide for Policy Makers and Practitioners*. Leeds, UK: University of Leeds.

Choi, B.-D. (2012). Gyeongbu expressway: Political economic geography of mobility and demarcation. *Korean Social Sciences Review* 2 (2): 181–218.

Cicmil, S., Williams, T., Thomas, J., and Hodgson, D. (2006). Rethinking project management: Researching the actuality of projects. *International Journal of Project Management* 24 (8): 675–686.

Comfort, L.K., Boin, A., and Demchak, C.C. (2010). *Designing Resilience: Preparing for Extreme Events*. University of Pittsburgh Press.

Conlon, E.J. and Parks, J.M. (1987). Information requests in the context of escalation. *Journal of Applied Psychology* 72: 344.

Cooke-Davies, T. (2002). The "real" success factors on projects. *International journal of project management* 20 (3): 185–190.

Cornelio, J.R.J., Sainati, T., and Locatelli, G. (2021). Infrastructure megaprojects governance escalation of commitment sunk cost. *International Journal of Project Management* 39 (7): 774–787.

Cuppen, E., Bosch-Rekveldt, M.G.C., Pikaar, E., and Mehos, D.C. (2016). Stakeholder engagement in large-scale energy infrastructure projects: Revealing perspectives using Q methodology. *International Journal of Project Management* 34 (7): 1347–1359.

Dimitriou, H.T., Ward, E.J., and Wright, P.G. (2012). *Lessons for Decision-Makers: An Analysis of Selected International Large-Scale Transport Infrastructure Project*. London: OMEGA Project 2, OMEGA Centre and Volvo Research and Educational Foundations, Bartlett School of Planning, University College.

Dörfler, I. and Baumann, O. (2014). Learning from a drastic failure: The case of the airbus A380 program. *Industry and Innovation* 21 (3): 197–214. https://doi.org/10.1080/13662716.2014.910891.

Drummond, H. (2017). Megaproject Escalation of Commitment: An Update and Appraisal. In: *The Oxford Handbook of Megaproject Management* (ed. B. Flyvbjerg), 194–216. Oxford: Oxford University Press.

Economic Development Research Group, Inc. (EDRG) (2006). *Economic Impact of the Massachusetts Turnpike Authority and Related Projects, Volume I: The Turnpike Authority as a Transportation Provider*. Boston, MA: EDRG.

Edmondson, A.C. (2011). Strategies for learning from failure. *Harvard Business Review* 89 (4): 48–55.

Esty, B. (2004). *Modern Project Finance*. Hoboken, NJ: John Wiley & Sons.

Feldman, G. and Wong, K.F.E. (2018). When action-inaction framing leads to higher escalation of commitment: A new inaction-effect perspective on the sunk-cost fallacy. *Psychological Science* 29 (4): 537–548.

Finkelstein, S. and Sanford, S.H. (2000). Learning from corporate mistakes: The rise and fall of iridium. *Organizational Dynamics* 29 (2): 138–148.

Flyvbjerg, B. (2011). Over Budget, over Time, Over and Over Again: Managing Major Projects. In: *The Oxford Handbook of Project Management* (ed. P.W.G. Morris, J.K. Pinto and J. Söderlund), 321–344. Oxford, UK: Oxford University Press.

Flyvbjerg, B. (2014). What you should know about megaprojects and why: An overview. *Project Management Journal* 45 (2): 6–19.

Flyvbjerg, B. (2021). Top ten behavioral biases in project management: An overview. *Project Management Journal* 52 (6): 531–546. https://doi.org/10.1177/87569728211 049046.

Fox, F.V. and Staw, B.M. (1979). The trapped administrator: Effects of a job insecurity and policy resistance upon commitment to a course of action. *Administrative Science Quarterly* 24: 449–471.

Gao, R. and Liu, J. (2020). Value of investors' escalation of commitment in PPP projects using real option thinking. *International Journal of Strategic Property Management* 24 (5): 348–364.

Goodman, R.J., Bazerman, M., and Conlon, E. (1980). *Research in Organizational Behavior* (ed. B.M. Staw and L.L. Cummings). Greenwich, CT: JAI Press.

Greiman, V.A. (2015). *Evaluating Megaprojects: What Constitutes Success?* Rethinking Infrastructure: Voices from the Global Infrastructure Initiative, 2, 14–17. London: McKinsey and Company.

Greiman, V.A. and Sclar, E. (2019). Meg-infrastructure as a dynamic ecosystem: Lessons from America's interstate highway system and Boston's Big Dig. *Journal of Mega Infrastructure and Sustainable Development* 1 (2): 188–200. https://doi.org/1 0.1080/24724718.2020.1742624.

Hafenbrädl, S. and Woike, J.K. (2018). Competitive escalation and interventions. *Journal of Behavioral Decision Making* 31 (3): 695–714. https://doi.org/10.1002/ bdm.2084.

Hertogh, M., Baker, S., Staal-Ong, P.L., and Westerveld, E. (2008). *Managing Large Infrastructure Projects*. The Netherlands: AT Osborne BV.

Hong Kong (1998). *Report of the Legislative Council Select Committee to inquire into the circumstances leading to the problems surrounding the commencement of the operation of the new Hong Kong International Airport at Chek Lap Kok*, July 6.

Iftikhar, R. and Müller, R. (2019). Taxonomy among triplets: Opening the black box. *International Journal of Management* 10 (2): 63–85.

Ika, L.A., Diallo, A., and Thuillier, D. (2012). Critical success factors for World Bank projects: An empirical investigation. *International Journal of Project Management* 30 (1): 105–116.

Flyvbjerg, B., Holm, M.S., and Buhl, S. (2002). Underestimating costs in public works projects: error or lie? *Journal of the American Planning Association* 68 (3): 279–295.

Keil, M., Tan, B., Wei, K. et al. (2000). A cross-cultural study on escalation of commitment behavior in software projects. *MIS Quarterly* 24: 299–325. https://doi.org/10.2307/3250940.

Kelman, S. (2009). *A Different Kind of Partnership: the UK's Ministry of Defence and EDS Develop the Joint Personnel Administration Program (JPA)*. Cambridge, MA: Harvard Kennedy School, Case Program.

Linnenluecke, M.K. (2015). Resilience in business and management research: A review of influential publications and a research agenda. *International Journal of Management Reviews* 19 (1): 4–30.

Loosemore, M. (1998). The three ironies of crisis management in construction projects. *International Journal of Project Management* 16 (3): 139–144.

Magnussen, O.M., and Samset, K. (2005). *Successful Megaprojects: Ensuring Quality at Entry*. [Paper Presentation] European Academy of Management (EURAM) Responsible Management in an Uncertain World, 1–12. May 4–7th 2005, TUM Business School, Munich, Germany.

Matinheikki, J., Aaltonen, K., and Derek, W. (2019). Politics, public servants, and profits: Institutional complexity and temporary hybridization in a public infrastructure alliance project. *International Journal of Project Management* 37 (2): 298–317.

McCain, B.E. (1986). Continuing investment under conditions of failure: A laboratory study of the limits to escalation. *Journal of Applied Psychology* 71 (2): 280–284. http://doi.org/10.1037/0021-9010.71.2.280.

Merrow, E.W. (2011). *Industrial Megaprojects: Concepts, Strategies and Practices for Success*. Hoboken, NJ: John Wiley & Sons.

Miller, R. and Hobbs, B. (2009). The complexity of decision- making in large projects with multiple partners: Be prepared to change. In: *Making Essential Choices with Scant Information* (ed. T. Williams, K. Samset and K. Sunnevåg), 375–389. Basingstoke, UK: Palgrave Macmillan.

Miller, R. and Lessard, D. (2001). *The Strategic Management of Large Engineering Projects: Shaping Institutions, Risks and Governance*. Cambridge, MA: MIT Press.

Mišić, S., and Radujković, M. (2015). *Critical drivers of megaprojects success and failure*. ScienceDirect Operational Research in Sustainable Development and Civil Engineering, Meeting of EURO working group and 15th German-Lithuanian-Polish colloquium (ORSDCE). https://doi.org/10.1016/j.proeng.2015.10.009.

Mitroff, I., Shrivastava, P., and Udwadia, F.E. (1987). Effective crisis management. *Academy of Management Executive* 1 (4): 283–292.

Montealegre, R. and Keil, M. (2000). De-escalating information technology projects: Lessons from the Denver International Airport. *MIS Quarterly* 24 (3): 417–447.

Morris, P.W.G. (2013). *Reconstructing Project Management*. Hoboken, NJ: Wiley-Blackwell.

Müller-Mahn, D., Mkutu, K., and Kioko, E. (2021). Megaprojects—Mega failures? The politics of aspiration and the transformation of rural Kenya. *European Journal of Development Research* 33: 1069–1090. https://doi.org/10.1057/s41287-021-00397-x.

National Aeronautics Space Administration (NASA) (2009, July 8). *Apollo 13*. Washington, D.C: NASA https://www.nasa.gov/mission_pages/apollo/missions/apollo13.html.

Oeij, P.R.A., Vuuren, T.V., Dhondt, S. et al. (2018). Mindful infrastructure as antecedent of innovation resilience behaviour of project teams. *Team Performance Management* 24 (7/8): 435–456.

Ogliastri, E. and Zúñiga, R. (2016). An introduction to mindfulness and sensemaking by highly reliable organizations. In *Latin America. Journal of Business Research* 69 (10): 4429–4434.

Olsson, N.O.E., and Samset, K. (2006). *Front-end management, flexibility, and project success*. [Paper presentation] PMI Research Conference: New Directions in Project Management, Montréal, Québec, Canada.

Organization for Economic Cooperation and Development (OECD) (2015). *Towards a Framework for the Governance of Infrastructure*. Paris: Public Governance and Territorial Development Directorate Public Governance Committee, OECD.

Organization for Economic Cooperation and Development (OECD) (2016). *Country Case: Governance Failures in the Management of the Berlin Brandenburg International Airport*. Paris: OECD.

Pearson, C.M. and Clair, J.A. (1998). Reframing crisis management. *Academy of Management Review* 23 (1): 59–76.

Pfeffer, J. (1981). *Power in Organizations*. Marshfield, MA: Pitman.

Pinto, J.K. and Winch, G. (2016). The unsettling of 'settled science: The past and future of the management of projects. *International Journal of Project Management* 34 (2): 237–245.

Platt, J. (1973). Social traps. *American Psychologist* 28: 641–651.

Rangan, V.K., Palepu, K.G., Bhasin, A. et al. (2004). *Enron Development Corporation: The Dabhol Power Project in Maharashtra, India (A)*. Boston, MA: Harvard Business School Publishing.

Rose, L., Savage, C., Jenkins, A., and Fransman, L. (2017). The Failure of Transport Megaprojects: Lessons from Developed and Developing Countries. In: *Pan Pacific Conference XXXIV: Designing New Business Models in Developing Economies* (ed. S.M. Lee and V. Charles), 133–136. Pan-Pacific Business Association http://eprints.hud.ac.uk/id/eprint/31178/.

Ross, J. and Staw, B.M. (1993). Organizational escalation and exit: Lessons from the Shoreham Nuclear Power Plant. *The Academy of Management Journal* 36 (4): 701–732. https://doi.org/10.2307/256756.

Samset, K. (2013). Strategic and Tactical Performance of Mega-Projects – Between Successful Failures and Inefficient Successes. In: *International Handbook on Mega-Projects* (ed. H. Priemus and B. van Wee), 11–33. Cheltenham, UK: Edward Elgar Publishing.

Serrador, P. and Turner, R. (2015). The relationship between project success and project efficiency. *Project Management Journal* 46 (1): 30–39. https://doi.org/10.1002/pmj.21468.

Shahid, M., Mehmood, R.K., and Anjum, M.N. (2021). Empirical analysis of poverty reduction and governance in Pakistan: A qualitative comparison with sub-continent. *Global Economics Review* VI (II): 87–102. https://doi.org/10.31703/ger.2021(VI-II).08.

Shenhar, A. and Holzmann, V. (2017). The three secrets of megaproject success: Clear strategic vision, total alignment, and adapting to complexity. *Project Management Journal* 48 (6): 29–46.

Smith, C. and Winter, M. (2010). The craft of project shaping. *International Journal of Managing Projects in Business* 3 (1): 46–60.

Song, J., Hu, Y., and Feng, Z. (2018). Factors influencing early termination of PPP projects in China. *Journal of Management in Engineering* 34 (1): 05017008. https://doi.org/10.1061/(ASCE)ME.1943-5479.0000572.

Staal-Ongm, P.L., Kremers, T., Karlsson, P.-O., and Stuart Baker, S. (2016). *10 Years of Managing Large Infrastructure Projects in Europe: Lessons Learnt and Challenges Ahead*. The Netherlands: NETLIPSE.

Staw, B.M. (1981). The escalation of commitment to a course of action. *Academy of Management Review* 6 (4): 577–587.

Staw, B.M. and Fox, F.V. (1977). Escalation: The determinants of commitment to a chosen course of action. *Human Relations 30* (5): 431–450. https://doi.org/10.1177/001872677703000503.

Staw, B.M. and Ross, J. (1980). Commitment in an experimenting society: A study of the attribution of leadership from administrative scenarios. *Journal of Applied Psychology* 65: 249–260.

Staw, B.M. and Ross, J. (1987). Behavior in escalation situations: Antecedents, prototypes, and solutions. *Research in Organizational Behavior 9*: 39–78.

Tabish, S. and Jha, K. (2011). Identification and evaluation of success factors for public construction projects. *Construction Management and Economics* 29 (8): 809–823.

Thomas, D.P. (2013). The Gautrain project in South Africa: A cautionary tale. *Journal of Contemporary African Studies* 31 (1): 77–94. https://doi.org/10.1080/02589001.2013.747292.

Transparency International (2021). Transparency International Corruption Index. Berlin, Germany. https://www.transparency.org/en/cpi/2021.

Transportation Research Board (TRB) (2014). *ABCs of Bridge Renewal*, TR News, Transportation Research Board of the National Academy of Sciences, January–February.

Trope, Y. and Liberman, N. (2010). Construal-level theory of psychological distance. *Psychological Review 117* (2): 440–463. https://doi.org/10.1037/a0018963.

Turner, J.R. and Yan, X. (2018). On the success of megaprojects. *International Journal of Managing Projects in Business* 11 (3): 783–805.

U.S. Congress (2011). Final report of the President's National Commission on the BP Deepwater Horizon Oil Spill and Offshore Drilling: Oversight hearing before the Committee on Natural Resources, U.S. House of Representatives. One Hundred Twelfth Congress, first session, January 26, 2011. United States. Congress. House. Committee on Natural Resources. Washington, D.C.

van Marrewijk, A. and Smits, K. (2016). Cultural practices of governance in the Panama Canal Expansion Megaproject. *International Journal of Project Management* 34 (3): 533–544.

van Marrewijk, A., Clegg, S.R., Pitsis, T.S., and Veenswijk, M. (2008). Managing public-private megaprojects: Paradoxes, complexity, and project design. *International Journal of Project Management* 26 (6): 591–600.

Wang, A. and Pitsis, T.S. (2019). Identifying the antecedents of megaproject crises in China. *International Journal of Project Management* 38 (6): 327–339.

Wang, L., Müller, R., Zhu, F., and Yang, X. (2021). Collective mindfulness: The key to organizational resilience in megaprojects. *Project Management Journal* 52 (6): 592–606. https://doi.org/10.1177/87569728211044908.

Weick, K.E. and Sutcliffe, K.M. (2006). Mindfulness and the quality of organizational attention. *Organization Science* 17 (4): 514–524.

Weick, K., Sutcliffe, K., and Obstfeld, D. (1999). Organizing for high reliability: Processes of collective mindfulness. *Research in Organizational Behavior* 21: 81–123.

Williams, T.A., Gruber, D.A., Sutcliffe, K.M. et al. (2017). Organizational response to adversity: Fusing crisis management and resilience research streams. *Academy of Management Annals* 11 (2): 733–769. https://doi.org/10.5465/annals.2015.0134.

Winch, G. (2013). Escalation in major projects: Lessons from the channel fixed link. *International Journal of Project Management* 31 (5): 724–734.

Xiong, W., Zhang, X., and Chen, H. (2015). Early-termination compensation in public-private partnership projects. *Journal of Construction Engineering and Management* 142 (4): 04015098. https://doi-org.ezproxy.bu.edu/10.1061/(ASCE) CO.1943-7862.0001084.

10

Laws and Contracts in Global Megaprojects

"No matter how many hundreds of pages of clauses the lawyers draft, the documentation simply cannot meet the test of reality." "Exemptions and waivers have to be sought, and the documentation itself has to be changed to satisfy the 200 banks" (p. 164).

— Eurotunnel Safeguard Procedures (2006)

Introduction

Contracts represent an important governance structure for megaprojects that brings together the thousands of participants including the banks, equity sponsor(s), owners, contractors, designers, engineers, subcontractors, local communities, and other project stakeholders. Contracts try to reduce uncertainty, mitigate risk, avoid ambiguity, and prevent conflict, yet as expressed in the above quote on the restructuring of Eurotunnel sometimes this task cannot always reach the "test of reality." Megaprojects require hundreds if not thousands of contracts between the numerous stakeholders. In this chapter we explore the development of a megaproject through the lens of the legal documentation and emerging issues in these epic projects. In Chapter 10, we will focus on the laws and contracts governing owners, designers, and contractors in megaprojects. We will look at legal contracting on a global scale from the perspective of the project participants as well as the ultimate beneficiaries. The following topics will be reviewed:

- The Legal Framework in Contract Procurement
- The Law Governing International Megaproject Contracting
- Challenging Negotiations in International Megaproject Contracts
- Emerging Issues in Global Development Laws
- Checklist for Development Contracts in Emerging Markets

Global Megaprojects: Lessons, Case Studies, and Expert Advice on International Megaproject Management, First Edition. Virginia A. Greiman.
© 2023 John Wiley & Sons, Inc. Published 2023 by John Wiley & Sons, Inc.

Planning for Procurement

All megaprojects begin with the recruitment of resources from diverse geographic locations and disciplines. Public Procurement laws can vary widely across countries and regions, so it is essential to refer to domestic procurement laws and in the case of Europe, international law. For example, members of the European Union must comply with the EU directives as incorporated into national legislation. The directives apply to public works such as civil engineering or construction and transport. They require compliance with general principles of freedom of access to public bidding, nondiscrimination, and equal treatment of bidders and to public–private partnership (PPP) projects. These Directives include:

Directive 2014/24/EC of 26 February 2014 on public procurement
Public procurement plays a key role in the Europe 2020 strategy, set out in the Commission Communication of 3 March 2010 entitled "Europe 2020, a strategy for smart, sustainable and inclusive growth" as one of the market-based instruments to be used to achieve smart, sustainable, and inclusive growth,

Directive 2014/25/EC of 26 February 2014 on procurement by entities operating in the water, energy, transport, and postal service sectors
In order to ensure the opening up to competition of procurement by entities operating in the water, energy, transport and postal services sectors, provisions should be drawn up coordinating procurement procedures in respect of contracts above a certain value. Such coordination is needed to ensure the effect of the principles of the Treaty on the Functioning of the European Union (TFEU) and in particular the free movement of goods, the freedom of establishment and the freedom to provide services as well as the principles deriving therefrom such as equal treatment, nondiscrimination, mutual recognition, proportionality, and transparency.

Directive 2014/23/EU of 26 February 2014 on the award of concession contracts.

An adequate, balanced, and flexible legal framework for the award of concessions would ensure effective and nondiscriminatory access to the market to all union economic operators and legal certainty, favoring public investments in infrastructures and strategic services to the citizen. Such a legal framework would also afford greater legal certainty to economic operators and could be a basis for and means of further opening up international public procurement markets and boosting world trade. The rules of the legislative framework applicable to the award of concessions should be clear and simple. They should duly reflect the specificity of concessions as compared to public contracts and should not create an excessive amount of bureaucracy.

Concession Procurement

Public authorities commonly use concession contracts to provide for the delivery of services or construction of infrastructure. Concessions involve a contractual arrangement between the public authority and an economic operator (the concession holder) whereby the latter provides services or carries out works and is substantially remunerated by being permitted to exploit the work or service. Concessions are a particularly attractive way of carrying out projects of public interest when state or local authorities need to mobilize private capital and know-how to supplement scarce public resources. They underpin a significant share of economic activity in the European Union and are especially common in network industries and for the delivery of services of general economic interest. Concession holders may, for example, build and manage motorways, provide airport services, or operate water distribution networks. Concessions involving private partners are a particular form of PPP. According to the World Bank, over 60% of all PPP contracts can be classed as concessions. As a model where, in remuneration, the private partner is given the right to exploit either the work or the service, with the inherent financial risks Concessions constitute a convenient legal framework for carrying out public tasks through PPPs and hence make it possible to deliver much needed public works and services, while keeping those assets off of the government balance sheet.

Directive 2016/2102/EU on procurement in the private sector

Private Sector Procurement

The Directive encourages Member States to extend the Directive to private entities that offer facilities and services which are open or provided to the public such as healthcare, social inclusion, social security, and electronic communication. In the private sector, the owner traditionally distinguishes the design phase from the construction phase of the project. During the design phase, the owner directly appoints the design team to develop the project in three stages: (i) the preliminary study followed by (ii) the detailed design, which then forms the basis of (iii) the tender documents. On the basis of these documents, the contractor(s) commit(s) to build the project for a fixed price (lump sum) or unit prices, in accordance with a defined schedule. A new type of contract known as a guaranteed maximum price contract has recently emerged in some countries such as France. The contractor undertakes to assist the design team and the owner during the design phase of the project to prepare tender documents in line with the maximum guaranteed price, which is then converted into a lump sum price. Procurement arrangements under which the contractor is responsible for the design and the construction of the project [design and build contracts and variants such as engineering, procurement, and construction (EPC) contracts] are unusual in the domestic private sector, with the exception of manufacturing facilities.

The International Federation of Consulting Engineers ("FIDIC")

International projects can use the standard form contracts described for local projects, but they most frequently use contracts specifically adapted from the standard form contracts provided by the International Federation of Consulting Engineers (Fédération Internationale des Ingénieurs-Conseils) (FIDIC). Depending on the type and complexity of the project and on the tasks and risks shared between the employer and the contractor, either the Red Book (Conditions of Contract for Construction for Building and Engineering Works Designed by the Employer) or the Yellow Book (Conditions of Contract for Plant and Design-Build) can be used.

The ("FIDIC") classifies construction contracts into three types: (i) turnkey or engineering, (ii) procurement and construction ("EPC"), and (iii) design-build and construction.

Other forms of international contracts include the Institute of Civil Engineers ("ICE") or the New Engineering Contract ("NEC") currently used in the Crossrail project in London. Since public procurement generally uses their own forms, standard forms are mostly left to private sector contracts. Usually, the application of these standard forms comes as a requirement from foreign investors or financing entities.

Turnkey Construction Contracts

The standard forms of construction contract for projects involving public works include Bill and Build (B & B); Build, Own, Operate, and Transfer (BOOT); and Turnkey Project contracts. In the case of B & B, interested contractors are normally required to send in their quotations, after which the successful contractor would then be paid some money (in advance) to commence the building project. With regard to BOOT, the successful contractor carries out the building work with its own funds, after which the said contractor will be entitled to own and operate the property for a certain duration of time for the purpose of recouping its capital. After recouping its capital within the agreed time, the next step will be for the contractor to transfer the property to the owner thereof. A Turnkey Project is a construction project in respect of which a contractor is employed to plan, design, and build a project and get the project ready for use at an agreed price and by a fixed date.

Megaproject Procurement Contracts in the United States

With 50 separate states, public and private entities in the United States utilize a wide variety of contract types with no standard form of construction agreement. Before the 1980s, there was a general preference for fixed-price design-bid-build contracts, based on a fully completed set of design documents. Time-and-materials arrangements have also long been used, especially where it is difficult to estimate a fixed price or when construction must begin before there is time to complete a design. In recent decades, however, statutory changes have facilitated an increased use of design-build contracting and its variations such as engineer procure construct (EPC) contracting often used internationally.

The Integrated Project Delivery (IPD) has been emerging as a promising method to avoid some of the traditional inefficiencies of traditional contracting systems and tends to maximize construction project success. However, IPD is struggling to spread in the industry and there is still the need to understand the value that it can unlock (De Marco and Karzouna 2018).

The contract forms provided by the International Federation of Consulting Engineers (FIDIC) are used infrequently in the United States because of the preference for other forms. Federal construction projects are generally governed by the Federal Acquisition Regulation (FAR), a book containing numerous clauses mandated on various types of jobs. At the state level, many states have adopted procurement codes that establish how public works contracts are awarded but leave the forms of the public works contracts up to the individual agency or municipality.

Therefore, the form for public works contracts can vary widely within a state and between states. The A-201 General Conditions and other forms published by the American Institute of Architects (AIA) are probably the most widely used. Other well-known contract forms are published by ConsensusDocs, the Engineers Joint Contract Documents Committee (EJCDC) and the Design-Build Institute of America (DBIA). All these forms aim to achieve a degree of balance between the various parties that typically participate in a complex construction project, though that balance can easily be lost when the forms are heavily modified to favor the interests of a particular party.

The form of construction contract to use for a specific project can be selected in a multitude of ways. An architect, engineer, or contractor preparing a proposal for an owner may present the owner with a form document to consider. In those cases, the owner then may ask its lawyer to review and revise the document as appropriate. A proactive owner, however, may prepare a form of contract, preferably before the owner seeks bids, and then include that contract with a request for proposals. Owners that construct projects regularly may have their own original manuscript forms of contract, rather than relying on published forms. Standard forms of contract are published and sold in the United States by organizations associated with the construction industry. As an example: AIA documents are developed by the American Institute of Architects. The AIA forms of agreement are the most widely used and the most heavily critiqued in the United States. Some perceive that the AIA forms allocate risk in favor of the architect and to the detriment of the owner. As a result, the AIA forms are generally modified by the parties before execution. It is common for construction companies to use their own model contract for both construction (which are typically lump-sum or unit price contracts) and engineering and design (typically services contracts).

Contract Formation and Delivery Methods

In addition to deciding the contractual standards and conditions for a global megaproject, delivery methods must also be determined that may include design-build contracts, design-bid-build contracts or engineering, procurement and construction (EPC) contracts. For example, the most common types of contracts used in Greece for private construction work are (i) measurement contracts; (ii) design-build contracts; (iii) design-bid-build contracts; and (iv) construction management contracts. In public works, design-build contracts are very commonly used, as it is common practice, in respect of construction agreements used for concession contracts, for the central government and/or local authorities to award them.

Megaproject International Competitive Bids

In megaprojects most contracts are awarded by the government or government agencies based on international competitive bids. Therefore, clauses of these contracts are not easily negotiated. Many contracts which are based on international standard-form contracts (e.g. the FIDIC conditions of contract), with some modifications, typically tend to favor the employer. It is difficult for the bidders or even the successful contractor to negotiate all the terms. However, parties do seek and get clarifications relating to some of the clauses before the bid submission date and some-times even manage to have a few of the clauses amended. However, this is rare. Private contracts between, for example, a developer or a special purpose vehicle (SPV) with a contractor or subcontractors are typically negotiated. Even in these contracts, the scope for negotiation is not large and the employer tends to retain their contract forms. Additionally, contracts with subcontracts can be on a back-to-back basis so that the provisions in the master agreement are consistent with the provisions in the subcontracts.

In build-operate-transfer (BOT), and build-own-operate-transfer (BOOT) projects and projects with developers, both debt and equity funding are used. The income stream is the security provided to the lenders (apart from guarantees granted by the parent companies). As a large number of projects do not permit the lender to take over the projects, and the project itself is not deemed to be a security. One of the recent issues with this model has been that in many projects (particularly road projects), even after the construction phase is over, the contractor or the SPV was not permitted to sell the equity. As a result, more complex arrangements were not entered into and only an escrow arrangement was worked out in which all earnings such as toll collection would be deposited.

However, it is expected that transfer of projects will be permitted more often in the future. This is because most lenders and banks are very reluctant to take over projects and run them, and, therefore, there are very few cases where projects have been taken over by the lenders. Where income streams form the guarantee to the lenders, there have been cases where the lenders have been forced to take up toll collection to recover their payments, which is not an ideal situation.

Laws Governing International Megaproject Contracting

> A nation that destroys its soils destroys itself. Forests are the lungs of our land, purifying the air and giving fresh strength to our people.
>
> *– Franklin D. Roosevelt*

There are so many laws that govern international contracting describing all these laws is beyond the scope of this book, however, it is important to include a brief description of some of the more important international laws as they are relevant to megaproject contracting (Table 10.1). One of these laws involves local building laws and regulations and is highlighted in Box 10.1.

Box 10.1   Codes and Regulations

In the United States, each of the 50 states and the District of Columbia have adopted the International Building Code. The International Fire Code and the International Energy Conservation Code have also been widely adopted. However, each jurisdiction has the authority to make alterations to the codes as it deems appropriate. There are several websites that include a compendium of the codes for each state, including the one found at https://www.buildingsguide.com/blog/resources-building-codes-state/. Federal public works projects must comply with these local codes in addition to any special requirements that the federal government may have for the project.

Table 10.1 Megaproject law and policy initiatives.

Country	Law and policy initiatives	Projects
Germany	The exit from nuclear and fossil-fuel energy means the focus will be more on sustainable finance.	The preservation and modernization of the German Rail System. The German Government and Deutsche Bahn, the German railway company, entered into an investment agreement with a total of €86 B for the preservation and modernization of the German railway networks within the next 10 years
		The acceptance of the 2.4 bn modernization of the Autobahn A3 in northern Bavaria.
Ghana	Projects approved in the power, maritime, and rail sectors	Eastern Railway Line Project This project involves the construction of a 340 km standard-gauge railway line, running from the Tema Port to Kumasi. It is being developed by the Ghana European Railway Consortium on a Build, Operate and Transfer (BOT) basis. ($2.2B)
Greece	Law 46-43/2019 National Energy and Climate Plan	Liberalization of the energy market, the modernization of the Public Power Corporation, and the privatization of the Public Natural Gas Company (DEPA)
Hungary	Renewable Energy	Development of the MET Dunai Solar Park (2018), and Airport Refinancing
France	Projects related to sustainable development for example "smart cities," energy efficiency and renewable energy as a result of France's commitment to climate finance, but also to its post-COVID-19 economic recovery plan	The Fécamp offshore windfarm (€2B): Expected to be completed in 2023, this windfarm is set to produce the equivalent of the annual electricity consumption of 770 000 inhabitants. A majority of the capital will be raised through non-recourse project-level debt. The Astérix project (€1.1B): This project consists of deploying the fibre network in medium-density areas in France. Crédit Agricole CIB is the lead arranger for financing

Ethiopia	In 2020, the PPP Board approved 25 projects (eight solar energy, eight wind energy, five hydropower, one housing development and three road projects) to be carried out through PPP arrangements. It was also reported recently that the government is considering PPP for the purpose of building a communication satellite that will cost over $350 million.	Two solar energy projects – Scaling Solar Dicheto in Afar Regional State and Scaling Solar Gad in Somali Regional State – were awarded to a Saudi Arabian firm named ACWA Power.
Malaysia	Among the major infrastructure and utility projects highlighted in the 2021 Economic Outlook are: (i) the Mass Rapid Transit 2 (MRT2); (ii) the Light Rail Transit 3 (LRT3); (iii) the West Coast Expressway (WCE); (iv) the Pan Borneo Highway; (v) the Langat 2 Water Treatment Plant; and (vi) the Sarawak Water Supply Grid Program.	Pengerang Refining and Petrochemical Integrated Refinery and Petrochemicals Complex: The US$21B petrochemical facility in the state of Johor, which is anticipated to have a combined capacity to produce 7.7 million tonnes per year of chemical products. The project was initially reported to be on course to begin commercial operations in March 2021, but this appears to have been pushed to the second half of 2021.
Japan	The global trend of decarbonization, in conjunction with the United Nations Sustainable Development Goals, has significantly discouraged the development of new thermal power plants. At the end of 2020, the Japanese government published the "Green Growth Strategies" and clarified its policy to increase renewable energy sources so that these constitute 50–60% of all electricity sources by 2050.	A substantial proportion of the existing Japanese social infrastructure was constructed during the 1960s and 1970s. To meet the need to restore or replace these facilities in the coming decades, the Japanese government is facilitating the use of public-private partnership (PPP) and private finance initiative (PFI) structures.
Indonesia	The most noteworthy project financings to have taken place in recent years in the public infrastructure sector have been for power plants, toll roads, and water treatment and supply.	In the private sector, such financing has mainly been used for smelting plants in relation to copper and nickel, and for oil-to-fuel processing plants.

(Continued)

Table 10.1 (Continued)

Country	Law and policy initiatives	Projects
Singapore	A Green Finance Industry Taskforce was convened to advance the green financing agenda, and its initiatives include developing a taxonomy based on international best practices to help Singapore-based financial institutions identify, classify, and transition to financing green activities, and offering guidance on best practices in environmental risk management.	A significant transaction in Singapore was the financing of the Jurong Island Desalination Plant. This modern seawater reverse-osmosis desalination plant development was completed in 2022 and is capable of desalinating 36 million gallons of seawater daily.
Turkey	There have been a significant number of ongoing public-private partnership (PPP) projects in the energy, highways, ports, healthcare, and railway sectors. The Turkish government promotes independent investment in these sectors and has also focused heavily on PPP projects in recent years. These PPPs are mainly centered around the Turkish energy, airport, road, railway, and infrastructure sectors.	Recent amendment made to the Capital Markets Law No. 6362, an alternative project financing method (i.e., project finance fund) and a new type of security (i.e., project-backed security) are also being introduced, with the purpose of contributing project financing for investments with long terms and high capital, such as infrastructure, energy, industry, technology, communications, and health projects.
		$4.5B financing and $1.6B additional financing of the Third Bosphorus Bridge and Northern Marmara Motorway. €125 m financing of the Third Airport of Istanbul. $1.2B financing of the road infrastructure project for a tunnel connecting Asia to Europe, namely Avrasya Tüneli. IC İçtaş's $1.2B expansion of the Tuz Gölü (Lake Tuz) Underground Natural Gas Storage

Source: Financial Institution and Country Project Databases.

Megaproject Risk and Regulation

The Grenfell Tower fire in West London on 14 June 2017, which caused 72 fatalities and further injuries, is a horrific example of why the world needs better contracts and regulations. The devastating fire was swiftly followed by industry-led consultations and public inquiries, some of which are ongoing. A Report to Parliament (2020) after the fire found the UK construction rules to be dangerously lax. However, "while some regulatory changes have been implemented, it is only recently that the first of the major statutory reforms concerning fire safety has come into effect. The reform included the Fire Safety Act 2021, which received Royal Assent in April 2021 – almost four years on from Grenfell." Meanwhile, the UK government has introduced various schemes to fund the replacement of defective cladding on the building sides, but eligibility and coverage are restricted, and, in most cases, funding is conditional on claimants also pursuing insurance claims and/or warranty claims against those involved in the installation of the cladding. Due to such factors, the English courts have seen a steady stream of claims relating to defective cladding and fire/building safety more generally.

Challenging Negotiations in Megaproject Contracts

Megaprojects are known for their complex networks, interfaces, and dependencies and must manage colossal risks, uncertainty, and ambiguity. All of this requires clarity in contracting and a road map that embodies the final deliverables which may take decades to complete. Because most megaprojects contain numerous work packages and subcontracts coordination among all contracts is a key challenge. For example, the Big Dig consisting of 135 major work packages and thousands of smaller subprojects was broken into geographic locations.

In Figure 10.1 you see examples of some of the most challenging clauses in large infrastructure construction projects, highlighting important yet contestable provisions of megaproject design and construction contracts and provisions that can impact owners, sponsors, lenders, designers, contractors, and the community at large. Each of these contractual provisions are analyzed from both a risk and sustainability perspective.

Clause 1: Social and Environmental Rights: Developing Sustainable Contracts

> What we are doing to the forests of the world is but a mirror reflection of what we are doing to ourselves and to one another.
>
> – *Mahatma Gandhi*

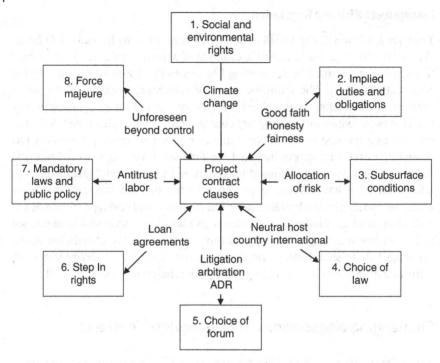

Figure 10.1 Project contract clauses.

In most countries today, environmental approval also considers sustainable development and, as a result, conditions are often imposed in this regard by the environmental authorities in the law and in contracts. Presently the requirements vary from country to country, but the movement is towards standardization on an international and national level. National courts have also considered environmental issues particularly relating to sustainable development and the right of the landowners whose lands have been acquired. Several hydroelectric projects have been postponed in view of lack of proper environmental studies relating to sustainable development.

"Sustainable development is about ensuring that legitimate human needs are met without sacrificing environmental resources in the process." If one is to believe that States have "adopted" sustainable development as the way forward, this needs to be reflected in what goes on within international organizations of which they are a part. International organizations must be more involved in global development if they are to be representative of their members' interests (Lorenzo 2018).

For example, the World Bank's Environmental and Social Framework (ESF) became effective on 1 October 2018, and applies to all Investment Policy Financing

(IPF) projects initiated after this date. Though the standards are not quite as extensive as the United Nations 2030 Agenda they respond to many of the Agenda's goals.

ESF	Standard
1	Assessment and Management of Environmental and Social Risks and Impacts
2	Labor and Working Conditions
3	Resource Efficiency and Pollution Prevention and Management
4	Community Health and Safety
5	Land Acquisition, Restrictions on Land Use, and Involuntary Resettlement
6	Biodiversity Conservation and Sustainable Management of Living Natural Resources
7	Indigenous Peoples/Sub-Saharan African Historically Underserved Traditional Local Communities
8	Cultural Heritage
9	Financial Intermediaries
10	Stakeholder Engagement and Information Disclosure

The goals of the ESF are critical to sustainability of all projects even outside the developing world and have been adopted by most multilateral organizations and even privately funded projects and can impact all design and construction contracts on a megaproject financing. Borrowers and projects are also required to apply the relevant requirements of the World Bank Group Environmental, Health and Safety Guidelines (EHSGs). These are technical reference documents, with general and industry specific examples of Good International Industry Practice (GIIP) that should be consulted. The ESF are designed to help borrowers manage the risks and impacts of a project, and improve their environmental and social performance, through a risk and outcomes-based approach.

In the context of the Belt and Road Initiative and the national strategy for sustainable development, China's major infrastructure projects (hereinafter referred to as megaprojects) are facing an unprecedented environmental sensitivity period, strategic opportunity period, and value reconstruction period (Malik et al. 2021). Therefore, in addition to paying attention to the progress, cost, and quality of the projects themselves, the stakeholders of megaprojects should also focus on project social responsibilities such as avoiding harm to public safety, maintaining ecological balance, and achieving sustainable development (Ma et al. 2017). Transforming our World: the 2030 Agenda for Sustainable Development. declares that "eradicating poverty in all its forms and dimensions, including extreme poverty, is the greatest global challenge and an indispensable requirement for sustainable development" (Pribrytkova 2020).

In reality, some project stakeholders excessively pursue their economic interests while neglecting social responsibilities, leading to frequent accidents. In 2013, the collapse of the Garment Factory, Sava District Building in Bangladesh, which killed 1127 people, shocked the world. The main reason was the owners' lack of a strict review mechanism, and the project was multi-level contracted. Subcontractors cut corners and built-in violation of regulations; government departments had information asymmetry in those mentioned above lacking social responsibility behaviors, leading to government supervision and management dereliction, which eventually led to tragic accidents (Brunet and Aubry 2016; Xue et al. 2022).

Clause 2: Implied Duties and Obligations Under International Contracts

> There are certain duties that are enforceable even though they are not part of the written contract.

This is often the case in international transactions. For instance, all contracts must comply with applicable laws, and the contractor shall be obliged to follow construction norms, health, safety and environmental rules, and any other law imposed by the State, at all times. Normally, general principles of contract apply in this regard.

However, certain terms, which necessarily arise from the contractual relationship between the project owner or employer and the contractor, or which are necessary in the business sense to give effect to the contract, will be implied. Implied terms may include that the employer will cooperate with the contractor, or that the employer will not deprive the contractor of possession of the construction site save in accordance with the terms of the agreement. It may also be an implied term that the contractor will do the work in a good and workmanlike manner, use suitable material, and perform his obligations in such a way as to conform to the applicable building regulations.

In the United States, every contract party owes an implied duty of good faith and fair dealing. For example, in a construction contract, parties are typically held to owe an implied duty that they will not hinder or delay each other. An owner who provides plans and specifications for use in construction impliedly warrants that they are suitable for use, and owners are typically held responsible for errors and omissions in their contract documents unless the contractor assumes responsibility for reviewing and completing the design as a design-builder (and even in that case the contractor may be able to place some degree of reliance on the employer's partial design and/or site information).

Many courts have held that an owner owes an implied duty to disclose any non-public "superior knowledge" about the site or the project that a court finds should have been disclosed to the contractor. When furnishing materials or equipment for a construction project, the seller typically owes implied duties that the goods will be of "merchantable" quality and that they will be reasonably fit for their intended purposes. See U.C.C. §2-314, 315. Such implied warranties may, however, effectively be disclaimed by contract. And in design contracts, architects and engineers may be held to an implied obligation to perform to the "prevailing standard" established for similar services in the area where the work is done. In multi-prime contracts, owners may owe an implied duty to coordinate their various prime contracts, and prime contractors will probably be held to an implied duty to coordinate their various subcontractors and suppliers.

As an example, Article 246 of the United Arab Emirates (UAE) Civil Code implies a duty to act in good faith upon the parties to a contract, and states that the contract must be performed in accordance with its contents. According to Article 265 of the Civil Code, a contract is to be interpreted in accordance with its clear wording. In the absence of clear wording, the mutual intention of the parties, the nature of the transaction, and the trust and confidence that should exist between the parties in accordance with the custom in such transactions, will be examined.

Every country in the world implies different obligations in their contracts but they also include basic rights to protect contracting parties. For example, In the Republic of Slovenia, construction contracts are concluded in accordance with the provisions of the Obligations Code (OZ), Special Construction Usages (PGU) and the International Federation of Consulting Engineers' (FIDIC) books.

The basic principles of the Slovenian Obligations Code include:
- The principles of conscientiousness and fairness,
- The duty to act in good faith, diligence, prohibition on abuse of rights,
- The duty to perform obligations, and prohibition on infliction of damage, shall also apply to construction contracts.

The term "fitness for purpose" does not apply in construction contracts in Slovenia. Contractors shall deliver what they have bargained for, and employers shall not expect contractors to guess what their future intentions might be.

Under PRC Contract Law, in addition to contractual obligations, the parties of a construction contract are generally required to:

- Observe the principle of fairness and reasonably allocate parties' rights and obligations.
- Observe the principle of good faith, be honest and fulfill their commitments.
- Observe not violate laws or the public order and good morals.
- Perform obligations including notification, assistance, confidentiality, and others based on the nature and purpose of the contract as well as the course of dealing; and
- Avoid wasting resources, polluting the environment, and damaging the ecology

As another example of implied conditions, the law of India recognizes the use of both express and implied terms in a construction contract. While express terms are easily identifiable, implied terms must be read into a contract while examining the intention of the contracting parties. However, such terms must not alter the intended commercial purpose of the contract as understood between the parties. While there is no agreed set of terms which can be implied in a construction contract, certain obligations are understood as impliedly binding on both the employer and the contractor. For example, a contractor is expected to perform its tasks while exercising a standard of care and must provide such materials as are fit to be used for the stipulated works.

> In summary, these implied obligations and rights exist in many countries and before entering into a contractual agreement with a host country or a project participant these obligations must be reviewed to make sure that they can be delivered.

Clause 3: Unforeseen Subsurface Conditions

One of the most complex and contentious provisions in international contracts for infrastructure and construction is drafting the allocation of risk for subsurface conditions or as they are sometimes referred to as differing site conditions. The reduction of subsurface risks requires the evaluation of the conditions to the fullest extent possible at the outset of the contract and then determining a proper allocation of risk that is clearly understood and defined in the contract documents.

There are no mandatory provisions under French law. Thus, it is the parties who contractually stipulate who bears the risk of unforeseen ground conditions. In practice, it is often the project owner who bears the risk. Indeed, this is what happens when the contract does not specify who bears the risk. The theory of unforeseen circumstances (théorie des sujétions imprévues) allows compensation to be obtained by the contractor for losses arising from unforeseen ground conditions in the context of public law construction contracts. In private law construction contracts, the theory of unforeseen circumstances (théorie de l'imprévision) may play an equivalent role, in that it enables the contract to be renegotiated (modified or terminated) in the event of new circumstances unforeseen at the time of execution of the contract which make the contract excessively onerous to perform.

Because of these unforeseen circumstances, it is important that the constructor and its engineer carefully review the relevant provisions of the concession agreement that involve design and construction risk assumption and, if at all possible, be involved in the negotiation of that agreement (Hatem and Cordum 2010).

America's largest inner-city project, Boston's Big Dig had major subsurface risk exposures that impacted schedule included design development, hazardous material removal, complexity of interfaces between contracts, unanticipated site conditions such as unchartered utilities, obstructions from old piles and seawalls, unexpected ground conditions, and archeological sites. Critical risk exposures were identified in the project's finance reports and in the project management monthly reports, as these exposures could have serious impacts on project schedule whether or not these impacts occurred on a critical path. Noncritical paths had to be carefully monitored for serious exposures that could change the status of that path to a critical path.

As described by one Bechtel engineer, "The soil beneath the city has a little bit of everything." The Big Dig tunnels had to be dug through four distinctly different types of soil – fill at the surface, followed by organics (silt, sand, etc.); a marine soil known as Boston "blue clay"; and, finally, a layer of boulders, gravel, and clay (glacial till) sitting atop the bedrock (Einstein 2012). A particularly large and deep deposit of the blue clay complicated proposed excavations to construct tunnels connecting the Ted Williams Tunnel, the Central Artery tunnels, and the Massachusetts Turnpike. Even small, shallow excavations would collapse without support. The solution was to combine the clay with cement, a "deep soil mixing" technique developed in Japan that makes the soil harder and thus easier to excavate, and makes it act as a buttress (Greiman 2013). Even that was not simple. To access the areas where soil needed to be improved, existing structures had to be moved. The tunnel ground freezing and jacking operations produced another level of unknown conditions and construction claims. Construction costs increased throughout the project and across all contracts as a result of these subsurface conditions (p. 219). As of 30 September 2004, the project reported that differing subsurface conditions represented about 19% of the almost $2B in requested claims and changes – a big number by any count (p. 219).

Differing site condition clauses represents a risk sharing mechanism commonly used in megaproject contracts. If the risk is not shared or allocated in the contract documents, then in the United States by default the common law applies which usually allocates the entire risk to the contractor. These clauses differ from one country to another and can vary widely. It is usually the obligation of the contractor to demonstrate that the conditions differ materially from those described in the contract. As demonstrated in the "Big Bertha" case study, subsurface conditions can create significant contractor and owner risk that includes cost, schedule, project finance, and relationship risk that may impede the project for years to come.

Big Bertha Case Study, The Washington State Department of Transportation (WSDOT)

Until April 2015, when engineers began hoisting its 2000-ton cutterhead to the surface, Bertha, the world's largest tunnel-boring machine, was stuck at a 120 feet below the streets of Seattle, Washington, too damaged to move far forward and incapable of going in reverse. The machine, which weighs some seven thousand tons and is about as long as a football field, is the centerpiece of a $2 billion project to build a stretch of underground highway two miles long, two lanes wide, and two levels high.

But, in December of 2013, after only four months and a 1000 feet of digging, Bertha overheated and was shut down. Attempts to fix it set off a cascade of other construction problems, helping to secure the tunnel's reputation as one of the biggest megaproject fiascoes in history. On 13 December 2019, after almost six years of disputes and litigation and a nine-week trial, the Washington State Department of Transportation (WSDOT) won a $57.2 million jury award against the contractors over delays in this project called the Alaska Way Viaduct project. There are no winners as the matter is now working its way through the appeal process which may take another several years of costly litigation and broken relations until a final decision is rendered.

Source: Adapted from AASHTO (2022).

Considerations for Allocation of Differing Site Conditions Risk

The allocation of risk for unknown subsurface conditions is an important provision in all megaprojects involving tunneling or other underground construction and can result in serious legal and other damages and years of litigation if allocation is not properly agreed upon by the project owner and its contractors For example, under the laws of Denmark and some other countries, the employer will, as a general rule, bear the risk of unforeseen ground conditions, as the employer must provide adequate information on hindrances with respect to ground conditions (Denmark 2016). Major sources of law in Denmark include the *Constitutional Act, statutory legislation, regulatory statute, precedent,* and *customary law.*

If an owner has knowledge of a differing site condition potential before the contract is signed, this should be disclosed, and an equitable adjustment clause should be included in the contract for additional costs that the contractor may anticipate. The additional costs may include the hiring of additional resources including people and equipment or allowing an extension of time to complete the contract or changing the scope of the contract to avoid the conditions entirely.

Most U.S. commercial construction contracts assign the risk of unforeseen subsurface conditions to the employer, assuming that the employer normally has more time and opportunity to study the ground that will be excavated. This is typically handled by a Differing Site Conditions clause. See, e.g., FAR 52.236-2. Such clauses typically allow compensation if a contractor encounters latent site conditions that differ materially from those indicated in the contract documents ("type 1") or latent conditions of an unusual nature that would not normally be expected at the site ("type 2"). Contractors must usually give prompt notice of such conditions and should take reasonable steps to mitigate their impacts. Contractor rights to claim Differing Site Conditions may be limited if the contractor is paid to conduct its own independent pre-bid investigation of subsurface conditions or if a particular contractor had knowledge of the condition based on prior work at the site.

One approach to mitigating but not necessarily avoiding the risk is through a baseline approach. The baseline approach is a method objectively defining, describing and/or quantifying physical conditions reasonably expected to be encountered during the performance of construction work (Hatem and Cordum 2010). The baseline approach primarily is intended to assist the project owner and contractor in more clearly defining and allocating risk for differing site conditions at the time of contract formation, thereby reducing misunderstandings, disputes and claims and project delays and disruptions during the construction process (p. 509).

One example of contract drafting is shown in Box 10.2 from the construction contract used on Boston's Big Dig governing differing site conditions.

Box 10.2   SPECIFICATION Section 4.04 Differing Site Conditions

If during the progress of the Work, the Contractor discovers that the actual subsurface conditions encountered at the site differ substantially or materially from those shown on the Plans or indicated in the Contract Documents, ... the contractor ... may request an Equitable Adjustment in the Contract Price ...

Problem: Disputes arising from this specification were quite common due to the differing interpretations of "substantially or materially" and whether the designer should be liable for not identifying the subsurface conditions in the drawing or the owner should be liable because the contractor was not at fault.

Solution: (i) the designer and the owner should work together to resolve any differences; (ii) the meaning of "differ substantially or materially" should be defined to a greater extent in the specification; (iii) the method of calculating the equitable adjustment should be included in the contract documents; and (iv) the risks and exposures should be identified and included in the contractor's baseline, or through a construction contingency.

Risk Allocation

Specifications are used to allocate risk among the various project participants. Serious problems can arise when these clauses contain errors, omissions, or inconsistencies. A review of the extensive case law that has arisen in construction contracting shows four areas of potential conflict that can seriously impact construction cost, schedule, quality, and risk. These clauses fall under the following three categories: excusable performance (force majeure); claims and changes; and project control and responsibility.

Force majeure clauses, as covered in Clause 8 Force Majeure, define the types of risks that will excuse performance such as severe weather, unforeseen conditions that were not contemplated at the time the contract was entered into, or a change in the law that impacts the investment. These risks must be carefully defined so the owner is protected from a risk that is within the control of the contractor, and the contractor is protected from a risk that is within the control of the owner. For instance, in a public construction project, the risk of differing site conditions is typically allocated to the contractor, while regulatory risk is allocated to the owner. These clauses should also define the obligations for mitigating an event once it occurs, as well as the time frame during which performance will be excused.

Claims and changes provisions detail the procedures that are to be followed when there is a change to the project scope, the process that will be followed to resolve the changes, as well as any dispute resolution process that is required of the involved parties.

Project control clauses generally define responsibilities of the parties and procedures for various aspects of the design and construction process. For example, the Spearin Doctrine, a landmark U.S. Supreme Court decision, is a legal principle that holds that when a contractor follows the plans and specifications furnished by the owner, and those plans and specifications turn out to be defective or insufficient, the contractor is not liable to the owner for any loss or damage resulting from the defective plans and specifications due to an implied warranty given by the owner to the contractor (*United States v. Spearin* 1918). Thus, efforts must be made to ensure that project participants understand their differing roles and responsibilities that are assigned in the contract clauses and to comply with these provisions. Prior to determining which law will govern the project contracts, it is important to have a complete legal analysis of the possible applicable laws by the organization's general counsel.

To the extent the contract does not address all issues, the local statutory and common law will govern the contract. Owners need to include in all project contracts, indemnity, liquidated damages, consequential damages, and differing site condition clauses to shift liability to the contractor for increased costs and delays in the event of the breach of the contract by the contractor. While the contractor and designer must negotiate their own protections for materially different conditions than were anticipated.

Recommended Practices for Subsurface Conditions:
1) Project owners should undertake an investigation of subsurface conditions prior to commencement of the project design and construction.
2) The Project contractors and designers should be involved early on in decisions regarding allocation of risk for materially different conditions.
3) Experts in the industry should be employed to review the conditions and recommend mitigation efforts.
4) The process for resolving disputes expeditiously should be determined early in the project planning phase to avoid unnecessary delays.
5) Allocation of risk should be based on a fair and equitable process considering the geo technical complexity of the project work site.
6) Insurance, warranties, and indemnification provisions should be clearly specified in each contract and should apply back-to-back to the subcontractors.

Clause 4: Choice of Law

Party Autonomy is one of the most important concepts in international law and is demonstrated best in the parties' freedom to choose the law of their contract. Virtually all legal systems and many treaties, and international legal agreements recognize the right to choose the law of your contract. Yet, there is variation in the enforceability and interpretation of these provisions. Enhanced predictability and efficiency are especially important in international settings because of the significant differences between national laws and decision-makers and the availability of multiple potential forums (Born and Kalelioglu 2021). Some of the key questions in selecting a choice of law to govern your contracts include the following:

1) Is there an applicable treaty, convention, or international law? These may include bilateral or multilateral investment treatments or enforcement of arbitration or judicial decisions.
2) Does the country have a constitution with provisions that may be relevant to international and foreign direct investment?
3) Do the national or local laws include provisions on environmental, labor, finance (investment limits), procurement processes, or other regulatory requirements?
4) Are there criminal laws that may be relevant to an investment such as bribery, financial crimes, fraud, or antitrust violations?
5) Who does the law favor most? The owners, the sponsors, the contractors, others?
6) Who will interpret the law? A court, an arbitrator, a mediator, an expert appointed by the parties.
7) Can the law be modified in the agreement?
8) Is there public policy or mandatory law that is applicable?
9) Will the law be enforced? In what jurisdictions can I bring a cause of action?

It is common for the concessionaire and its contractors to choose a law that is familiar to them and that in their view adequately governs the issues addressed in their contracts. Depending upon the type of contract, different issues concerning the governing law clause will arise. For example, equipment supplies, and other contracts may be entered into with foreign companies and the parties may wish to choose a law known to them as providing, for example, an adequate warranty regime for equipment failure or nonconformity of equipment. In turn, the concessionaire may agree to the application of the laws of the host country in connection with contracts entered into with local customers.

For some megaprojects there may be different choice of law provisions depending on the type of contract. There may be one provision for the financing agreements with the lenders, and another provision for the financing agreement with the equity sponsors. Sometimes, the host country's legal system will mandate a particular law or will have mandatory provisions that cover part of a contract but not the entire contract. For contracts between construction contractors and the project owner or employer a different law is usually negotiated that will provide more flexibility for the contractors. Which particular jurisdiction's laws govern a project is a critical question for all parties to a project finance or development agreement.

As an example, In Thailand, the parties may agree on the governing law provision in the project agreements and financing agreements. If the project agreement or financing agreement is governed by foreign law (law other than Thai law), it will be recognized and applied, but only to the extent such law is proven to the satisfaction of the courts of Thailand not to be contrary to the public order or good morals of the people of Thailand.

In the English Chunnel a compromise was agreed upon between the parties to the construction contract calling for ICC arbitration in Belgium and designating not one, nor two, but three different legal systems to govern the relationship. The arbitrators were to apply "principles common to both English law and French law." Absent such common principles, the agreement was to be construed according to "general principles of international trade law as have been applied by national and international tribunals" (Park 2015).

The concessionaire sought judicial relief outside the arbitral process, asking English courts to restrain the contractors from suspending work pending resolution of the dispute. Ultimately the matter reached the House of Lords. Although admitting that an English court had power to grant an injunction, the Arbitrator reasoned that such power should not be exercised under the circumstances of the case. Noting the heavily negotiated arbitration and choice-of-law provisions, rather than a routine standard-form clause, he wrote as follows.

> The parties chose an indeterminate "law" to govern their substantive rights; an elaborate process for ascertaining those rights; and a location for

that process outside the territories of the participants. This conspicuously neutral "anational" and extra-judicial structure may well have been the right choice for the special needs of the Channel Tunnel venture. But whether it was right or wrong, it is the choice which the parties have made. The appellants now regret that choice. . . .. Notwithstanding that the court can and should in the right case provide reinforcement for the arbitral process by granting interim relief. I am quite satisfied that this is not such a case and that to order an injunction here would be to act contrary both to the general tenor of the construction contract and to the spirit of international arbitration.

Thus, as shown by the Arbitral decision in the English Chunnel, courts and arbitrators will tend to uphold the chosen law of the parties even if it is seriously detrimental to one of the parties. Choice of law is an internationally recognized right that all parties have and protect party autonomy which is crucial to the encouragement of foreign investment in all countries. In the next section we will look at choice of forum and dispute resolution clauses and will see how these two provisions are intricately linked and must be considered together in drafting contracts and financial documents.

Clause 5: Choice of Forum and Dispute Resolution

> The whole duty of government is to prevent crime and to preserve contracts.
> — *William Lamb*

Disputes – Crossrail

Differences of opinion are a reality of construction contracts, especially of those as complex as the Crossrail Project in London, Europe's largest megaproject. Contracts encourage parties to reach agreement, or at least a decision, or state a final position, at the time the issue is encountered. This approach is seen as the antidote to long-drawn-out disputes running for months or years after contracts finish. The approach requires firm agreement to be made on the basis of a prospective forecast rather than a retrospective analysis of actual events. For complex issues this can be difficult to do, especially when so much is at stake and many issues are interwoven. The client encountered several disputes with contractors, mostly revolving around issues of time, but generally engaged successfully in a process of managerial discussions to reach agreement. To date, the client has had just one formal adjudication in relation to the main works contracts. Full and open managerial discussions at an appropriate level are essential and are more likely to result in a positive outcome for both parties than formal dispute routes.

The parties need to persevere through sometimes very difficult issues and relationships (Morrice and Hands 2017).

The United States

The United States is a signatory nation to the Convention on the Recognition and Enforcement of Foreign Arbitral Awards (the "New York Convention"), and the U.S. complies with this agreement by enforcing arbitral agreements and awards issued in signatory countries. Consistent with Article V of the New York Convention, defenses to enforcement of a foreign award include lack of due process, a conclusion that enforcement would be contrary to public policy, and other listed defenses. Chapter 2 of the Federal Arbitration Act provides terms under which courts of the United States shall enforce foreign arbitration awards in accordance with the New York Convention.

Under a federal government contract, disputes are ultimately subject to resolution in a federal court or agency board, without a jury. Under state, local, or private contracts, disputes are ultimately subject to resolution in a court (often with a right to jury) unless the parties have agreed otherwise. Although some federal agency boards have significant experience in construction disputes, most courts have little such experience, and parties often agree by contract that disputes will be referred to mediation and ultimately to binding arbitration by a person with construction industry expertise.

Dispute Review Boards

Many contracts initially refer disputes to high-level executives of the parties before they may be submitted to a court or arbitration panel. Many large complex projects also establish a Dispute Review Board that helps to resolve issues while the job is being performed. Nonbinding mediation is also very commonly used to resolve complex construction disputes. A few companies provide mediation and arbitration services. Two of the largest providers are the American Arbitration Association (AAA) and JAMS, each of which publishes detailed rules for mediation and arbitration.

Figure 10.2 shows the steps used in a dispute resolution process utilizing the Dispute Review Board model. At the Big Dig, every effort was made to resolve disputes internally, which required the parties to agree to waive any claim against each other except in the case of willful conduct or default. The Big Dig's dispute resolution process required a progression from field office determination through various stages, including senior-level partnering, executive-level partnering, mediation before the dispute review board up to litigation, as shown in Figure 10.2. At any point along the way, the dispute could be resolved. As noted in the figure, the longer a dispute took to resolve, the greater the expense; thus, early resolution was always encouraged. The internal resolution of disputes allowed the integrated and collaborative process to continue throughout the life of the project. During

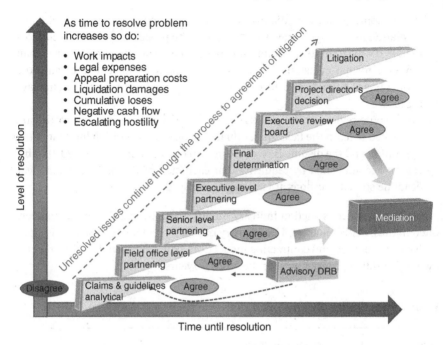

Figure 10.2 Steps in claims dispute resolution at the Big Dig. *Source:* Greiman (2013)/ with permission of John Wiley & Sons.

the project, more than 25 000 claims were generated between the contractors and the project owner.

Lessons learned from dispute resolution under a Review Board Model include the following:

1) To prevent large backlogs of claims, dispute resolution programs "must have robust and experienced staffs in both the contractor and owner organizations so that claims can be addressed as they arise and do not migrate to the end of the project as part of an omnibus claim" (Dettman et al. 2010).

2) Dispute resolution methodologies should be designed to encourage conflict avoidance and dispute prevention before controversies escalate and cause serious communication breakdowns among the key stakeholders.

3) Creating proactive, innovative dispute resolution processes with independent oversight is essential in large, complex projects where schedule delays can cause serious financial losses.

4) Constantly assess the effectiveness of the dispute review board framework and revise the processes and procedures as the requirements and priorities of the project change. As an example, if the mediation process proves

cumbersome because mediators are issuing opinions through an evaluative mediation process, you may want to change the process so that the mediators merely try to facilitate a resolution between the disputing parties rather than render a formal decision. Another change might be to reduce the dispute review board from five to three members to expedite the process and save on costs.

5) The dispute review board process can be effective if the disputing parties are compelled to learn the facts and evaluate not only their position but that of the opposition and the dispute is resolved in a timely manner (Harmon 2009). This can be accomplished by requiring that each step of the process has a clear deadline so that decisions are not unduly delayed.

According to the construction industry reports, partnering also occurs relatively often in the Swedish market. The industry has seen an increased use of this alternative work method and contracting model from both private and public property owners/ building proprietors over the last 7–10 years.

Examples of Global Dispute Resolution Provisions

Country	Dispute resolution mechanisms
Slovenia	Slovenian courts offer the possibility to solve disputes amicably with mediation in front of the court. In rare cases, when the parties do not resolve the dispute in such a way and conclude a court settlement (or out-of-court settlement, for that matter), parties shall proceed with the litigation. However, there has been an increase in the number of disputes that have been settled through arbitration.
Mexico	The most common way of resolving disputes is litigation; however, arbitration and negotiation are gaining ground, especially for complex projects in sectors such as large-scale construction, energy, and oil. In public contracts, the parties can agree on arbitration by way of dispute resolution (as a so-called conciliation that is undertaken before the Ministry of Public Function) but termination of a contract by the Contracting Entity is excluded from the subjects eligible for arbitration (article 98 of the Public Works and Related Services Law). It is also common to see different kinds of ad hoc alternative dispute resolution in contracts executed with Public Companies [like PEMEX (Mexican Petroleum) or CFE (Federal Commission of Electricity)] in order to resolve technical or administrative disputes.
Nigeria	Arbitration as a method of dispute resolution is the most preferred mechanism for the resolution of disputes in Nigeria's construction industry. Arbitral proceedings are initiated in accordance with the Arbitration and Conciliation Act, Cap. A18, LFN, 2004, and the rules made pursuant thereto.

Country	Dispute resolution mechanisms
India	Arbitration is the most common method for resolution of construction disputes. Section 48 of the Act provides for conditions which must be satisfied for enforcement of a foreign arbitral award in India under the New York Convention. Public Policy is narrowly construed. The Indian Arbitration and Conciliation Act 1996 also provides for conciliation though it is rarely adopted. India also uses mediation and a Dispute Review Board to resolve construction disputes.
Denmark	According to the dispute resolution ladder prescribed in the AB Standards, disputes are finally resolved by arbitration before the Danish Building and Construction Arbitration Board. The arbitral procedure is governed by the Danish Arbitration Act and rules prepared by the Board.
France	The French Arbitration Act (Art. 1520 of the Civil Procedure Code) follows closely the limited grounds for Annulment of an Award under the New York Convention.
China	Most disputes are resolved out of court through settlement. For disputes that cannot be settled litigation or arbitration are used for dispute settlement.

Allocation of Risk

The types of risk and their allocation vary depending on the nature of the construction project. The main risks common to most projects are the unforeseen events or circumstances relating to:

Errors, omissions, or contradictions in the design specifications on which the contractor has relied to prepare his offer. Such risks are generally borne by the contractor, who represents and warrants to the owner that he has fully reviewed the design specifications and conducted all necessary additional studies. However, in public works contracts, the contractor may, in certain circumstances, have the right to an adjustment of the contract price due to unforeseen events or circumstances.

Ground conditions. The owner provides the contractor with any and all surveys he has undertaken. As above, the contractor reviews them and conducts all additional surveys that he considers necessary to submit his offer to the owner. As a result, the contractor bears the associated risks.

Cost escalations. These are borne by the contractor in international construction contracts, and in domestic, short-term contracts. Domestic long-term construction contracts generally provide for limited price revision mechanisms linked to common relevant market indexes.

The surrounding area, such as difficult or constrained access to the building site or damage to nearby buildings and infrastructure. The contractor typically bears these risks, although insurance may be put in place in some cases.

Clause 6: Step-in Rights

It is important that Contractors are aware in global projects of the right of a lender to step-in and take over a project when the project company is not performing in accordance with the project contract. Some national laws expressly authorize the contracting authority to take over temporarily the operation of the project, normally in case of failure to perform by the concessionaire, in particular where the contracting authority has a statutory duty to ensure the effective delivery at all times of the project concerned. In some legal systems, such a prerogative is considered to be inherent in most government contracts and may be presumed to exist even without being expressly mentioned in legislation or in the project agreement.

Step-In clauses are typically negotiated between the project lender(s) and the project owners or sponsors. A typical step in clause is shown in Box 10.3.

Box 10.3   Step-In Clause

---

1) (a) At any time on or after the occurrence of an Event of Default (as defined in the Senior Loan Facilities Agreement) and while it remains un-waived and unremedied, but prior to the Termination Date if a Termination Notice has been served, the Agent or Representative may give notice on behalf of the Banks (a Proposed Novation Notice) to the Authority that it wishes another person (which may (but need not necessarily) be a company owned by the Banks or any of them or which has an interest in the shares of the Contractor) (a) Proposed Substitute to assume by way of novation all of the rights and obligations of the Contractor under the Concession. (b) The Proposed Novation Notice shall specify a date, falling not less than 30 days nor later than 120 days after the date of the Proposed Novation Notice, and in any event not later than the Termination Date (if any), on which such assumption is to be effective (the Novation Effective Date).

2) A novation in accordance with a Proposed Novation Notice shall only be effective if the Authority consents to that novation in writing and the Proposed Substitute assumes the rights and obligations of the Contractor under the Concession and such rights and obligations of the Contractor under the Financing Agreements as the Banks may require. The Agent shall (as soon as practicable) supply the Authority with such information as the Authority reasonably requires to enable it to decide whether to grant such consent, including without limitation in relation to the Proposed Substitute: (a) its name and registered address; (b) the names of its shareholders and the share capital held by each of them; (c) the names of its directors and secretary; (d) the manner in which it is proposed to finance the Proposed Substitute and the extent to which such finance is committed.

Clause 7: Mandatory Law and Public Policy

All countries have laws that cannot be overridden in a contract and must be clearly understood as a requirement of a contract even if the choice of law is that of another country. For example, in Belgium indemnifications must be paid to distributors for termination of a contract that has not concluded or is not subject to grounds for breach of contract. Pursuant to an EU Directive termination of agency contracts before the legal termination date may also result in required indemnity payments.

Although contractual freedom is the rule, in France, some clauses may be unenforceable (deemed unwritten – réputée non écrite). This is particularly the case if the clause is contrary to French public order (ordre public) or mandatory rules (lois de police) within the meaning of Art. 3 of the French Civil Code. Examples include the decennial liability regime (obligations of contractors for up to ten years after completion of a project) and the protection of subcontractors in accordance with laws on subcontracting.

Many countries have enacted laws that apply extraterritorially regardless of the choice of law of the parties or the location of the contract. These mandatory laws include some criminal acts such as Bribery, Tax Fraud, Securities violations, and Antitrust laws.

In the United States criminal law is usually territorial. It is a matter of the law of the place where it occurs. Nevertheless, a surprising number of criminal laws apply extraterritorially outside of the United States. Application is generally a question of legislative intent, express or implied. There are two exceptions. First, the statute must come within Congress's constitutional authority to enact. Second, neither the statute nor its application may violate due process or any other constitutional prohibition (Doyle 2016). Although the crimes over which the United States has extraterritorial jurisdiction may be many, so are the obstacles to their enforcement. For both practical and diplomatic reasons, criminal investigations within another country require the acquiescence, consent, or preferably the assistance, of the authorities of the host country. The United States has mutual legal assistance treaties with several countries designed to formalize such cooperative law enforcement assistance. It has agreements for the same purpose in many other instances. Cooperation, however, may introduce new obstacles. Searches and interrogations carried out jointly with foreign officials, certainly if they involve Americans, must be conducted within the confines of the Fourth and Fifth Amendments. And the Sixth Amendment imposes limits upon the use in American criminal trials of depositions taken abroad (p. 2).

Clause 8: Force Majeure

The definition of "force majeure" will vary from project to project and in relation to the country in which the project is to be located. The definition of "force majeure" generally includes "risks beyond the reasonable control of a party, incurred not as a product or result of the negligence of the afflicted party, which have a materially adverse effect on the ability of such party to perform its obligations." Sometimes an even stricter requirement, requiring impossibility of fulfillment, is imposed. This is a very difficult fact to prove and could result in the operator bearing an unacceptable level of risk. Parties should also consider whether it is appropriate to exclude consequences which could reasonably be avoidable by either party (World Bank 2021).

The concept of force majeure is well known in U.S. construction contracting. Although some contracts allow compensation to a contractor whose progress is interrupted by such unexpected events, most contracts merely allow only an uncompensated extension of time. Force majeure generally refers to a clause that is included in contracts to remove liability for natural and unavoidable catastrophes that interrupt the expected course of events and prevent participants from fulfilling obligations typically referred to as "Acts of God."

Force majeure events are typically defined to include two common provisions:

Only unforeseen events

Events outside the control of both contract parties.

As an example, labor strikes may be excluded from this category if they arise from specific acts or omissions of the affected contractor (i.e., to be distinguished from national or regional strikes). The fact that a project becomes more difficult or costly due to rising costs of labor or materials is generally not regarded as a force majeure condition, although courts may regard sudden and extreme price escalations in this category. Many U.S. employers have argued that the COVID-19 pandemic is a force majeure event for which no resulting costs are recoverable, but contractors counter that claims for changes in work and lost efficiency are compensable even if pure delay costs are not. This area of U.S. construction law is likely to undergo substantial development as a result of economic burdens imposed by the pandemic. In some limited cases, performance may also be excused (i.e., effectively allowing its termination) on force majeure principles if the contractor can show that the specified work was frustrated by impossibility or commercial impracticability.

Force majeure events may widely be defined as extraordinary, unpreventable, and unforeseeable circumstances caused by a natural phenomenon (such as an earthquake, landslide, hurricane, drought, and others) or social and economic circumstances (such as war, blockade, import and export bans in the State interest and others) which are not controllable by the will or action of either party and due to which the parties cannot perform their contractual obligations. The mere fact that a contract has become uneconomic is not sufficient grounds for a claim for force majeure. The parties are also free to agree on the events or circumstances constituting force majeure and how such events or circumstances shall be treated.

In French law, force majeure is provided for in Art. 1218 of the Civil code and is defined as an event beyond the aggrieved party's control which could not be foreseen at the time of the conclusion of the contract and the effects of which cannot be avoided, preventing the performance of the obligation arising from the contract. The aggrieved party can claim for an extension of time, and either party can terminate in case of extended force majeure. A contract which has become uneconomic might lead to a claim under the imprévision theory, under public or private law (Art. 1195 of the Civil Code), not under force majeure.

In India, exclusion of liability for force majeure is provided for in Article 1148 of the Civil Code (Code Civil). Liability can be entirely excluded for any loss arising from events that are: unforeseeable. unavoidable, and external (that is, completely outside a party's control). The effect of a force majeure event on the contract varies according to the nature of the event. It may: suspend performance of the contract, vary the contract by altering its scope, or terminate the contract where performance has been rendered impossible.

The parties can choose to apportion the risks of force majeure between them by agreeing that the contractor will either bear all risks associated with force majeure events, or on the contrary exclude his liability entirely. In practical terms, parties should carefully draft their force majeure clauses to: Define which events amount to force majeure. Strikes, for example, may be excluded, or the parties may provide that only strikes affecting a particular sector amount to force majeure. The parties may decide that the consequences of the event will be apportioned between them according to a certain formula or criteria (e.g. the contractor will be entitled to an extension of time but will bear the disruption costs).

Additional Contractual Considerations

Emerging Contract Issues During the outbreak of COVID-19, the People's Republic of China (PRC) Supreme People's Court issued three Guiding Opinions on Several Issues Concerning the Lawful and Proper Trial of Civil Cases Involving COVID-19 ("Guiding Opinions"), which provide guidance on legal applications for various matters that are possibly influenced by COVID-19. As far as construction contracts are concerned, paragraph 7 of the second Guiding Opinion applies the rules of force majeure and change in circumstances respectively for different situations: (i) If a contractor fails to complete the construction within the agreed time limit due to the epidemic or the epidemic prevention and control measures, the People's Court will not uphold the employer's claim against the contractor for breaches. If the contractor requests to extend time for completion, the People's Court shall uphold such request considering the degree of impact of the epidemic situation or the epidemic prevention and control measures on the performance of the contract. (ii) Where the epidemic situation or epidemic prevention and control measures cause a sharp rise in the costs of labor and construction materials, etc., or cause contractor's losses such as labor costs and equipment rental fees, etc., and the continued performance of the contract is obviously unfair to the contractor, if the contractor requests an adjustment of the contract price, the People's Court shall, in consideration of the actual circumstances of the case, make adjustments in accordance with the principle of fairness.

Ambiguity   Ambiguity is resolved through principles similar to those that apply to contracts generally. While the subject is too broad to be properly summarized here, extrinsic evidence can be considered as an aid to interpretation, and various presumptions and canons of construction may be employed by the courts to discern the meaning of the contractual clauses in question. In construction contracts in particular, a well-drafted contract will often include a hierarchy or precedence clause, providing for the order of priority in interpretation between documents. The aim of such clauses is to provide a framework for addressing inconsistencies and ambiguities in the contract, and would generally be given effect to, where necessary, to resolve a clear and irreconcilable discrepancy (although courts would generally be slow to resort to such a clause).

In interpreting ambiguous terms in a contract in Taiwan, Article 98 of the Civil Code prioritizes the parties' real intent over the literal meaning of the terms. According to the prevailing court view, the purpose and context of the contract, and earlier drafts of the terms, shall be considered to understand the parties' real intent.

Collaborative Approaches to Contracting

Megaproject sponsors have gradually moved away from traditional, adversarial forms of contracting and focused more on collaborative approaches that encourage mutual cooperation (Brady and Davies 2014; Davis and Love 2011; Galvin et al. 2021). The result has been an increasing reliance on various forms of non-traditional contracts to deliver megaprojects, including alliances, relational contracts, or public-private partnerships (Bygballe et al. 2010; Clifton and Duffield 2006) – as these contractual forms enable participants in the contract to "share the gain and share the pain" (Lloyd-Walker et al. 2014).

Galvin et al. (2021) in their recent studies on PPPs found that establishing the right governance, trust, and culture were critical. On its own, each of these is necessary but not sufficient to support collaboration becoming the default behavior within the alliance.

Previous studies had identified the contribution of these three enablers to the successful establishment of collaborative behaviors in alliances (e.g., Clegg et al. 2002; Kadefors 2004; Pinto et al. 2009; Xue et al. 2016).

Alliance contracts have been introduced in megaprojects to improve the alignment of objectives, risk and reward between client and contractor. However, the relational norms of alliances are not sufficient on their own to eliminate opportunistic behaviors (Galvin et al. 2021).

Joint Ventures

In projects involving international competitive bidding, particularly where foreign lenders are involved [for example, the Asian Development Bank (ADB) and the Japan International Co-operation Agency (JICA)], the preferred structures are Joint Ventures (JVs) or Special Purpose Vehicles (SPVs). The form of a JV typically used in India and many other countries is an unincorporated JV, referred to as an "association of persons." An "association of persons" is recognized by the Income Tax Act 1961. It enables various parties to combine their qualification requirements and skills without actually having to start or incorporate an SPV. However, for build-operate-transfer (BOT) projects, the preferred structures involve setting up of an SPV. In some projects, incorporation of an SPV is required by the project owner. Typically, in these projects, the SPV does not carry out all the work and subcontracts some of the work (either to their own principals or to other subcontractors). In many large projects (for example, airport projects), the SPVs are set up by the developers with the government entity and in turn subcontract the entire construction work to an EPC contractor (who further subcontracts specialized work). Most large projects involve international competitive bidding.

It has become increasingly prevalent for contracting companies to form joint venture partnerships in the UAE, especially in respect of the so-called "mega projects" which can have a contract value above one billion. Given that this often involves major global contractors, these joint ventures can be unincorporated, meaning that any liability is joint and several.

Another example of collaborative contracting was used on London's Crossrail Project. Many of the contractors have formed unincorporated joint ventures for the Crossrail contracts of two, three, or even four parties. This obviously reduces the risk for individual contractors and gives a greater pool of expertise and experience to draw from. From Crossrail Limited's perspective as client, joint ventures can take longer to "bed down" and establish common processes and systems. Some joint ventures can also struggle with internal governance and decision-making (Morrice and Hands 2017).

Collaborative contracting is relatively common in Portugal, especially in complex projects or in projects that involve a combination of different construction and/or engineering capabilities. The most common form of association is the consortium, in which two or more contractors agree to perform one or more contracts together without incorporating a new legal person for that purpose.

There are different forms of consortium but usually employers require that contractors form external consortia in which all members assume joint and several liability towards the employer for completion of the works. The complementary group of companies ("ACE") is another form of association, usually for major projects requiring a longer performing period and in which the contractors incorporate a new legal person (the ACE) to assume the role of main contractor. In public concessions, contractors are usually required to incorporate corporate special purpose vehicles ("SPVs") to act as concessionaires.

Collaborative Contracting

Collaborative contracting often used in public private partnerships is prevalent in Nigeria. It is an agreement between two or more establishments that are desirous of coming together to undertake construction work. The forms of collaborative contracts that are commonly used are: Joint Venture Agreements; Partnership Agreements; and Merger Agreements.

It is important to know whether a country or a municipality has a statutory framework for collaborative arrangements. For example, in India, there is no specific legal or statutory framework for PPPs at the central level. However, a few states (e.g. Punjab) have State Acts that govern construction contracts and have their own statutory tribunal. There is no typical procurement/tender process in PPP transactions in India. All government tendering is governed by the guidelines laid down by the Central Vigilance Commission to ensure transparency and avoid corrupt practices. Tender procedures are also guided by the lending agencies such as the World Bank, the Asian Development Bank (ADB), European Investment Bank, and the Japan International Co-operation Agency (JICA), among others. Generally, there are no standard forms for PPP projects. However, various authorities do have approved formats, which are regularly amended. For example, the National Highways Authority of India (NHAI) has its own format which is regularly updated and amended.

Some of the significant PPP transactions that have recently been completed in the United States include the sale and refinancing of the Chicago Skyway in Illinois, the Pennsylvania Rapid Bridge Replacement Project, the Purple Line Light Rail Project in Maryland, the LaGuardia Central Terminal Building Replacement Project in New York, the Denver Airport Great Hall, the Ohio State University Energy Project, the Virginia Outside the Beltway I-66 Project, the Colorado I-70 East Project, the LAX Consolidated Rent-A-Car, the LAX Automated People Mover and the Newark Liberty International Airport Consolidated Rent-A-Car.

Emerging Issues in Global Development Laws and Contracts

Despite the recent market volatility arising from the COVID-19 pandemic, there remains a pressing need throughout the world for large-scale investment in infrastructure across a broad spectrum of industries (in particular, in emerging markets such as Africa). Large-scale project finance typically focuses on "green-field" projects in sectors ranging from power generation (conventional, nuclear, and renewables) to transmission, oil and gas, petrochemicals, infrastructure, mining, and telecoms. With the heightened focus on climate change, there has been an increased focus on projects which will facilitate the "energy transition" away from our dependence on hydrocarbons. Global economic growth and demand for energy and commodities is a major driver for capital investment in these sectors and notwithstanding recent market volatility, the economies of fast-growing countries such as China, India and Brazil have underpinned the upward trend in energy and commodity prices.

Some of the largest projects in the world are currently being developed in emerging markets: projects involving capital expenditures of $10–50B are moving forward in countries such as Saudi Arabia, the United Arab Emirates and Malaysia (UNCTAD 2021). The European Union (EU) alone decided to make a total of €1.8 trillion available for Europe's recovery in the wake of the health crisis. These funds are also designated to support the fight against climate change and the forging of a more digitally oriented Europe. It is therefore to be expected that there will be a great deal of activity around renewable energy and the digital transformation. In this context, it can also be expected that sustainable finance, though currently still a new product, will gain more importance.

The changes in emerging market law will also have an impact on international megaproject contracting. All contracts must be reviewed to ensure they comply with these new laws.

Bilateral Investment Treaties

An important consideration for a private investor in an emerging market or in a less developed country is the existence of the Bilateral Investment Treaty (BIT). These Treaties are entered into by countries seeking protection for investments made by investors from their own state. For example, State A (the host state) agrees to provide certain protections to investors coming from State B (the home state) and vice versa. If the investor considers that these protections have been violated, investment treaties also grant the investor permission to bring an arbitral claim directly against the host state. As a result, the agreement is entered into by the home and host state (collectively, the treaty parties) but the protections are created for the benefit of, and are typically enforced by, an investor from one state against the other state.

Investment treaties expressly protect investors against certain unilateral actions by host states, such as expropriation without compensation, nationalization of an industry, and discriminatory treatment, and permit investor-state arbitration to enforce these obligations. For home states, an increase in efficient foreign investments will not just benefit their investing nationals; it is also likely to benefit their non-investing nationals because the state's development is likely to be enhanced through increased tax revenues. For host states, promoting such investments means increasing development through the creation of new jobs, the development of new infrastructure, and the enhancement of tax revenues. The U.S. negotiates investment treaties on the basis of a model text as do most countries. The Bilateral Investment Treaty Program administered by the U.S. Department of State and the U.S. Trade Representative provides investors with six core benefits:

- U.S. BITs require that investors and their "covered investments" (that is, investments of a national or company of one BIT party in the territory of the other party) be treated as favorably as the host party treats its own investors and their investments or investors and investments from any third country. The BIT generally affords the better of national treatment or most-favored-nation treatment for the full life cycle of investment – from establishment or acquisition, through management, operation, and expansion, to disposition.
- BITs establish clear limits on the expropriation of investments and provide for payment of prompt, adequate, and effective compensation when expropriation takes place.
- BITs provide for the transferability of investment-related funds into and out of a host country without delay and using a market rate of exchange.
- BITs restrict the imposition of performance requirements, such as local content targets or export quotas, as a condition for the establishment, acquisition, expansion, management, conduct, or operation of an investment.
- BITs give covered investors the right to engage the top managerial personnel of their choice, regardless of nationality.
- BITs give investors from each party the right to submit an investment dispute with the government of the other party to international arbitration. There is no requirement to use that country's domestic courts.

The United States presently has 47 Bilateral Investment Treaties, though two of those treaties will terminate after the covered investments are completed. As of 2022 more than 2228 investment treaties are in place around the world. Some of the countries with the largest number of BITS in force include China at 110, the United Kingdom at a 98, Turkey at 100, Germany at 125, and India with 86 (UNCTAD 2022). All foreign investors should be aware of their rights under these Bilateral Investment Treaties, and to the extent a Treaty is not available the contract or Concession Agreement between the foreign investor and the host country should include at a minimum full compensation for expropriation or nationalization and protection against discrimination. A complete listing of Bilateral Investment Treaties for all countries can be found at the Investment HUB at the

United Nations Commission on Trade and Development (UNCTAD 2022) as well as at the U.S. Department of State for US. Treaties.

Emerging Markets Impacting Contracting in Megaprojects

As presented in previous chapters emerging markets and the developing world constituting about 60% of the GDP and 80% of the population creates opportunities for development projects that focus on energy, transport, agriculture productivity, and social and environmental initiatives. This opportunity also creates numerous legal issues not only for the financing of these projects but also for the implementation of these projects in difficult terrains with serious economic and finance risk. These risks range from environmental, social and governance issues (ESG) to supply chain and procurement risk to offtake purchasing agreements, investment treaties, corporate responsibility and liability and green procurement in contracts among many other issues.

Oil and gas companies struggle to find new growth opportunities in mature basins. Remote and undeveloped regions of the world offer the potential for significant value creation with their abundant and often under-exploited oil and gas resources. However, along with these opportunities and potential rewards, emerging markets present substantial obstacles and risks, as these regions can be plagued with numerous problems, including unstable political systems, weak rule of law, corruption, lack of transparency, and human rights issues. Companies are exposed to significant legal and reputational risks if they do not effectively identify and manage the challenges and risks in these regions. In recent years, directors and officers of companies operating in these regions have been subject to increased scrutiny.

The questions below are based on a compilation of issues that have arisen in the developing world through the author's professional experience and the experience of the global financial institutions as reflected in the numerous documents and case studies that have been issued by these institutions over many decades. Checklists should never be seen as a comprehensive list of all provisions that should be in a contract. Some contracts may have laws that impact the project and must be included in the project contract, while other contracts may require less detail but more a broad summary of the applicable law.

Checklist for contract development in emerging markets:

1) How transparent is the host government and the political system where the project will be located?
2) Are Special Purpose Vehicles permitted as a form of corporate organization or are other options available?
3) What is the ability of the Public to question and protest project plans if they are averse to the building of the project?

4) What are the cultural issues concerning large scale megaproject development?
   a) Are there important social requirements that must be considered in the project framework?
   b) Are there particular issues regarding green development, climate change and other environmental concerns?
   c) Will there be community involvement in development decisions?
   d) What are the expectations of the local community?
5) How lengthy is the project regulatory approval process?
   a) To whom is the risk of permitting allocated?
   b) Will there be one stop shopping for all permits?
   c) Will the government assist with the obtaining of permits?
6) What are the Labor Customs and Practice?
   a) The viability and roles of the trade unions
   b) The ability to employ managers and workers outside the host country
   c) The application of international labor standards
   d) The right to strike
7) What are the procurement requirements?
   a) How are projects advertised and noticed to potential bidders?
   b) Are home country bidders preferred over foreign bidders?
   c) What are the general project criteria for selecting the project designers and contractors?
   d) Is the project low bid or subject to weighting criteria?
   e) How are projects scored or evaluated?
   f) Will the considerations for selection be shared with the public?
   g) Can suppliers come from other countries? Are there any exceptions?
8) What are the Key Environmental, Social and Governance Considerations?
   a) What are the sustainability requirements required by the countries laws or regulations and who oversees that these requirements are met?
   b) What are the rights of the members of the local communities impacted by the project?
   c) How are climate change and environmental impacts assessed?
   d) What is the availability of skilled labor and how will shortfalls be remedied?
   e) What are the standards for health, safety, and security under the host country laws?
   f) What are the implications if sponsors fail to meet ESG requirements?
9) Is there a Bilateral Investment Treaty with the host country?
   a) Has the Host Country implemented a bilateral investment treaty?
   b) What are the protections under this Treaty? Are investors protected from expropriation, nationalism, discrimination, compensation for violations?

10) Does the country have a Public-Private Partnership Law or Framework?
   a) What are the minimum and maximum investment provisions?
   b) How are these arrangements structured legally?
   c) What are the potential liabilities for the private partners?
11) Does the country have security rights for creditors?
   a) If so, what are the terms and formalities for security creation and are these liens enforced?
   b) In what assets can security be obtained?
   c) Are there any restrictions on enforcement of security? Can a blanket lien be obtained over all assets?
12) Employment Laws
   a) Are there any restrictions on foreign workers, technicians, engineers, or executives being employed by a project company?
   b) Will workers from other countries be able to retain their home country rights including protections from discrimination or termination?
13) How is Force Majeure Applied?
   a) Does the country have a legal definition for force majeure?
   b) Is it stricter or more flexible than the stakeholder's home countries laws?
   c) Can a force majeure provision be agreed upon by the parties that may create lesser or greater liability than the home country law?
14) Bribery and Corruption
   a) Are there laws that protect against corruption, fraud, and other related crimes?
   b) What are the penalties for bribery and corruption?
   c) Are the laws enforced?
   d) Is the host country a member of any treaty such as the Budapest Convention?
15) Arbitration
   a) Is arbitration of disputes permissible and may the parties choose the arbitral forum where disputes will be resolved. May the parties select their own arbitrators?
   b) Are there any matters that are not arbitrable?

Summary

The goal of this chapter was to provide an overview of the major risks and issues in international contracting as well as the related laws. As discussed in earlier chapters the financial documents required for project financing are extensive and complex and involve the participation of multiple lenders, equity sponsors, the borrowing entity, and the obligations of the Concessionaire. In Chapter 10 the focus moved to the engineers, developers, and contractors in these massive projects and the challenges and opportunities available to these participants to prevent potential liability and ensure successful outcomes by reducing uncertainty, mitigating risk, avoiding ambiguity, and preventing conflict. The types and benefits of collaborative efforts were explored to encourage movement away from the more rigid, bureaucratic contracting methods of the past.

Lessons Learned and Best Practices

1) Contracts represent an important governance structure for megaprojects
2) Public procurement plays a key role in the Europe 2020 Strategy
3) Guaranteed maximum price contracts are a useful structure for megaproject contracts
4) The World Bank's Environmental and Social Framework (ESF) applies to almost all Multilateral Bank contracts
5) There are certain duties that are enforceable even though they are not part of the written contract
6) In the United States and most countries, every contract party owes an implied duty of good faith
7) One of the most complex and contentious provisions in international contracts for infrastructure is drafting the allocation of risk for subsurface conditions
8) Choice of law and forum can be selected by the parties, but both must be clearly stated in the contract so there is no ambiguity
9) In global projects the right of a lender to step-in and take over the project when a contractor defaults is a commonly recognized provision
10) As shown by the choice of law provision in the English Chunnel, courts and arbitrators will tend to uphold the chosen law of the parties even if it is seriously detrimental to one of the parties
11) Dispute Review Boards have been successfully used in many megaprojects
12) Collaborative contracting is a mechanism used to reduce disputes and align parties' interests and goals and is a move away from the hierarchical contracts of the past

---

Discussion Questions

1 What are the different types of contracts that can be used in a global infrastructure project?

2 If you are a contractor or consulting firm what provisions will be most important to you in protecting your company from the damage that may occur from a force majeure event?

3 What are the employment and labor provisions you will be most concerned about in a global infrastructure contract?

4 What are the advantages of arbitrating a dispute between parties rather than litigating in a foreign court?

5 Why is choice of law important in a global contract?

6 What are some important internationally recognized standards and how will they be beneficial in an international infrastructure project contract?

7 What is the role of a concession agreement and who are the typical parties to the agreement?

8 What is a mandatory law and how does it impact a written contract in a global transaction?

9 What is the purpose and benefit of a bilateral investment treaty?

10 What is the purpose of the U.S. Foreign Corrupt Practices Act and who does it benefit in a global infrastructure contract?

References

American Association of State Highway and Transportation Officials (AASHTO) (2022, Dec. 20). Jury Awards WSDOT $52.7M for tunnel construction delays. *AASHTO Journal.* https://aashtojournal.org/2019/12/20/jury-awards-wsdot-52-7m-for-tunnel-construction-delays/.

Born, G. and Kalelioglu, C. (2021). Choice-of-law agreements in international contracts. *Georgia Journal of International and Comparative Law* 50: 44–118.

Brady, T. and Davies, A. (2014). Managing structural and dynamic complexity: A tale of two projects. *Project Management Journal* 45 (4): 21–38.

Brunet, M. and Aubry, M. (2016). The three dimensions of a governance framework for major public projects. *International Journal of Project Management* 34 (8): 1596–1607.

Bygballe, L.E., Jahre, M., and Swärd, A. (2010). Partnering relationships in construction: A literature review. *Journal of Purchasing and Supply Management* 16 (4): 239–253.

Clegg, S.R., Pitsis, T.S., Rura-Polley, T., and Marosszeky, M. (2002). Governmentality matters: Designing an alliance culture of inter-organizational collaboration for managing projects. *Organization Studies* 23 (3): 317–337.

Clifton, C. and Duffield, C.F. (2006). Improved PFI/PPP service outcomes through the integration of alliance principles. *International Journal of Project Management* 24 (7): 573–586.

Davis, P. and Love, P. (2011). Alliance contracting: Adding value through relationship development. *Engineering, Construction and Architectural Management* 18 (5): 444–461.

De Marco, A. and Karzouna, A. (2018). Assessing the benefits of the integrated project delivery method: A survey of expert opinions. *Procedia Computer Science* 138: 823–828. https://doi.org/10.1016/j.procs.2018.10.107.

Denmark Contract Law (2016). AB 18 and ABT 18.

Dettman, K., Harty, M.J., and Lewin, J. (2010). Resolving megaproject claims: Lessons from Boston's 'Big Dig. *The Construction Lawyer* 30 (2): 1–13. American Bar Association, Washington, D.C.

Doyle, C. (2016). *Extraterritorial Application of American Criminal Law: An Abbreviated Sketch* Updated. Congressional Research Service (CRS) October 31. https://crsreports.congress.gov RS22497.

Einstein, D. (2012). *Laying the Groundwork: How Bechtel Geotechnical Experts Help Projects Succeed by Overcoming Everything from Boston Blue Clay to Athenian Schist*. San Francisco: Bechtel Corporation.

Eurotunnel Safeguard Procedures (2006). *Eurotunnel obtains protection from Paris Commercial Court*. Paris, France: GETLINK, formerly Groupe Eurotunnel https://press.getlinkgroup.com/news/eurotunnel-obtains-protection-from-paris-commercial-court-fee4-0791e.html.

Galvin, P., Tywoniak, S., and Sutherland, J. (2021). Collaboration and opportunism in megaproject alliance contracts: The interplay between governance, trust. and culture. *International Journal of Project Management* 39 (4): 394–405.

Greiman, V.A. (2013). *Megaproject Management: Lessons on Risk and Project Management from the Big Dig*. Hoboken, NJ: John Wiley & Sons.

Harmon, K.M.J. (2009). Case study as to the effectiveness of dispute review boards on the central artery/tunnel project. *Journal of Legal Affairs and Dispute Resolution in Engineering and Construction* 1 (1): 18–31. American Society of Civil Engineers (ASCE), Reston, VA.

Hatem, D. and Cordum, M. (ed.) (2010). *Megaprojects: Challenges and Recommended Practices*. Washington, D.C: American Council of Engineering Companies (ACEC).

Kadefors, A. (2004). Trust in project relationships: Inside the black box. *International Journal of Project Management* 22 (3): 175–182.

Lloyd-Walker, B.M., Mills, A.J., and Walker, D.H.T. (2014). Enabling construction innovation: The role of a no-blame culture as a collaboration behavioral driver in project alliances. *Construction Management and Economics* 32 (3): 229–245.

Lorenzo, J.A.P. (2018). "Development" versus "sustainable development"? (re-) constructing the international bank for sustainable development, 51. *Vanderbilt Journal of Transnational Law* 399: 399–476.

Ma, H.Y., Zeng, S.X., and Lin, H. (2017). The societal governance of megaproject social responsibility. *European Journal of Operations Research* 35: 1365–1377.

Malik, A., Parks, B., Russell, B. et al. (2021). *Banking on the Belt and Road: Insights from a New Global Dataset of 13,427 Chinese Development Projects.* Williamsburg, VA: AidData at William & Mary.

Morrice, D. and Hands, M. (2017). Crossrail project: Commercial aspects of works contracts for London's Elizabeth line. *Civil Engineering* 170 (6): 42–47. Special issue on Crossrail Project: Programme managing the Elizabeth line, London, UK. ISSN 0965-089X | E-ISSN 1751-7672.

Park, W. (2015). *Lord Mustill and the Channel Tunnel Case.* Boston, MA: Scholarly Commons at Boston University.

Pinto, J.K., Slevin, D.P., and English, B. (2009). Trust in projects: An empirical assessment of owner/contractor relationships. *International Journal of Project Management* 27 (6): 638–648.

Pribrytkova, E. (2020). Global obligations for sustainable development: Harmonizing the 2030 Agenda for Sustainable Development and International Human Right Law, 41. *University of Pennsylvania Journal of International Law* 1030: 1031–1136.

The American Institute of Architects (AIA) (n.d.). The A-201 General Conditions and Other Forms. https://www.aiacontracts.org/contract-documents/25131-general-conditions-of-the-contract-for-construction.

The World Bank (2018). *Environmental and Social Framework (ESG).* Washington, D.C: The World Bank Group.

The World Bank (2021). *Force Majeure Clauses Checklist and Sample Wording.* The Public-Private Partnership (PPP) Legal Resource Center (PPPLRC), The World Bank.

U.S. Department of Labor (DOL) (2020). *Bureau of Labor Statistics.* https://www.bls.gov/news.release/archives/osh_11032021.pdf.

UK Parliament (2020, Jan. 20). *Grenfell Tower Fire: Background. Research Briefing.* UK Parliament: House of Commons.

United Nations Commission on Trade and Development (UNCTAD) (2022). *Investment Policy Hub.* https://investmentpolicy.unctad.org/international-investment-agreements.

United Nations Conference on Trade and Development (UNCTAD) (2021). *World Development Report.* Geneva, Switzerland: UNCTAD.

United States v. Spearin. 1918. 248 U.S. 132.

Xue, J., Yuan, H., and Shi, B. (2016). Impact of contextual variables on effectiveness of partnership governance mechanisms in megaprojects: Case of Guanxi. *Journal of Management in Engineering* 33 (1): 04016034.

Xue, F., Chen, G., Huang, S., and Xie, H. (2022). Design of social responsibility incentive contracts for stakeholders of megaprojects under information asymmetry. *Sustainability* 14 (3): 1465. https://doi.org/10.3390/su14031465.

11

Megaproject Innovation and Resilience

Never tell people how to do things. Tell them what to do and they will surprise you with their ingenuity.
— George Patton (1885–1945), General of the U.S. Army

Introduction

In this chapter, we explore megaproject innovation to understand better not only what are the innovations that megaprojects have created but also more importantly how are innovations in megaprojects initiated, supported, and developed? What inspires innovation and how does it succeed? The Danish Architect Jørn Utzon, famously found inspiration in monumental Mayan and Aztec architecture for his design of the Sydney Opera House, a UNESCO World Heritage site described as a masterpiece of the twentieth century.

While early societies made extensive use of stone, bronze, and iron, it was steel that fueled the Industrial Revolution and built modern cities. Evidence of steel tools dates back 4000 years to 1800 BCE, but the alloy was not mass-produced until Henry Bessemer developed an effective way to use oxygen to reduce the carbon content in iron in the 1850s. Steel then exploded into one of the biggest industries on the planet and was used in the creation of everything from building bridges and railroads to skyscrapers and engines. It proved particularly influential in North America, where massive iron ore deposits helped the United States become one of the world's biggest economies.

Throughout time, innovation in megaprojects has come from great ideas created by many of history's civil engineers. Without civil engineering, our tunnels would not be constructed, our bridges would not be designed, and our railways

Global Megaprojects: Lessons, Case Studies, and Expert Advice on International Megaproject Management, First Edition. Virginia A. Greiman.

would not have existed. They can be found in the aerospace industry, designing jetliners and space stations, the ship building industry, designing submarines and tankers, and the power industry, fulfilling the great energy needs of the world.

Every infrastructure throughout civilization is the result of a civil engineering mastermind. A few of the greatest civil engineers in history include:

Archimedes of Syracuse (287–212 BCE)
Archimedes from the ancient Greek City of Syracuse invented the compound pulley which turned out to be one of the greatest inventions in history. He also invented the Archimedes' Screw (a device for raising water) that is still used in crop irrigation and sewage treatment plants today.

John Smeaton (1724–1792)
John Smeaton was a British civil engineer responsible for the design of bridges, canals, harbors and lighthouses during the Industrial Revolution. Smeaton was the first self-proclaimed "civil engineer" and is often regarded as the "father of civil engineering."

Benjamin Wright (1770–1842)
Benjamin Wright was an American civil engineer who was chief engineer of the Erie, the Chesapeake and the Ohio Canals. In 1969, the American Society of Civil Engineers declared him the "Father of American Civil Engineering."

George Stephenson (1781–1848)
George Stephenson revolutionized transportation and urban infrastructure by creating the world's first public inter-city railway line that used steam locomotives. This British Engineer, often referred to as "The Father of Railways," is also credited with devising the historic measurement of the rail gauge at four feet eight-and-a-half inches, which became the standard railway gauge measurement worldwide.

Squire Whipple (1804–1888)
Squire Whipple's work on iron bridges contributed toward the development of the railroad industry in the United States. Although his designs became known around the world, he's perhaps best known for his work on the Erie Canal.

Henry Bessemer (1813–1898)
An Englishman, Sir Henry Bessemer's steel-making process would become the most important technique for making steel in the nineteenth century for almost one hundred years from 1856 to 1950. He is credited for inventing the first mass-producing steel process that led to the development of the modern skyscraper. He was knighted in 1879 for his contribution to the scientific community.

Henry Larcom Abbot (1831–1922)
Henry was a military engineer and career officer in the United States Army. He served in the Union Army during the American Civil War and was appointed brevet brigadier general of volunteers for his contributions in engineering and artillery. He conducted several scientific studies of the Mississippi River and served as a consultant for the locks on the Panama Canal. He was elected a Fellow of the American Academy of Arts and Sciences in 1863.

Duff A. Abrams (1880–1965)
A professor with the Lewis Institute, he studied the component materials of concrete in the early twentieth century. He developed the basic methods for testing concrete characteristics still in use today.

Othmar Hermann Ammann (1879–1965)
Othmar was a Swiss-American structural engineer whose bridge designs include the George Washington Bridge, Verrazzano-Narrows Bridge, and Bayonne Bridge. He also directed the planning and construction of the Lincoln Tunnel. Ammann designed more than half of the eleven bridges that connect New York City to the rest of the United States. His talent and ingenuity helped him create the two longest suspension bridges of his time.

Arthur Casagrande (1902–1981)
A former Harvard professor, Arthur Casagrande developed the design principles used in the construction of earth and rock dams throughout the world. While working as a consultant to the U.S. Army Corps of Engineers, he contributed to theoretical work in the field of soil behavior and the construction of many dams around the world.

Where Does Innovation Come From?

The decision to start Microsoft, for example, wasn't based on a momentous flash of insight. It was based on incremental developments in a nascent personal computing industry, the fact that Paul Allen and I had access to mainframe computers at the high school we attended, and our hunch about what people could do with computers in the future.

— *Bill Gates*

The popular perception of innovation is that it comes from big breakthroughs. However, the reality is that most discoveries emanate from discoveries in the "near field" of the known. Steven Johnson (2010) in his book, *Where Good Ideas Come From* compares one-off ideas with innovative ideas that incubate and develop over a longer period of time. The author details the hunches of Darwin and Newton revealed in their personal writings, which existed long before their ideas were published formally. The book reveals that during the early days of the enlightenment in Europe, scholars maintained a "commonplace book," where they noted down various ideas which came to them and also took notes from their readings. According to Johnson, "most hunches that turn into important innovations unfold over much longer time frames. They start with a vague, hard-to-describe sense that there is an interesting solution to a problem that has not yet been proposed, and they linger in the shadows of the mind, sometimes for decades, assembling new connections and gaining strength. . . . But that long incubation period is also their strength because true insights require you to think something that no one has thought before in quite the same way" (Johnson 2010, p. 75).

Scientific discovery, by and large, illustrates how a "hunch" (or more formally conceived as a "hypothesis") guides the pursuit of understanding and is tested over time. These hunches either gather support or are discarded. In this way, there is a certain trial and error to our thinking, where the strongest ideas survive, both in our own minds and in conversation with others (see Weick 1989). Johnson's theory of the adjacent possible can also be used to explain the theory of "multiples," which is the event where more than one researcher discovers a fact at about the same time, while working independently of each other. This fairly common phenomenon occurs because the accumulated knowledgebase obtained by other researchers reaches a point where a move into the adjacent possible by anyone will yield a new idea.

Earlier in the nineteenth century, a British inventor, Charles Babbage, tinkered with two ideas – a "Difference Engine" to calculate polynomial functions, and an "Analytical Engine," which would have been the world's first programmable computer (CBI n.d.) Neither machine was built at the time, but many of the ideas underlying the "Difference Engine" took hold quickly, leading to the mass production of mechanical calculators. Although the Analytical Engine was a brilliant idea that included many of the key concepts in today's computers, it was, Johnson suggests, beyond the adjacent possible of the day. Babbage's design would have required a huge number of mechanical gears and switches, which probably would have made the machine too slow to operate effectively. It took another 100 years for researchers to independently rediscover Babbage's ideas and apply them using newer technology – vacuum tubes and eventually integrated circuits (Gates 2011).

The power of networks as explained by Johnson (2010) comes not only from their numbers, but also from their interconnections with each other. He describes innovative systems as existing at the "edge of chaos," which separates the orderly state (akin to solids) from the chaotic state (akin to gases). The liquid network thus facilitates existence at the edge of chaos where exploring of the adjacent possible is the easiest (Johnson 2010).

Notably, innovation can also come from failure. For example, as Cantarelli and Genovese (2021) found in their systematic literature review of innovation that frequent failures occur in megaprojects when transferring from implementation to operations. This led to innovations in the organizational design of Crossrail (Davies et al. 2014).

What Is an Innovation Megaproject?

> The value of an idea lies in the using of it.
>
> *Thomas Edison (1847–1931), Inventor*

Innovation has been defined in the context of megaprojects as a new product, process, or service that has a step change and creates value including financial value, environmental value, societal value, and job creation (Sergeeva and Zanello 2018). Innovation is generally considered as "the successful commercial exploitation of new ideas." It includes the scientific, technological, organizational, financial, and business activities leading to the introduction of a new (or improved) product or service' (Dodgson et al. 2008, p. 3).

In the first systematic literature review on innovation in megaprojects, Cantarelli and Genovese (2021) observed that new ideas in megaprojects may come from both the permanent and temporary organization of the megaproject itself, or external resources. Technological innovation rarely occurs through the activities of single firms. It is more commonly a result of inputs from a variety of organizations, working together as customers and suppliers, in various forms of communities and networks, and in formal technological collaborations, alliances, and networks (Dodgson et al. 2008, pp. 3–4). Centrally important is the management of research and development (R & D) which provides an organized source of idea generation and improves the ability of firms to absorb useful information from ·outside (p. 4).

Innovation is part of the nature of megaprojects due to their growth in size, cost, and impact with iconic designs and advanced technologies (Flyvbjerg 2014). Technology innovation takes time and a return on investment may take years and it may not benefit the specific project where the technology was developed but could benefit projects in the future. Appropriating value involves intellectual property rights (IPRs), licensing, the creation of standards, speed, and secrecy, and the ownership of "complementary assets" (Dodgson et al. 2008). To the extent new technology or processes were developed on a megaproject, it may have involved ideas generated by multiple project engineers, architects, contractors and may have been financed by public funds. Without a clear technology development policy and legal agreements, it may take years to determine the rightful owner(s) of the technology.

It may be new to a megaproject but not necessarily new to the world. The innovation potential for megaprojects is subjected to a fundamental tension: on the one hand, they offer a one-off opportunity to invest in cutting-edge innovative solutions; on the other hand, innovations are risky and, in some cases, could lead to failure and budget and time loss (Davies et al. 2009; Gil et al. 2012). In these settings, championing innovation is increasingly recognized as important for successful delivery of megaprojects that impacts on the policymaking, economy, and society as a whole (Sergeeva and Zanello 2018).

Project management has a long history of innovation including technical, product, and process innovation (Severo et al. 2019). Innovation in project management has also been studied across industries including construction, energy, medical, and agriculture. Engineering is a key driver of innovation. For example, at its core, innovation engineering is the result of using the approaches, processes, behaviors, and mindsets of entrepreneurs/innovators within the context of engineering projects.

Although there have been many innovations developed over the years through megaprojects, research, and studies on innovation in megaprojects is still in its early stages. According to some scholars, innovation focuses more on innovation as a program within the megaproject and less on the driver of innovation and how innovation is fostered within a megaproject (Brockmann et al. 2016; Davies et al. 2014; Sergeeva and Zanello 2018).

> A lack of innovation may also contribute to project failure. Davies and Gann (2017) argue that one of the reasons megaprojects fail is due to the inability of their delivery model to innovate and adapt to changing and unexpected circumstances.

The complexity of megaprojects drives the need for innovation, while at the same time making it important to integrate these innovations throughout the projects long-life. Chen et al. (2018) described the complexity associated with several technological innovations in the case of the Hong Kong–Zhuhai–Macau bridge in China, which was nonlinear, and due to a lack of integration between inventors (many teams and industries), one innovation change had a major impact on others. Technological innovations are critical to megaprojects; however, a lack of regulatory synchronization between regions in developing countries hinders the integration of such innovations (Damayanti et al. 2021). Ozorhon (2013) studied four construction projects and found that environmental sustainability is the main driver of innovation, and that collaborative working among team members and strong commitment can enable innovation, but that reluctance, inexperience, and cost are the main barriers to innovation.

China's Hong Kong-Zhuhai-Macao Bridge complex. *Source:* Peri/World Highways.

Megaproject Innovation Programs

As shown in Table 11.1, formal innovation programs have been established in some of the world's largest megaprojects. These programs have assisted other megaprojects in creating a structure for innovation as well as incentivizing innovation among the many sponsors, designers and contractors involved in these programs. Importantly, every megaproject should place a high priority on formalizing innovation early in the project life. Recognizing that megaprojects throughout their long life create many innovations some of which are never recognized, it is necessary to develop policies that will protect innovation and provide attribution and ownership to the real inventor.

Boston's Central Artery Tunnel Project (Big Dig) Innovations and Advancement Program

Since the Big Dig, in the United States innovation knowledge transfer has been incorporated into every large-scale project that receives federal funding and is also required under many state statutes and regulations (Allen and Barnes 2004). The basis of the program was used to develop the existing Federal Highway Administration Center for Accelerating Innovation (CAI). The CAI is one of four

Table 11.1 Crossrail and the central artery tunnel project innovation programs.

Project: Crossrail Innovation Program (CRL): Innovate 18 Programme: Joint endeavor with Imperial College	*Project:* Central Artery Tunnel Project (CA/T): Innovation and Advancement Program: Joint endeavor with U.S. Department of Transportation, Federal Highway Administration
Approach: CRL created a procurement approach called "Optimized Contractor Involvement" (OCI) where each individual contractor, joint venture, and supplier can bring new ideas and practices to the project, while sharing the risk and reward.	*Approach:* The CA/T Project created an Innovation and Advancement Program in collaboration with the Federal Highway Administration that focused on technological, social and process innovation
Lessons learned: Although the Innovate 18 governance process allowed for top level support from various directorates through the Crossrail Innovation Forum, this impact was at times limited as the Crossrail Program was too far along to introduce independent directorate ownership of innovation activities. If innovation was built into a major infrastructure project from the beginning, and relevant KPIs were developed in collaboration with different directorates, the impact of the innovation program and the ability to get buy in at all levels would be increased (Gilmour 2016).	*Lessons learned:* The Innovations and Advancements Program was established by the U.S. Department of Transportation to promote innovation on Megaprojects in the United States. The innovations on the Big Dig were numerous and included environmentally enhancing context-sensitive design, safety incentive programs, innovative ground-freezing techniques, and the largest use of urban slurry wall modules, which are now being used in other projects around the world (Greengard 2007; Greiman 2015). Formal innovation programs can add considerable value to a project organization, but it is critical to have contractual agreement as to ownership of the innovation and ownership of any improvements to existing intellectual property.

innovation-focused centers within the FHWA Office of Innovative Program Delivery (OIPD). The CAI works with FHWA Division Office EDC Coordinators and Innovation Deployment Teams to compile information on the state-of-the-practice, as well as the innovation implementation goals, of states and other partners.

This knowledge transfer includes innovations in managerial, operational, and technological tools and processes. Environmental improvements including conservation initiatives, health, and safety improvements, and the development of urban parks and islands in Boston's nearby harbor have added much to the ecosystem of not only Boston, but the region in general (EDRG 2006a, 2006b). More entrepreneurial endeavors will be needed to create value in the future (R. Paaswell, personal interview, 26 October 2017). A twenty-first century element of major infrastructure in the center of a metropolis like Boston could not be a singular element of highway construction. It had to serve a range of important urban needs. Moving traffic more quickly through the central business district is, in many ways, of less importance. The replacement of a concrete infrastructure with a two-mile contemporary public park and increased housing and commercial opportunities stimulated economic and social growth for the long-term future including 20 million visitors per year to Boston's landmark Faneuil Hall (USNPS 2019).

Crossrail Innovate 18

By comparison Crossrail, the largest construction project in Europe, broke new ground on 15 May 2009 in the UK construction industry by being the first major project in Europe to develop and deliver a strategy and process for systematically managing innovation in a major project environment (Crossrail 2022).

The Innovate 18 program was developed by Crossrail in conjunction with Imperial College to support Crossrail's mission to deliver a world-class railway that fast-tracks the progress of London. The program has been transferred to the Knowledge Transfer Network to become the basis for the Infrastructure Industry Innovation Portal I3P. Established in 2016, the Infrastructure Industry Innovation Partnership (I3P) is a community of client and supply chain organizations that have made a commitment to delivering collaborative innovation through projects supported by a large network of experts and innovators and world leading industry knowledge that will drive the future transformation of the infrastructure and construction industry (I3P 2021).

Innovate 18 provided the opportunity to capture and explore pioneering ideas from all of those involved in the project. The techniques, products, and methods used on Crossrail are providing a benchmark for other construction projects. It is important to note that people were central to Innovate 18. The network of innovation champions and innovators, drawn from Crossrail, its supply chain and its partners, provided energy and inspiration.

As the program evolved, the project team reviewed and revised the strategy, to ensure focus and that innovation strategic themes (health and safety, sustainability, and efficiency and digital integration) were aligned with desired project outcomes. This approach saw a change in the nature of innovations being generated, moving from primarily health and safety related through the construction phase, to a trend of digital and efficiency related innovations focused on supporting M&E fit-out and handover in the later stages of the program.

> At some stages in the programme there has been a lot of focus on whether innovation has provided benefits. However, many of the benefits were difficult to quantify, either because they were soft benefits, such as improvements in behaviors; safety, etc., or because there was no baseline data available against which to compare. Careful thought should be given to objectives and KPIs for an innovation programme to avoid retrospective assessments.
>
> *(Gilmour 2016)*

Despite the challenges, the Network of European Foundations for Innovative Cooperation (NEF) Innovation Institute awarded Crossrail's innovation program, first place for "Best Example of Innovation in the Public or Voluntary Sector" in December 2014.

Both projects relied on global supply chains, the preferences of stakeholders, the financial support of investors and lenders, the trustworthiness of their customers and users. This was in addition to a host of other project participants to achieve its goals of delivering a first of its kind project to millions of users around the globe. To meet the needs of a diverse group of stakeholders' innovation is an essential component. However, questions of what to innovate, why to innovate, when to innovate, where to innovate, and how to innovate are not easy to answer.

Systems Innovation in the Railway Industry in India

A third example of the development of a formal innovation program can be found in the railway system in India. The design and expansion of the railway system in India marked the transformation from one social technological system to another Kumpf et al. (2021). The railway system fundamentally altered the infrastructure and economic activity of India by connecting important ports to industrial and commercial centers in the country. Its advent highlights the dynamic interplay between innovations at the micro-level, such as steam engine technology and railway construction design, with significant societal and economic transformations at the macro-scale, notably industrialization and the advent of capitalism. Such a transformation from one social technological system to another can be called "systems innovation." It includes more than just optimize the system but goes beyond the field of engineering and includes innovation along the entire value chain. For example, startups designing digital systems to simplify bookings are now being widely used by passengers. The Indian Railway company is also testing and scaling innovative ideas proposed by its own employees. Most of the ideas are technical solutions to boost safety, whereas others seek to improve the passenger's experience such as low-cost water coolers with zero electric consumption. Overall, in international development, considerable innovation efforts are targeted toward strengthening national health or education systems. In this case, there is less emphasis on the deliberate support to bring new systems into being.

Recent Industrial Innovations

Researchers at VW have recognized the importance of collaboration with competitors and innovative AI companies to advance their electric cars and autonomous vehicles in the coming years (Su et al. 2019). Wal-Mart uses sophisticated data analytics and data mining technology to understand its customers and its supply chain. Many companies have focused on open innovation to include as many disciplines as possible because innovation requires an open and transparent process including working with its competitors. Of course, openness has to be balanced with the dangers of losing market share or sharing the market with unreliable trespassers that can deflate the value of an intellectual property portfolio. Peter Drucker (2002) through his extensive research on innovation points out that most innovations result from a conscious, purposeful search for innovation opportunities rather than just letting innovation take a natural course of evolution.

What Is an Innovation Strategy?

We build too many walls and not enough bridges.

— *Sir Isaac Newton*

Sir Isaac Newton has been recognized as one of the greatest minds in history, and his contributions have shaped the world as we know it. Sir Isaac Newton recognized that innovation does not just happen but must be planned and strategized. In military terms, strategy refers to how wars are won, while tactics refers to the specific means by which wars are won. An innovation strategy can be measured by a firm's investment in innovation. Strategy entails judgment about which kinds of innovation processes are most appropriate for the firm's circumstances and ambitions (Dodgson et al. 2008).

The two ends of the spectrum of innovation models can be defined as incremental innovation and radical innovation. Incremental is a small change, based upon current knowledge and experience. In contrast a radical innovation is a breakthrough in science or technology that often changes the character and nature of the industry (Slaughter 1998). Radical change means a transformation of an industry such as moving from self-driving to autonomous vehicles. Radical changes are rare and unpredictable in their impacts. As mentioned earlier, an example of a radical change was the introduction of structural steel from which entirely new infrastructures could be built.

Aversa et al. (2017) in their study of business model portfolios stress the importance of exploiting new opportunities to create additional value, such as cross-selling, differentiation, reputation, user data, and capability development to advance the purpose of the business enterprise. This approach is applicable to megaprojects where business models are critical to linking resources, technology, and innovation across projects to monetize the project's value through technical and process improvements. In Table 11.2, an overview of the different types of

Table 11.2 Types of innovation.

Type	Description	Case studies
Technical	The multistage process whereby organizations transform ideas into new/improved products, service, or processes	Technical innovation has arisen on megaprojects in the form of ground freezing, slurry wall tunnel wall construction, integrated project control systems, ventilations systems, and inner-city mining operations (Greiman 2015)
Architectural	Requires change in the interacting set of components or systems	Apple used existing cell phone technology and repackaged it from a phone to a watch
Process	The "introduction of new production methods, delivery models, new management approaches, and new technology that can be used to improve production and management processes" (Wang and Ahmed 2004, p. 305)	Process innovations have included the protocols for better management of health and safety including incentive programs, and sustainable innovation processes that encourage better environmental practice and focus on data collection and analytics to provide essential information for risk and uncertainty decision making. More recent process innovations include the use of Artificial Intelligence (AI), the Cloud, Machine Learning (ML) and Robotics to provide more efficient processes to reduce or eliminate risk, and control cost, schedule, and quality, risk and uncertainty
System	Integration of multiple independent innovations (Slaughter 1998). Such a transformation from one social technological system to another can be called "systems innovation." It includes more than just optimize the system it goes beyond the field of engineering and includes innovation along the entire value chain	The India Railway Freight Operations Information System (FOIS) was introduced as an application to track and monitor the movement of wagons, locomotives, and unit trains comprising both customer services and train operations

Systems innovation is widely used in the construction industry because systems are reconfigured in every project |

Business model	Transforming the delivery of the value such as the bundling of different entities for design and construction	Business model innovation is related to creating new value for customers (Cantarelli and Genovese 2021). Transforming the delivery of the value, or delivering the value to new customers (Davila et al. 2006). An organization's innovation strategy includes the processes that create and capture value by combining and coordinating resources (Aversa et al. 2017; Dodgson et al. 2015). Other innovations related to creating value to customers range from value engineering methods (Husin et al. 2015) to innovative financing (Johnston 2011)
Modular	A significant change within a component but does not require change in the interacting set of components	A new machine that ties the wires for reinforcing bar, or updates to mobile phones and computers that eliminates the requirement of buying a whole new phone as technology evolves. Tesla's model X is another example of modular innovation
Incremental	A small gradual change based upon current knowledge and experience (Slaughter 1998)	Changes made in construction techniques or equipment, or the functionality and features of existing products and services such as automobiles, mobile phones or computers
Radical	Can occur outside the industry and based on engineering or scientific breakthroughs (Koseoglu et al. 2019)	Space manned missions, and the introduction of structural steel. Radical innovation can entail financial or project risk or technical risk. Building information modeling (BIM) has been described as a radical innovation and a digital transformation in the architecture, engineering, and construction industry

(Continued)

Table 11.2 (Continued)

Type	Description	Case studies
Management referents	Adopted or adapted from other models	Heathrow Terminal 5 (Cost plus contract) (Davies et al. 2009); Integrated project teams
Administrative	Organizational structure, administrative processes and human resources, communities of practice	Collaborative and networking approaches to procurement, contracts, and other areas for innovation (Arena et al. 2017)
Sustainability	Incudes socio-environmental innovation that can take various forms such as poverty alleviation, climate change, green projects, digitization, and quality of life initiatives	The $35B Songdo International Business District, Icheon, South Korea built on reclaimed land near the Yellow Sea is widely recognized as a model of sustainable city-scale development for smart cities around the globe
		The Smart City approach of the Municipality of The Hague is reflected in the LivingLab Scheveningen, a coastal town that is part of the municipality. Since 2020 Scheveningen's new boulevard has been a special learning area for the city of the future. The municipality is testing smart, digital inventions here that solve social problems

innovations employed in projects is described along with examples and case studies of how innovation can be advanced within a megaproject. Although not a comprehensive list it provides important examples of how innovation is developed and implemented in different businesses and industries.

Impediments to Innovation

The inability of megaprojects to innovate their delivery and adapt to changing and unexpected circumstances is one of the reasons for project failure (Davies and Gann 2017). Due to the high risk and uncertainty with megaprojects, the level of innovation is low with clients and contractors relying on proven and standardized techniques, technologies, and approaches (Davies et al. 2009; Maghsoudi et al. 2016; van Marrewijk et al. 2008). Besides risk and uncertainty, other inherent features of megaprojects impede their ability to be innovative, including their temporary nature (Davies et al. 2014), size, complexity, the separation of design and construction, and risk attitudes (Brockmann et al. 2016). Recently, there seems to be a drive toward increasing levels of innovation in megaprojects (Holzmann et al. 2017) with project complexity (Ozorhon et al. 2016) being one of the reasons for these innovations.

If consortia are free to choose their methods, materials, and techniques, it motivates them to develop innovative solutions if this has an economic value (Davies et al. 2014; Parrado and Reynaers 2020). Targets set by the Government, changes in regulations or the political environment (Sergeeva and Zanello 2018), and globalization of markets and economic conditions are also key drivers for innovation in megaprojects. Besides, job creation was an economic motivation on the French side to make process improving innovations in the Channel Tunnel (Winch 2000).

Internal and External Drivers for Sustainability Innovations

Innovation can come from within or outside the organization. It can be based on project risk, uncertainty, complexity, discovery, regulatory requirements, and financing limitations. Table 11.3 shows a comparison of sources that can advance innovation both intentionally and sometimes accidentally in the course of discovery or problem solving.

Enablers and Challenges to Innovation

> The reasonable man adapts himself to the world; the unreasonable one persists in trying to adapt the world to himself. Therefore, all progress depends on the unreasonable man.
>
> — *George Bernard Shaw (1856–1950), Playwright*

Table 11.3 Internal and external drivers of sustainable innovation.

Internal drivers	External drivers
• Socio-environmental risk drives sustainability innovations	• Supply chain expertise
• Disciplines within the project	• Offtake purchaser requirements
• Geotechnical risk analysis	• Value Construction and Value Engineering
• Engineering and contractor discovery of new technologies and construction methods	• Project community resistance
	• Local Indigenous people rights
• Stakeholders who support innovation initiatives	• Demands of Special interest groups – conservationists, developers, scientists, businesses, labor, residents
• Innovation champions	• Project audits and oversight
• Group innovation and team innovation	• Project finance challenges
• Project research and development laboratories	• Adoption and adaptation of methods from other projects
	• Cultural adaptation
• Material testing laboratories	• Changes in the law
• Onsite inspections	
• Critical safety factors	

Recently, there seems to be a drive toward increasing the levels of innovation in megaprojects (Holzmann et al. 2017), devising mechanisms to foster innovation (Worsnop et al. 2016), and making "significant efforts to create a more innovative and flexible delivery model" (Cantarelli and Genovese 2021; Davies and Gann 2017) however the dimensions of innovation in megaprojects are not sufficiently explored. For example, there are many important questions that could help spur innovation if they were better understood. These include:

- How is innovation developed differently when innovation is viewed as an outcome versus innovation as a process?
- How does innovation start and develop within megaprojects (top-down v. bottom-up, internal v. external)?
- What type of innovation is appropriate during the various life phases?
- What are the enablers and challenges to innovation?

Anthony et al. (2019) in their research on breaking down the barriers in innovation found there are certain habits and routines that lead to growth and that organizations must build these routines into the daily work environment. Drawing from the innovation, entrepreneurial, managerial, and project management literature and practice the enablers and challenges to innovation as shown in Table 11.4 were referenced. Though the list is far from complete it is intended to stimulate discussion on ways of better encouraging innovation frameworks and process in large scale projects.

Table 11.4 Twenty-six enablers and challenges to innovation in megaprojects.

Enabler of innovation	Challenges to innovation
Local, regional, national, and international investment in innovation	Limited public investment in innovation
Government initiated tax credits such as France's R&D tax credit, in order to address the decline in public funding of R&D in construction. Other enablers include cooperative research among project participants, and direct funding for innovation through grants, venture capital, accelerators, and crowd funding	An absence of incentives to stimulate investment to encourage countries to consider dedicated engineering and construction R&D funding sources
Development of an innovation strategy is nothing more than a commitment to a set of coherent, mutually reinforcing policies or behaviors aimed at achieving a specific competitive goal. Good strategies promote alignment among diverse groups within an organization, clarify objectives and priorities, and help focus efforts on them (Pisano 2015)	The failure to implement a strategy for innovation development and implementation at the enterprise level and then ensure that the strategy and policies are strategically aligned and understood and implemented at the portfolio, program, and project levels
Strategic alignment of enterprise, organizational, portfolio, and project goals	Lack of strategic alignment of goals among enterprise components
Support from the public and project sponsors for innovative sustainable development	The inability to recognize the benefits of innovation for the long term
Promoting megaprojects as laboratories of R&D	The failure to recognize that megaprojects are by their very nature opportunities for R&D innovation.
Organizational goals of sustainability and resilience as a measurement of success	Adherence to conventional measurements of success
Outcomes-based specification in contracts as opposed to overly detailed specifications	Lack of flexibility and collaboration in developing and negotiating contracts
Accepting failure as a risk of innovation and being open and transparent about successes and failures	Overcoming the adversity to risk taking is essential to increasing the likelihood that project managers and teams will embrace innovation
Innovation program and laboratory of learning with dedicated resources	Absence of dedicated resources to project innovation

(Continued)

Table 11.4 (Continued)

Enabler of innovation	Challenges to innovation
Integrated regulatory processes to share knowledge, develop best practices, and avoid duplication of efforts	Multiple regulatory processes without continuity or integration
Stakeholder engagement during all project phases	Disengagement or lack of interest in stakeholders concerns or influence
Open and transparent procurement process that encourages innovation in proposals	Top-down procurement process that rewards the low bidder
Focus on value creation and stewardship	Focus limited to cost and schedule adherence
Open working environment that rewards Innovation and encourages communication	Micromanagement that encourages obedience to standardized processes and fixed criteria
High-level support and R&D investment	Lack of investment or support
Innovative approaches to contracting	Use of standard contracts and boilerplate language
Space and time built into schedules for reflection	Little time for reflection due to cost, schedule, and other pressures
Collaboration among partners, alliances, organizations, teams, and disciplines	Collaboration is either not encouraged or difficult to structure
Performance-based criteria for measuring success	Failure to incentivize project participants through performance-based criteria
Integrated project teams (IPT) and integrated project organizations (IPO)	Isolation of project components including lack of integration of disciplines such as integration of design, construction, quality assurance, risk, and program management
Understanding drivers of uncertainty	Uncertainty is connected with risk rather than treated as a separate but important discipline
Capabilities and talent management	Hiring and retaining talent that demonstrates creative and critical thinking, and showing through your recruitment and management processes, and showing that creativity and strategic risk-taking are essential values of your project
Risk taking encouraged	Focus on negative risks rather than opportunity
Competitive environment that seeks new ideas	Acceptance of the status quo
Public–private partnerships that support innovation and cooperative research	Public–private partnerships with little or no shared programs for innovation

Best Practices for Innovation

> If I had asked the public what they wanted, they would have said a faster horse.
>
> — *Henry Ford*

1) Innovation Strategy

Megaprojects by their very nature are innovative and most megaprojects create and build something that has never been done before. Despite the natural innovation that occurs in megaprojects there is a desperate need to see that megaprojects develop and implement an innovation strategy. Yet, this is rarely the case and innovation in megaprojects if it happens at all often occurs by accident. Paul E. Gray, a former president of the Massachusetts Institute of Technology (MIT), stated in 1989 that "furthering technological and economic development in a socially and environmentally responsible manner is not only feasible, but also the great challenge we face as engineers, as engineering institutions, and as a society" (Gray 1989).

2) Value Engineering

Value engineering (VE) is an effective technique for reducing costs, increasing productivity, and improving quality. It can be applied to hardware and software; development, production, and manufacturing; specifications, standards, contract requirements, and other acquisition program documentation; facilities design and construction. It may be successfully introduced at any point in the life cycle of products, systems, or procedures. VE is a technique directed toward analyzing the functions of an item or process to determine "best value," or the best relationship between worth and cost. In other words, "best value" is represented by an item or process that consistently performs the required basic function and has the lowest total cost. In this context, the application of VE in facilities construction can yield a better value when construction is approached in a manner that incorporates environmentally sound and energy-efficient practices and materials.

Lee and Wang (2021) studied value engineering from the perspective of the Cruise Lines supply chain. The sustainability and competitiveness are just like two sides of a coin. On one hand, reducing the pollution and raw material waste is not only to protect our precious natural resources, but also to bring the cost down. On the other hand, a good global sourcing decision could deliver the cost saving and also keep the cruise ship industry heading toward sustainability.

> The secret of change is to focus all of your energy, not on fighting the old, but building on the new.
>
> — *Socrates (470–399 BCE), Philosopher*

3) **What is a Blue Ocean Strategy and how does it impact innovation in megaprojects?**

The cornerstone of Blue Ocean Strategy is – "value Innovation." Value Innovation puts equal emphasis on both value and innovation. Megaproject processes need to not only redefine the value proposition it is providing to existing customers (clients) but also needs to create a new value proposition for target segments (customers).

Just as large organizations need to innovate to survive so do megaprojects. Innovation strategy is not new and has been articulated frequently with various recommendations in the economic and commercial literature. Despite massive investments of management, time, and money, innovation remains a frustrating pursuit in many organizations (Pisano 2015). Innovation initiatives frequently fail, and successful innovators have a hard time sustaining their performance – as Polaroid, Nokia, Sun Microsystems, Yahoo, Hewlett-Packard, and countless others have found. Why is it so hard to build and maintain the capacity to innovate? The reasons go much deeper than the commonly cited cause: a failure to execute. The problem with innovation improvement efforts is rooted in the lack of an innovation strategy (Pisano 2015).

A strategy is nothing more than a commitment to a set of coherent, mutually reinforcing policies or behaviors aimed at achieving a specific competitive goal. Good strategies promote alignment among diverse groups within an organization, clarify objectives and priorities, and help focus efforts on them. Companies regularly define their overall business strategy (their scope and positioning) and specify how various functions – such as marketing, operations, finance, and R&D – will support it (Pisano 2015).

A great example of an innovation strategy can be found in the R&D programs in the automobile industry (Su et al. 2019). Within the next decade at the current rate of innovation we will experience transition from the fuel driven auto, to the electronic vehicle, to the autonomous vehicle. Automobile manufacturers around the world, in particular Tesla, Volkswagen, Daimler AG, General Motors, and Ford Motor Company are partnering to innovate in the development of the electric vehicle, and also working on battery cell development. At the same time automotive manufacturers are partnering with AI Companies to develop the tools necessary for the safety and reliability for these innovations.

> If I have a thousand ideas and only one turns out to be good, I am satisfied.
> — *Alfred Nobel (1833–1896) Chemist, Engineer, and Inventor*

The Red v. Blue Ocean Strategy

Chan Kim and Renée Mauborgne (2015) coined the terms red and blue oceans to denote the market universe. Red oceans are all the industries in existence today – the known market space, where industry boundaries are defined, and companies try to outperform their rivals to grab a greater share of the existing market. Cutthroat competition turns the ocean bloody red. Hence, the term "red" oceans.

Blue oceans denote all the industries not in existence today – the unknown market space, unexplored and untainted by competition. Like the "blue" ocean, it is vast, deep and powerful – in terms of opportunity and profitable growth.

The blue ocean strategy seeks to bring differentiation to organizations and brands to create awareness and presence in a new marketplace and create demand amongst its users and customers. The blue ocean strategy focuses on creating demand in the uncontested market space, and by doing this, it makes the competition unrelated and irrelevant (Kim and Mauborgne 2017). All of the successful companies in the world have a blue ocean strategy which is why they remain in the list of the top 100 brands in the world year after year (Interbrand 2021).

Tesla and Apple are two examples of leading companies that created their blue oceans by pursuing high product differentiation at a relatively low cost, which also raised the barriers for competition. They also were paradigmatic of burgeoning industries at the time that were later exemplified and emulated by others.

The red and the blue ocean strategy for Innovation in megaprojects can be described as follows:

Red ocean strategy	Blue ocean strategy
Innovation competes with existing players in existing market spaces and defined industry barriers	Creates a new market space through innovation rendering the competition irrelevant and redefines industry barriers
Takes customers from the existing pool of competitors	Creates new and innovative demand because it has created a new product altogether
Makes the value-cost tradeoff	Innovation breaks the value-cost tradeoff because it is a unique product or project that has never been done before
Focus on performance by implementing better marketing and a lower cost base.	Creates new market space diffusing the impact of the competition

4) Cracking the Code of Megaproject Innovation

The Boeing 787 Dreamliner brought aviation into a new next-generation era when it took its first flight on 15 December 2009, and represented the future for Boeing Commercial Airplanes.

The Dreamliner was designed to be the most advanced commercial aircraft ever built and the most efficient to operate, however, its eight long years in development and early technical problems were troubling for one of the world's most experienced aircraft manufacturers. Holzmann et al. (2017) in their research on the Dreamliner's innovation challenges, used the Diamond of innovation analysis developed by Shenhar and Dvir (2007) to determine that the Dreamliner should be classified as an innovation because of its:

Novelty. The build-to-performance business model constituted an unknown experience requiring pilot testing and iterative model modifications.

Technology. It was considered a high-tech innovation because the technology of composite materials and electronic controls was new to the industry.

Complexity. The exceptional amount of outsourcing to a global network of 700 local and foreign suppliers without the requisite experience (MacPherson and Pritchard 2005).

Pace. The Dreamliner was a fast competitive product aiming to get to market in record time.

To avoid innovation challenge pitfalls, Boeing offered the following advice:
1) Past performance does not guarantee future success
2) Unknown–unknowns are potential pitfalls
3) Assessing the degree of innovation means estimating how long it will take to get it right
4) Do not rush to execution
5) The most successful projects spent a long time in up front planning
6) Assess your capabilities for the emergence of new challenges
7) Use a framework to analyze the new and unknown parts

Though the Dreamliner had a long phase of success, in 2022, due to prolonged delays in making and delivering its 787 Dreamliner jet, the company was driven to a dramatic $4.2B loss. The Dreamliner costs were caused in part by a realization that the fixes Boeing needed to make to win Federal Aviation Administration approval for the twin-aisle plane would take longer than expected. Thus, innovation can extract a serious cost through delays and compliance with regulatory requirements that might not occur, but for the introduction of new more innovative products. As of December 2021, Boeing has built 1006 Dreamliners, and estimates vary between 1300 and 2000 will be needed to break even.

1) Innovation in the Developing World

The 2021 Global Innovation Index (GII) showed that investment in innovation showed great resilience during the COVID-19 pandemic, often reaching new peaks, but that it varies across sectors and regions. Investment in innovation reached an all-time high prior to the pandemic, with R&D having grown an exceptional 8.5% in 2019.

The ultimate goal of the Global Innovation Index (GII) a partnership of the World Intellectual Property Organization (WIPO) and the Portulans Institute is to discover what works best in producing an ecosystem where people can achieve their highest potential, innovating and creating to improve lives everywhere. The GII 2021 finds that governments and enterprises in many parts of the world have scaled up their investments in innovation during the COVID-19 pandemic. Meantime, scientific output, expenditures in research and development, intellectual property filings and venture capital deals continued to grow in 2020, building on strong peak pre-crisis performance (Dutta et al. 2021).

Innovation growth policies have played an important part in the successful implementation of new science, technology, and innovation projects throughout the world but particularly in the developing economies. The policies have helped advance new product innovation but has also been recognized as a tool in developing innovation in large-scale megaproject innovation development. We will take a look at a few of these national policies to understand the impact on national economic and social development in selected countries.

The World Intellectual Property Organization has been instrumental in tracking the development of innovation through its Global Innovation Index (GII). The seven pillars of this Index are shown in Figure 11.1.

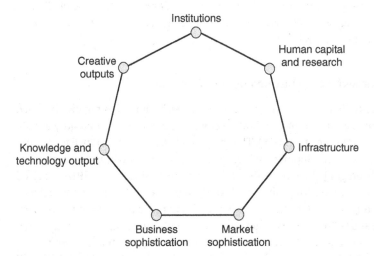

Figure 11.1 The Seven Global Innovation Index Pillars. *Source:* Adapted from the Global Innovation Index, The World Intellectual Property Organization.

Innovation and the Economic Growth of Asia

The economic growth of East Asian and Southeast Asian economies including, India, Korea, and Taiwan have been based on the development of their technological capabilities (Unger et al. 2017). The lessons from technological innovation can be learned by studying an economy such as Taiwan, which has been successful through support from a national innovation system (NIS). Through understanding the triple helix (integration of technology, business, and academia) many innovations were created that would otherwise still be a dream (Unger et al. 2017). The multilateral development banks have recognized the importance of knowledge in gaining strategic advantage. The importance of innovation technology is critical for future economic growth of all countries.

Developing Innovation from Science Parks to Industrial Districts

In recent years, we have seen a migration of innovation from the urban science parks such as Hsinchu City to the urban areas of Taipei and other cities (Chen et al. 2018) due to the desire of the younger generations to relocate to the urban cities (Chen and Huang 2014). To address the challenges of globalization in terms of enhancing market accessibility and market size, as well as maintaining technological and institutional set up, a city must create a competitive environment (Mauro and Forster 2008; Unger et al. 2015). They also must develop a diversity of products and markets. This is costly financially and in terms of other essential resources, such as human capital, as well as in terms of cultural, intellectual, social, and environmental aspects (Friedmann 2005). Some countries respond to these challenges by creating new agglomerations, ranging from industrial districts to competitiveness poles (Ganne and Lecler 2009).

Innovation Growth Policy Development in Thailand

A study of methodologies of innovation in countries in the Asian region identified Thailand as an example of how policy development can fuel sustainability as well as innovation (Greiman et al. 2018). Thailand attaches great importance to the concept of sustainable development which has long taken root in the country. The country has been guided by the Sufficiency Economy Philosophy (SEP), conceived by His Majesty the Late King Bhumibol Adulyadej. The philosophy of sufficiency economy highlights a balanced way of living. Three principles – moderation, reasonableness, and self-immunity – along with the conditions of morality and knowledge can be applied to any level of the society – from an individual to a country.

Thailand's 13th National Economic and Social Development Plan, which will run from 2023 to 2027, intends to transform the country to achieve five goals. To begin, Thailand must restructure its manufacturing to adapt to the digital economy. It must place a high priority on research and development. Second, the country must invest in human resources to keep up with the rapid pace of digital development. Third, the plan emphasizes providing everyone with equal chances and justice. Fourth, the country must place a greater emphasis on environmental protection and long-term solutions to the effects of global climate change. Finally, as development accelerates, Thailand must prepare for any global concerns and issues, such as connectivity with neighboring nations and the creation of economic corridors in each region of the world (Greiman et al. 2018).

Table 11.5 highlights the national economic development goals and plans looking forward of Asia's major economies.

Table 11.5 Innovation in Asia.

Thailand's 13th National Economic and Social Development Goals (2023–2027)	*Vietnam's Five-Year Economic Development Plan (2021–2025)*
1) Importance on research and development	1) Improving the efficiency of resources
2) Development of human resources to keep up with rapid digital changes	2) Focus on self-reliance and adapting capacity of the economy
3) Emphasis on equal opportunities and being fair to people	3) Renewable energy sources to be further prioritized in electricity generation
4) Focus more on environmental conservation and ways of handling the effects of global climate change in a comprehensive and sustainable manner	4) The formation of a science-technology and innovation-based economy, ensuring fast and sustainable growth
5) Preparation for any future global challenges and issues	5) The public–private partnership law which came into effect January 2021 encourages private sector participation in infrastructure development
(Thailand 2021)	(MOIT 2021)
Malaysia National Policy on Science, Technology, and Innovation	*People's Republic of China 14th Five-Year Plan for National Economic and Social Development*
1) Advancing Scientific and Social Research	1) Urbanization, infrastructure, and regional cooperation
2) Development and Commercialization	2) Innovation and industrial modernization
3) Developing, Harnessing and Intensifying Talent	3) Digitization
4) Energizing Industries	4) Sustainability
5) Transforming Science, Technology, and Innovation (STI) Governance	5) Culture, education, and ideology
6) Promoting and Sensitizing	6) Agriculture and rural development
7) Enhancing Strategic International Alliances	7) Governance and public sector
(NPSTI 2013; Malaysia 2021)	(China 2021)

Kazakhstan

Basically, in a broad sense, in Kazakhstan innovation is synonymous with the successful development of social, economic, educational, managerial, and other areas. Based on innovation, scientific and scientific–technical development, the invention becomes embodied in the goods, services as well as methods. Consequently, the innovation cycle is preceded by research, development, or design work. Their results create the groundwork on the basis of which innovation begins (Sadyrova et al. 2021).

The Creative Economy for Sustainable Development and Recovery

> We urge Governments to recognize culture and creativity as an integral part of wider policy agendas (. . .). We recommend including culture, cultural heritage and the creative sector in national and international post-pandemic recovery strategies, recognizing that international cultural exchanges depend on strong cultural and creative actors in all countries.
> — *G20 Italy, Rome Declaration of the G20 Ministers of Culture 30 July 2021*

The contention of the G20 is not that the creative economy needs public subsidies to resume its previous growth. It does however require governments and multilateral organizations to recognize and address constraints and regulatory structures that have failed to keep pace with the rapid evolution of the sector (G20 2021).

As the race for recovery (COVID-19) focuses on infrastructure investments, including long neglected and underfunded social infrastructure (Fransen et al. 2018) to revive the 2030 Agenda, the creative economy could play a significant macroeconomic role in support of people-centered sustainability policies, as recognized by the doubling of the Creative Europe funding program by the European Commission in May 2021, up to €2.3B. A key issue therefore arises as to whether and how such programs could be expanded globally, starting with a G20 wide perspective.

Summary

In this chapter, we explored megaproject innovation by researching how innovations in megaprojects are initiated, supported, and developed and how innovation rather than being a sudden flash of insight, evolves over time. The chapter provided insights on both enablers and challenges of innovation and how to break down barriers to innovation. Innovation was explained from the perspective of formalized developed country frameworks to strategies and policies on innovation and entrepreneurship in the developing world. The internal and external drivers of innovation in megaprojects were examined and the role of risk, project failure, and regulation were recognized as potential drivers of innovation. Finally, various strategies for innovation were reviewed to determine the role of strategy in developing successful innovation models. Building a culture of innovation at all levels is a core component of megaprojects.

Lessons and Best Practices

1) Innovation requires support from a megaproject's sponsors and a commitment to invest in innovation at the critical upfront stages of the project.
2) Innovation comes in many forms including process innovation, systems innovation, radical innovation, administrative and managerial innovation, and technological innovation.
3) Innovation enablers and barriers should be assessed through a structured risk management process to determine how barriers to innovation can be reduced and opportunities for innovation can be enhanced.
4) Blue Ocean strategies can be efficient enablers of innovation on megaprojects.
5) An innovation program should be developed at a high level in the sponsoring organization(s) and optimized to capture the full benefits and return on investment that innovation can provide.
6) Recognizing the role of a champion for innovation will assist in addressing resistance and other barriers to innovation.
7) The relationship between innovation and sustainability should be developed so that new initiatives include specific requirements for sustainability.

Discussion Questions

1 Why are innovative initiatives so hard to develop in megaprojects?

2 What are some ways in which barriers to innovation can be reduced?

3 What do you think are the most important enablers of innovation?

4 How would you reduce the barriers to innovation in your project?

5 Which types of innovations are most used in megaprojects and why?

6 How would you convince your project sponsor that an investment in innovation is critical to megaproject success?

7 How do risk and system failure contribute to the development of innovation strategies?

8 Provide a few examples of how Blue Ocean strategies can be deployed on a megaproject.

References

Allen, C. and Barnes, P.E. (2004). Sharing experiences and lessons learned. *Public Roads*, July–August 68 (1): 54–64. Federal Highway Administration, Washington, D.C.

Anthony, S.D., Cobban, P., Nair, R., and Painchaud, N. (2019). Breaking down the barriers to innovation: Build the habits and routines that lead to growth. *Harvard Business Review* (November/December): 92–101.

Arena, M., Cross, R., Sims, J., and Uhl-Bien, M. (2017). How to catalyze innovation in your organization. *MIT Sloan Management Review* 58 (4): 39–47.

Aversa, P., Haefliger, S., and Reza, D.G. (2017). Building a winning business model portfolio. *MIT Sloan Management Review* 58 (4): 49–54.

Brockmann, C., Brezinski, H., and Erbe, A. (2016). Innovation in construction megaprojects. *Journal of Construction Engineering and Management* 142 (11): 1–59. https://doi.org/10.1061/(ASCE)CO.1943-7862.0001168.

Cantarelli, C.C., and Genovese, A. (2021). Innovation potential of megaprojects: A systematic literature review. *Production Planning & Control*, 1–21. https://doi.org/10.1080/09537287.2021.2011462.

Charles Babbage Institute (CBI) (n.d.). University of Minnesota, College of Science and Engineering and University Libraries. https://cse.umn.edu/cbi/who-was-charles-babbage.

Chen, L., and Huang, E. (2014). *Leaving Hsinchu Science Park Behind: Taiwan's Changing High-Tech Center of Gravity*, Commonwealth Magazine, February 27, No. 541.

Chen, H., Su, Q., Zeng, S. et al. (2018). Avoiding the innovation island in infrastructure mega-project. *Frontiers of Engineering Management* 5 (1): 109–124.

China (2021). *People's Republic of China 14th Five-Year Plan for National Economic and Social Development*. Chinese Parliament, the National People's Congress.

Crossrail (2022). *Crossrail Learning Legacy: Innovation*. London: Crossrail Ltd.

Damayanti, R.W., Hartono, B., and Wijaya, A.R. (2021). Clarifying megaproject complexity in developing countries: A literature review and conceptual study. *International Journal of Engineering Business Management* 13: 1–25. https://doi.org/10.1177/18479790211027414.

Davies, A. and Gann, D.M. (2017). Innovation and Flexibility in Megaprojects: A New Delivery Model. In: *The Oxford Handbook of Megaproject Management* (ed. B. Flyvbjerg), 313–338. Oxford, UK: Oxford University Press.

Davies, A., Gann, D., and Douglas, T. (2009). Innovation in megaprojects: Systems integration at London Heathrow Terminal 5. *California Management Review* 51 (2): 101–126.

Davies, A., MacAulay, S., DeBarro, T., and Thurston, M. (2014). Making innovation happen in a megaproject: London's Crossrail suburban railway system. *Project Management Journal* 45 (6): 25–37. https://doi.org/10.1002/pmj.21461.

Davila, T., Epstein, M.J., and Shelton, R. (2006). *Making Innovation Work: How to Manage It, Measure It, and Profit from It*. Upper Saddle River, NJ: Wharton School Publishing, Pearson Education Inc.

Dodgson, M., Gann, D., and Salter, A. (2008). *The Management of Technological Innovation: Strategy and Practice*. Oxford, UK: Oxford University Press.

Dodgson, M., Gann, D., MacAulay, S., and Davies, A. (2015). Innovation strategy in new transportation systems: The case of Crossrail. *Transportation Research Part A: Policy and Practice* 77: 261–275. https://doi.org/10.1016/j.tra.2015.04.019.

Drucker, P. (2002). The discipline of innovation. *Harvard Business Review* 80 (8): 95–102.

Dutta, S., Lanvin, B., León, L.R., and Wunsch-Vincent, S. (ed.) (2021). *Global Innovation Index (GII): Tracking Innovation Through the COVID-19 Crisis*, 14the. Geneva, Switzerland: World Intellectual Property Organization (WIPO).

EDRG (Economic Development Research Group, Inc.) (2006a). *Economic Impact of the Massachusetts Turnpike Authority and Related Projects, Volume I: The Turnpike Authority as a Transportation Provider*. Boston, MA: Prepared for Massachusetts Turnpike Authority.

EDRG (Economic Development Research Group, Inc.) (2006b). *Economic Impact of the Massachusetts Turnpike Authority and Related Projects, Volume II: Real Estate Impacts of the Massachusetts Turnpike Authority and the Central Artery/Third Harbor Tunnel Project*. Boston, MA: Prepared for Massachusetts Turnpike Authority.

Flyvbjerg, B. (2014). What you should know about megaprojects and why: An overview. *Project Management Journal* 45 (2): 6–19. https://doi.org/10.1002/pmj.21409.

Fransen, L., del Bufalo, G., and Reviglio, E. (2018). *Boosting Investment in Social Infrastructure in Europe: Report of the High-Level Task Force on Investing in Social Infrastructure in Europe*. Luxembourg: European Commission, European Association of Long-Term Investors (ELTI).

Friedmann, J. (2005). Globalization and the emerging culture of planning. *Progress in Planning* 64 (3): 183–234. https://doi.org/10.1016/j.progress.2005.05.001.

G20 Declaration (2021). Rome Declaration of the G20 Ministers of Culture. July 30, 2021.

Ganne, B. and Lecler, Y. (2009). *Asian Industrial Clusters, Global Competitiveness and New Policy Initiatives*. World Scientific Pub. Co. *SSRN Electronic Journal*. https://doi.org/10.2139/ssrn.1469382.

Gates, B. (2011). *It's not always a eureka moment*. The Blog of Bill Gates. September 29, 2011. https://www.gatesnotes.com/books/where-good-ideas-come-from.

Gil, N., Miozzo, M., and Massini, S. (2012). The innovation potential of new infrastructure development: An Empirical Study of Heathrow Airport's T5 Project. *Research Policy* 41 (2): 452–466. https://doi.org/10.1016/j.respol.2011.10.011.

Gilmour, M. (2016). *The Innovator's Way: Essential Practices for Successful Innovation*. Cambridge, MA: MIT Press.

Gray, P.E. (1989). The Paradox of Technological Development. In: *Technology and Environment* (ed. J.H. Ausubel and H.E. Sladovich), 192–204. Washington, D.C: National Academies Press.

Greengard, S. (2007). *How the Big Dig Is Transforming Boston*. Washington, D.C: Engineering, Inc., The American Council of Engineering Companies.

Greiman, V.A. (2015). *Evaluating Megaprojects: What Constitutes Success? Rethinking Infrastructure* 2: 14–17.

Greiman, V.A., Unger, B., and Wan, W. (2018) *Thailand's Young Innovators and Their Perspectives on the Challenges Facing the New Creative Economy*. In Conference Proceedings, 27th World Business Congress, June 11–13. Transformation, Coopetition, and Sustainability in the Era of Globalization, Engagement, and Disruptive Technology (CPCI-SSH). Heng Seng University, Hong Kong, Thompson Reuters. https://imda-usa.org/advances-in-global-business/.

Holzmann, V., Shenhar, A., Zhao, Y., and Melamed, B. (2017). Cracking the Code of Megaproject Innovation. In: *The Oxford Handbook of Megaproject Management* (ed. B. Flyvbjerg), 453–457. Oxford: Oxford University Press.

Husin, A.E., Berawi, M.A., Dikun, S. et al. (2015). Forecasting demand on mega infrastructure projects: Increasing financial feasibility. *International Journal of Technology* 6 (1): 73–83. https://doi.org/10.14716/ijtech.v6i1.782.

Interbrand (2021). *Best Global Brands Report 2021*. New York, NY: Interbrand https://www.rankingthebrands.com/PDF/Interbrand%20Best%20Global%20Brands%202021.pdf.

Johnson, S. (2010). *Where Good Ideas Come from: The Natural History of Innovation*. New York, Riverhead: Penguin Publishing Group, Riverhead Books.

Johnston, V.R. (2011). Entrepreneurial megaproject leadership, innovation, and accountability: Denver's international airport, T-REX, and FasTracks. *Public Works Management & Policy* 16 (3): 199–227. https://doi.org/10.1177/1087724X11409949.

Kim, W.C. and Mauborgne, R. (2015). *Blue Ocean Strategy, Expanded Edition: How to Create Uncontested Market Space and Make the Competition Irrelevant*. Brighton, MA: Harvard Business Review Press.

Kim, W.C. and Mauborgne, R. (2017). *Blue Ocean Shift: Beyond Competing - Proven Steps to Inspire Confidence and Seize New Growth*. Boston, New York: Hachette Books.

Koseoglu, O., Basak Keskin, B., and Ozorhon, B. (2019). Challenges and enablers in BIM-enabled digital transformation in mega projects: The Istanbul New Airport Project Case Study. *Buildings* 9 (5): 115. https://doi.org/10.3390/buildings9050115.

Kumpf, B., Strandberg, N., and Barkell, R. (2021). *Part One: Systems Innovation: An Inside-Look at Diversifying Innovation Efforts in International Development Organizations*. New York, NY: International Development Innovation Alliance (IDIA).

Lee, C.-W. and Wang, C.-H. (2021). Value analysis and value engineering on the sustainability of global sourcing competitiveness. *Journal of Applied Finance and Banking* 11 (6): 1–19.

MacPherson, A. and Pritchard, D. (2005). *Boeing's Diffusion of Commercial Aircraft Design and Manufacturing Technology to Japan: Surrendering the US Aircraft Industry for Foreign Financial Support*, Canada-United States Trade Center Occasional Paper No. 30. Buffalo: State University of New York.

Maghsoudi, S., Duffield, C., and Wilson, D. (2016). Innovation in infrastructure projects: An Australian perspective. *International Journal of Innovation Science* 8 (2): 113–132. https://doi.org/10.1108/IJIS-06-2016-008.

Malaysia National Policy on Science, Technology & Innovation (NPSTI) (2013). *2013–2020 Harnessing STI for Socio- Economic Transformation and Inclusive Growth*, July 2013.

Malaysia (2021). *The National Policy on Science, Technology and Innovation (DSTIN) 2021–2030*. Ministry of Science, Technology and Innovation (MOSTI), Kuala Lumpur, Malaysia. https://www.mosti.gov.my/en/category/berita/.

van Marrewijk, A., Clegg, S.R., Pitsis, T.S., and Veenswijk, M. (2008). Managing public–private megaprojects: Paradoxes, complexity, and project design. *International Journal of Project Management* 26 (6): 591–600. https://doi.org/10.1016/j.ijproman.2007.09.007.

Mauro, F. and Forster, K. (2008). *Globalization and the Competitiveness of the Euro Area*. [Occasional Paper]. Frankfurt, Germany: European Central Bank.

Ministry of Industry and Trade (MOIT) (2021). *Vietnam's Five-Year Economic Development Plan (2021–2025)*. Vietnam: MOIT.

Ozorhon, B. (2013). Analysis of construction innovation process at project level. *Journal of Management Engineering* 29 (4): 455–463. https://doi.org/10.1061/(ASCE)ME.1943-5479.0000157.

Ozorhon, B., Oral, K., and Demirkesen, S. (2016). Investigating the components of innovation in construction projects. *Journal of Management in Engineering* 32 (3): 04015052. https://doi.org/10.1061/(ASCE)ME.1943-5479.0000419.

Parrado, S. and Reynaers, A.M. (2020). Agents never become Stewards: Explaining the lack of innovation in public–private partnerships. *International Review of Administrative Sciences* 86 (3): 427–417. https://doi.org/10.1177/0020852318785024.

Pisano, G.P. (2015). You need an innovation strategy. *Harvard Business Review* 93 (6): 44–54.

Sadyrova, M., Yusupov, K., and Imanbekova, B. (2021). Innovation processes in Kazakhstan: Development factors. *Journal of Innovation and Entrepreneurship* 10 (36): 1–13. https://doi.org/10.1186/s13731-021-00183-3.

Sergeeva, N. and Zanello, C. (2018). Championing and promoting innovation in UK megaprojects. *International Journal of Project Management* 36 (8): 1068–1081. https://doi.org/10.1016/j.ijproman.2018.09.002.

Severo, E.A., Sbardelotto, B., de Guimarães, J.C.F., and de Vasconcelos, C.R.M. (2019). Project management and innovation practices: Backgrounds of the sustainable competitive advantage in Southern Brazil enterprises. *Production Planning and Control* 31 (15): 1276–1290. https://doi.org/10.1080/09537287.2019.1702734.

Shenhar, A.J. and Dvir, D. (2007). Project management research: Challenge and opportunity. *Project Management Journal* 38 (2): 93–99.

Slaughter, S.E. (1998). Models of construction innovation. *Journal of Construction Engineering and Management* 124 (3): 226–231. https://doi.org/10.1061/(ASCE)0733-9364(1998)124:3(226).

Su, N., Fang, Y., and Yang, D. (2019). *Volkswagen Group: Adapting in the Age of AI*. London, Canada: Harvard Business School Case Study, IVEY Publishing.

Thailand (2021). *Thailand's Thirteenth National Economic and Social Development Plan (2021–2025)*. Bangkok, Thailand: Office of the National Economic and Social Development Board, Office of the Prime Minister.

The I3P Institute for Information Infrastructure Protection (2021). https://www.thei3p.org/.

U.S. National Park Service (USNPS) (2019). *Faneuil Hall*. Boston, MA: Boston National Historical Park.

Unger, B., Greiman, V.A., Tu, T.-T., and Wan, W.W.N. (2015). *Science Parks in Taiwan and Their Challenges in the Era of the Creative Economy*. [Proceeding Paper]. In: *Advances in Global Management Development*, vol. 24 (ed. E. Kaynak and T.D. Harcar), 195–204. ISBN 1-888624-14-0. https://imda-usa.org/advances-in-global-business/.

Unger, B., Greiman, V.A., and Wan, W.W.N. (2017). *Can Taiwan's New Government Respond to the Changing Character and Needs of Innovation and Entrepreneurship*, 265–267. [Proceeding Paper] International Conference on Business and Economics (ICBE 2017), the Korea Distribution Science Association/Ho Chi Minh City, Vietnam. Web of Science.

Wang, C.L. and Ahmed, P.K. (2004). The development and validation of the organisational innovativeness construct using confirmatory factor analysis. *European Journal of Innovation Management* 7 (4): 303–313.

Weick, K.E. (1989). Theory construction as disciplined imagination. *The Academy of Management Review* 14 (4): 516–531. https://doi.org/10.2307/258556.

Winch, G.M. (2000). Innovativeness in British and French construction: The evidence from Transmanche-Link. *Construction Management and Economics* 18 (7): 807–817. https://doi.org/10.1080/014461900433096.

Worsnop, T., Miraglia, S., and Davies, A. (2016). Balancing open and closed innovation in megaprojects: Insights from Crossrail. *Project Management Journal* 47 (4): 79–94. https://doi.org/10.1177/875697281604700407.

12

The Future of Global Megaprojects

The future depends upon what we do in the present

— Mahatma Gandhi

Introduction

The megaprojects of the future will have a vastly different focus than the current and past megaprojects. They not only will have a new engineering and technological focus, but they will also be implemented using new frameworks, systems, and infrastructure architecture. While current practice is focused on cost and schedule compliance to maintain budgets, this will be weighed against the emerging principle-based standards that place attention more on outcomes than deliverables (PMI 2021).

This chapter presents the challenges of the next generation megaprojects including the rise of digitization and big data, the need for blue ocean, and blue economy megaprojects, the relocation of people from rural to urban megacities, human machine transformations, space exploration, curing the world's cancers, pandemics, and disease, developing more viable climate friendly energy resources, water transfer, connecting the world through tunnels and bridges, and meeting the new social requirements of sustainability. These new megaprojects of the twenty-first century will impact all industries and all regions of the world, and will require different competencies, a greater understanding of complex adaptive systems, technological innovation including artificial intelligence (AI) and machine learning, sustainability and resilience, and creative leadership to advance and succeed.

The goal of this final chapter is to understand the capabilities we will need and several of the emerging issues we will face in the megaprojects of the future.

Global Megaprojects: Lessons, Case Studies, and Expert Advice on International Megaproject Management, First Edition. Virginia A. Greiman.
© 2023 John Wiley & Sons, Inc. Published 2023 by John Wiley & Sons, Inc.

Table 12.1 Projects of the future.

Focus	Description
Opportunities	A risk or uncertainty turned into an opportunity that has a positive effect on a project's mission and goals
Outcomes	The long-term results based on benefits realization versus short-term budget and schedule adherence
Openness	Transparency in project planning, processes, policies, and communication channels
Optimization	Managed change and continuous improvement in project operations including governance, finance, technology, economic, and social dimensions
Blue Ocean Innovation Projects	A focus on differentiating the product or project from its competitors through innovation
Blue Economy Projects	The blue economy is a globally emerging concept for oceans governance that seeks to tap the economic potential of the oceans in environmentally sustainable ways

The Changing Practice of Megaprojects

> The consequences of our actions are so complicated, so diverse, that predicting the future is a very difficult business indeed.
> — *J. K. Rowling, Harry Potter, and the prisoner of Azkaban*

To ensure success, the megaprojects of the future will need to focus on opportunities, outcomes, openness, optimization, and blue ocean and blue economy projects as described in Table 12.1.

Blue Ocean Thinking

> Although the term "blue economy" has been used in different ways, according to the World Bank it is understood as comprising the range of economic sectors and related policies that together determine whether the use of oceanic resources is sustainable.
>
> (WB 2017)

In Chapter 11, we introduced the concept of Blue Oceans. In this chapter, we contrast Blue Ocean Thinking with the Blue Economy to better understand the challenges that the megaprojects of the future will face. The Blue Ocean Strategy Institute at Insead has compiled multiple academic studies that show Blue Ocean strategy has a higher ROI than traditional Red Ocean competition (INSEAD 2022). While you can be successful following a Red Ocean strategy (there are myriad examples of companies that have dominated their space by competing better than their peers, such as Disney, Exxon, and GE). Overall, the results from pursuing a Blue Ocean strategy are likely to surpass the results of competing in a Red Ocean (Kim and Mauborgne 2015, 2017). While organizations globally are implementing blue ocean strategies, at the same time, we are seeing a form of blue ocean thinking creeping into the megaproject arena that includes never-been-done-before projects like colonizing Mars, tube transportation networks, floating islands, and underwater cities (Frey 2016).

Examples of Global Megaproject Blue Ocean Thinking

1) Industrial Corridors
2) Global Tube Transportation Projects
3) Hadron Collider
4) Small Modular Reactors

Blue Economy Thinking

The blue economy is a globally emerging concept for oceans governance that seeks to tap the economic potential of the oceans in environmentally sustainable ways. Yet understanding and implementation of particular visions of the blue economy in specific regions diverge according to national and other contexts (ECA 2016).

While the very concept of the blue economy is still awaiting greater global consensus, the Nairobi Scientific and Technology Conference adopted and accepted a definition that has gained recognition. The blue economy is seen as:

> The productive pillar of the oceans, seas, lakes, and rivers, including fisheries, aquaculture, tourism, transport, shipbuilding, energy, bioprospecting, and underwater mining, as well as all related activities (UN 2018). Two conceptual pillars have been constructed including: (i) sustainability, climate change and controlling pollution; (ii) production, accelerated economic growth, jobs and poverty alleviation.

About 71% of the planet's surface is covered by water serving as our planet's largest life support system, and 96.5% of this water is found in the oceans. In addition, ocean currents govern the world's weather and its dependent biomes; they regulate global climate and temperature, and so support many diverse forms of life. The oceans are estimated to have absorbed 25% of human-induced carbon dioxide since the Industrial Revolution resulting in a 26% increase in ocean acidity (Visbeck 2018). The acidity has reduced the ability of oceans to absorb more carbon dioxide, therefore undermining marine ecosystems through changing ocean chemistry. Oceans currently provide only 2% of the world's food supply on a caloric basis. Aquaculture, though, is the fastest-growing food sector producing about 47% of the fish consumed globally. However, 87% of global fish stocks are already fully or overexploited. Globally, 350 million jobs are linked to marine fisheries. There is, therefore, little doubt that the blue economy holds huge untapped potential for eradicating poverty, providing full employment and decent jobs, and addressing environmental challenges (WB 2017).

An important challenge of the blue economy is thus to understand and better manage the many aspects of oceanic sustainability, ranging from sustainable fisheries to ecosystem health to pollution as well as how tomorrow's megaprojects can contribute to these new opportunities.

A second significant issue is the realization that the sustainable management of ocean resources requires collaboration across nation-states and across the public–private sectors, and on a scale that has not been previously achieved. This realization underscores the challenge facing the Small Island Developing States (SIDS) and Least Developed Countries (LDCs) as they turn to better managing their blue economies (Outa et al. 2021). The "blue economy" concept seeks to promote economic growth, social inclusion, and the preservation or improvement of livelihoods while at the same time ensuring environmental sustainability of the oceans and coastal areas. At its core it refers to the decoupling of socioeconomic development through oceans-related sectors and activities from environmental and ecosystems degradation. It draws from scientific findings that ocean resources are limited, and that the health of the oceans has drastically declined due to anthropogenic activities (Rodriques et al. 2020). These changes are already being profoundly felt, affecting human well-being and societies, and the impacts are likely to be amplified in the future, especially in view of projected population growth (WB 2017).

Grand Cancun Eco Island, Quintana Roo, Mexico. *Source:* Beautifullife.info.

Challenges of Future Megaprojects

To address the challenges of megaprojects in the future, it will require an understanding of the following concepts: sustainability; complex adaptive systems; resilience; and emergence.

Sustainability

Sustainability in the project profession is an approach that balances the environmental, social, economic aspects of project-based working to meet the current needs of stakeholders without compromising or overburdening future generations (APM 2019). The association of project management (APM) in the United Kingdom divides sustainability into four elements – the environment, social, economic, and administration.

Sustainability, defined in the United Nations 2030 Sustainable Development Goals as improving the quality of our lives, protecting our ecosystem, and preserving natural resources for future generations (UN 2015) will be advanced through a focus on resilience.

As we discussed in prior chapters, sustainability is essential for the success of all projects across economic and social dimensions including environmental, climate, and food security issues. Sustainability also matters in financial decisions. Multilateral institutions such as the World Bank and IMF play a critical role in improving sustainability in the financial realm. This is done by improving coverage in the databases they manage, providing technical assistance, performing assessments on a country's current debt management framework, designing international standards on debt, and building awareness. Lenders, along with borrowers, also have a responsibility to report debt information fully, accurately, and transparently. All these steps will help creditors to (i) assess accurately the debt sustainability of their potential borrowers, (ii) help citizens to hold their governments accountable for the debt they assume, and (iii) borrowers to design strategies based on a clear understanding of the level, cost, and risk profile of their debt portfolio. Increased debt transparency will also help many low and middle-income countries to assess and manage their external debt during and after the COVID-19 crisis and to work with policy makers toward achieving sustainable debt levels and terms (WB 2021).

> Trust, resilience, and adaptive capacity are built by inviting stakeholders into the decision-making process.

Complex Adaptive Systems

Complex adaptive systems are self-organizing and constantly evolving. "They are characterized by feedback loops, tipping points and emergent properties, and their future is impacted by their past, and surprise is inevitable" (Gillin 2019). Examples of complex adaptive systems include farms, forests, cities, infrastructure, and our immune systems. All complex adaptive systems share common characteristics (Folke et al. 2010), irrespective of whether the complex adaptive system is the economy, the internet, an ant colony, or the brain (Holland 2014; Johnson 2010). These characteristics include: first, a complex system where many diverse agents interact with each other. Second, it is adaptive, meaning it has the capacity to change based on experience and thus evolve, giving it resilience in the face of disturbance. Finally, the interaction can lead to a third stage called emergence (Holland 1999). The key issue is that you cannot really understand the whole system by simply looking at its individual parts (Sullivan 2011). Current project management practices are poorly equipped to deal with the nonlinear nature of complex adaptive systems. As such, alternative ways of working will need to be developed, to enable management of projects in a manner that is holistic, embraces uncertainty, and values adaptive capacity.

When we look toward the future in any project, we see a lot of uncertainty often brought on by these adaptive systems. Traditionally, projects have managed uncertainty as a risk rather than thinking of uncertainty as an opportunity to exploit. As discussed in the Chapter 12 the projects of the future will require a focus on innovation and entrepreneurship to remain resilient. They will require innovation that is sustainable.

Daniel and Daniel (2019) studied complex adaptive theory in the Hinkley Point C (HPC) nuclear power plant megaproject launched in 2012. The ambition was to build a Franco-British power station, which was to be commissioned in 2018 at an estimated cost of £15B. It is a very high-profile case with organizational complexity, and it has undergone performance changes during the front-end phase. The research highlighted the necessity to revise the conventional project management model, emphasizing the strategic role of collaborative planning and control, in a project management process that is built on adaptation. Based on the results from their theoretical analysis they proposed a revised project management model based on four findings:

1) Megaproject perception of success is multidimensional – The project's performance is the result of shared (limited) agents' visions and goals.
2) Megaproject self-organization is based on each stakeholder's source of control – the project's performance is the result of shared capacity to control the outputs during project implementation.

3) Megaproject coevolution is based on stakeholders' reactions towards new deliverables – the process of adaptation is a result of a collaborative interaction of agents because of other agents producing unexpected outputs.

4) Megaproject recombination reveals transformations in the output achieved and in the stakeholders' relations and roles in the project – the project agents make decisions of adaptation resulting in modifications of their role and action. This process of adaptation is a result of (involuntary collaboration). This is an important research contribution even with its limitations (being a single case study) as it enhances the understanding of how large megaprojects are planned and controlled by multiple agents under conditions of self-organization and coevolution.

Sustainable Repurposing

Taking care of our environment will be a major concern of all projects moving forward, while at the same time looking for new forms of sustainable energy. The nuclear industry will be facing many power-plant decommissioning's over the coming decades and how we manage these massive megaprojects and repurpose these sites in a sustainable way will be a tremendous challenge to countries globally. Aspects of what sustainable decommissioning entails have already been demonstrated to be successful, through projects such as the revitalization of the Fernald site in the United States, as well as the repurposing of the RI reactor hall in Stockholm, Sweden, and the Greifswald nuclear power plant site. Attributes of a sustainable nuclear decommissioning approach include (Gillin 2019):

- **Inclusive:** Enabling the public and other external stakeholders to actively participate in the decision making. Making room at the table for difficult decisions.
- **Asset Focused:** Considering every part of the system as being of value as a building block in a future use. Repurposing ought to be considered as well.
- **Integrated:** How to replace the power and jobs lost, and how to mitigate impact on surrounding communities.
- **Vision-based:** Determine what the site will be used for post-decommissioning prior to commencing the planning. Rather than viewing the end point as a release from regulatory control, it should be placed later beyond decommissioning and coincide with the point in time when new uses and reuses are fully operational on the site. This way, a shared vision is created, towards which both internal and external stakeholders can strive. Planning for decommissioning becomes planning for site transformation from current to future use.

Battersea Power Station Apple London Campus. *Source:* Ben Lovejoy/Twitter, Inc.

An example of sustainable repurposing is found in the London iconic Battersea Power Station, a decommissioned coal-fired power station, located on the south bank of the River Thames (see Photo). The power station lay derelict for 30 years before work began on its redevelopment in 2014. The redevelopment is London's biggest regeneration scheme and one that sets a model for other power station transitions (Lovejoy 2021). The 42-acre mega project is breathing new life into the landmark building (featured on a Pink Floyd LP album cover) and surrounding area of Nine Elms, and will include thousands of homes, 2.5 million square feet of office and retail space, a new tube station, eateries, cultural institutions, and entertainment venues. The Apple Inc. campus will fill up to 40% of the available office space when it is completed. The total cost of the mega project is estimated at $10.6B. These examples show that sustainability can be achieved even in the most difficult situations of repurposing nuclear brownfield sites. As of 2021, the building and the overall 42 acres (17 ha) site development is owned by a consortium of Malaysian investors.

Resilience

The third concept essential to managing future large scale and complex projects is resilience. Resilience was originally introduced by Holling (1973) as a concept to help understand the capacity of ecosystems with alternative attractors to persist in the original state subject to perturbations, as reviewed by Gunderson (2000), Folke

(2006), and Scheffer (2009). In some fields the term resilience has been technically used in a narrow sense to refer to the return rate to equilibrium upon a perturbation called engineering resilience by Holling (1996).

Walker et al. (2004) exposited the relationships among resilience, adaptability, and transformability by defining resilience as "the capacity of a system to absorb disturbance and reorganize while undergoing change so as to still retain essentially the same function, structure, identity, and feedbacks" (Walker et al. 2004, p. 4).

Resilience requires continually adapting to our changing environment through innovation and an entrepreneurial spirit. It means learning from the past, while at the same time pushing forward to a higher level of understanding. Examples of resilience are abundant in our world history, but the characteristics of resilience are often not applied in the course of megaproject development at least not intentionally. The key to managing such systems is to understand and attempt to influence their resilience (Gillin 2019).

Resilience thinking addresses the dynamics and development of complex social–ecological systems (SES). Three aspects are central: resilience, adaptability, and transformability (Folke et al. 2010). These aspects interrelate across multiple scales. Resilience in this context is the capacity of a SES to continually change and adapt yet remain within critical thresholds. Adaptability is part of resilience. It represents the capacity to adjust responses to changing external drivers and internal processes and thereby allow for development along the current trajectory (stability domain). Transformability is the capacity to cross thresholds into new development trajectories allowing transformational change at smaller scales, while enabling resilience at larger scales (Folke et al. 2010).

Given the impossibility of anticipating, planning, and defending against all the crises that a megaproject might face, the concept of resilience has been given a great amount of attention in the project management realm, especially since the ongoing COVID-19 pandemic (OECD 2020). Improving the resilience of projects, which refers to the ability to prepare, respond, and recover from the unexpected is not only a decisive factor for the project's survival, but also related to the interests of all participating stakeholder organizations. The complexity, internal and external risks, and significant social impact of megaprojects make their organizational resilience particularly important. To survive potential adversities, megaproject organizational resilience depends on collective mindfulness (Wang et al. 2021).

Infrastructure, built in large part through megaproject activity, is the backbone of our communities, providing not only critical services (such as water, transportation, electricity, and communications), but also the means for health, safety, and economic growth. These systems often extend beyond our communities providing service to entire regions and contributing to the delivery of national critical functions. Given the vital importance of infrastructure to our social and economic well-being, it is imperative we ensure our networks are strong, secure, and resilient. For communities to thrive in the face of uncontrollable circumstances and adapt to changing conditions (e.g. evolving security threats, impacts from extreme weather, technological development, and socio-economic shifts), we must work to make our infrastructure more resilient.

In the United States resilience in the cyber space has become a major factor in addressing the threat of cyber-attacks (Greiman and Bernardin 2021). The Cybersecurity and Infrastructure Security Agency (CISA) developed the Infrastructure Resilience Planning Framework (IRPF) to provide an approach for localities, regions, and the private sector to work together to plan for the security and resilience of critical infrastructure services in the face of multiple threats and changes. The National Institute of Standards and Technology's (NIST) Community Resilience Planning Guide (CRPG) provides tools and resources for integrating critical infrastructure into planning as well as a framework for working regionally and across systems and jurisdictions.

The Era of Emergence

> Tomorrow is the most important thing in life. Comes into us at midnight very clean. It's perfect when it arrives, and it puts itself in our hands. It hopes we've learned something from yesterday.
>
> — *John Wayne*

Because of the vast eco system they encompass, there is interest in megaprojects worldwide. Contemporary megaprojects consist of a multitude of stakeholders that have both an interest and often influence in the project. "They leverage cutting-edge technology to 'see' complex systems as legible and singular phenomena" (Schindler et al. 2019). "As a result, they are more ambitious, more pervasive and they have the potential to reconfigure longstanding relationships that have animated social and ecological systems" (p. 1).

Emergence is critically important to understand the complex and interrelated systems embedded in megaprojects. Emergence is the principle of self-organization that explains how complexity can arise from simplicity, and order can sometimes emerge from chaos. Emergence explains complex phenomena in diverse fields such as thermodynamics, biology, and digital technology. It has been used to explain how cities evolve, and what makes some institutions succeed in the long term while others fail. It is an underlying principle behind the complex collective behavior of natural systems, and it is part of why computers can achieve such useful complexity, even though they are built on a foundation of just a few simple processes.

According to Beth Comstock, former Vice-Chair of General Electric we are in an emergent era, characterized by a constant state of evolution. She defines the emergent organization as an adaptive organization where solutions to problems and opportunities will spontaneously emerge, before needs or demand exists. She has suggested six principles for leading in the emergent era:

1) Organize around information flows; ditch hierarchy and bureaucracy – this involves access to real-time data and open communication throughout the organization.
2) Empower individuals – encourage collaboration and localized decision-making.
3) Replace long lists of rules with a good *modus operandi* (m.o.), but more importantly mission objective and mindset orientation. This combined m.o. is intended to encourage creativity and speedy execution.
4) Get used to living in the "in between" – we have to abandon the concepts of total safety (risk aversion) and comprehensive knowledge and embrace uncertainty.
5) Open up new feedback loops – feedback needs to be open and honest. Failure, after honest effort, is seen as a mode of learning and should be communicated. It could lead to the next great idea.
6) Tap into the power of minds and machines – capitalizing on machine simulations and artificial intelligence to spark human creativity can multiply the independent strengths of each and lead to innovation.

Meeting the Grand Challenges of the Twenty-first Century

> Artificial intelligence will reach human levels by around 2029. Follow that out further to, say, 2045, we will have multiplied the intelligence, the human biological machine intelligence of our civilization a billion-fold.
>
> — *Ray Kurzweil*

In recent years, there has been a growing scientific and societal interest for grand challenges (e.g. climate change, global health, sustainable cities, justice and equality, and transformative technology). Such grand challenges are strongly dependent on the performance of megaprojects and on our human capacity to understand how collaboration can support project management teams in coping with emergence and instability. Technological advancements in sustainable transport, energy, and waste management, zero carbon building design, pollution abatement, urban food systems, wearable sensors, mobile applications, and big data analysis provide opportunities to improve the health of urban populations while reducing their pollution and carbon footprint.

However, technology solutions alone are unlikely to solve the persistent environmental, societal and health problems that twenty-first century cities are facing around the world. To achieve sustainable, long-term solutions we need concerted action from key stakeholders – government, planners, health professionals, researchers, businesses – as well as individual citizens. Importantly, we need to develop a better understanding of complex societal, technological, and behavioral barriers and enablers for effective change toward healthier and more sustainable cities. This requires holistic, system-based approaches that address multiple environmental and health hazards, taking into account co-benefits, and unintended consequences (Vardoulakis and Kinney 2019).

Grand challenges are ambitious but achievable goals that harness science, technology, and innovation to solve important national or global problems. They also have the wonderful potential to capture and enhance the public's imagination. Among the many benefits, grand challenges can:

- Help create the industries and jobs of the future.
- Expand the frontiers of human knowledge about ourselves and the world around us.
- Tackle important problems related to energy, health, education, the environment, national security, and global development; and
- Bring together private industry, governments, countries, and regions of the world to partner for social and economic good.

Grand challenges include the National Institute of Health (NIH), Defense Advanced Research Project Agency (DARPA) and the National Science Foundation's (NSF) Brain Initiative to revolutionize our understanding of the human mind and uncover new ways to treat, prevent, and cure brain disorders such as Alzheimer's schizophrenia, autism, epilepsy, and traumatic brain injury, DOE's SunShot Grand Challenge to make solar energy cost competitive with coal by the end of the decade or fill the roads with electric vehicles that are as affordable as gasoline powered vehicles within the next five years, or NASA's Planetary Defense Grand Challenge to find all asteroid and other threats to human populations and know what to do with them in the next 10 years. Recently, the Office of Science and Technology Policy (OSTP) of the White House, introduced a new strategic plan that aims to strengthen America's quantum ecosystem by developing a diverse, inclusive, and sustainable workforce (WH 2022).

Defining a National Ambition

To meet the grand challenges of the future requires a national ambition and a national strategy that clearly defines the opportunities and goals of the strategy and requires government and private sector buy-in and sponsorship.

Megaprojects are a completely different type of project in terms of their level of aspiration, lead times, complexity, and stakeholder involvement. Consequently, they are also a very different type of project to manage requiring high level government sponsorship and financial support from multiple sources. Albert O. Hirschman, the renown social scientist in the economics and politics of the developing world, (1995, p. vii, xi) calls such projects "privileged particles of the development process" and points out that often megaprojects are "trait making"; in other words, they are designed to ambitiously change the structure of society, as opposed to smaller and more conventional projects that are "trait taking"; that is, they fit into preexisting structures and do not attempt to modify them (Flyvbjerg 2014, p. 6). Megaprojects, therefore, are not just magnified versions of smaller projects. Defining a national ambition requires the following steps (Maher et al. 2021):

1) The first step in defining a national ambition is the development of a national strategy. The massive undertakings of the future will evolve around governments willingness to define a national ambition in light of the evolving opportunities in the emerging, digitally connected, global economy.
2) Governments should set key goals, such as new job creation, launching regional champions, boosting exports, or meeting urgent social needs.
3) Policymakers should then thoroughly assess their nation's existing competitive strengths and innovation capacity to deliver on the vision including targeting priority innovation sectors. Based on this analysis, policymakers should identify industrial sectors that are in the strongest position to achieve key national goals.

Renewable Energy Projects

Egypt possesses an abundance of land, sunny weather, and high wind speeds, making it a prime location for renewable energy projects. The renewable energy equipment market is potentially worth billions of dollars. The Egyptian government is cognizant of the need for a sustainable energy mix to both address increasing demand, and to move to a more environmentally sustainable and diverse electricity sector. The 2035 Integrated Sustainable Energy Strategy, which builds on previous strategies, emphasizes the importance of renewable energy. Egypt intends to increase the supply of electricity generated from renewable sources to 20% by 2022 and 42% by 2035, with wind providing 14%, hydro power 2%, photovoltaic (PV) 22%, and concentrating solar power (CSP) 3% by 2035. The private sector is expected to deliver most of this capacity (ITA 2021).

National Digital Health Strategy

South Africa's National Digital Health Strategy, launched in 2019, aims to build on its extensive health technology ecosystem to scale up innovative solutions for fighting disease and improving the quality of care.

Green Energy Initiatives

Some governments have taken a stronger environmental stance by removing subsidies for carbon-intensive sectors or making bailouts dependent upon environmental performance or commitments.

- Air France, for example, is required to reduce emissions by 50% and achieve a minimum standard of 2% renewable fuel by 2030 as conditions of its rescue by the French government (Atalia et al. 2021). Fleet renewal, with the integration of the latest generation of environmentally more efficient aircraft – Airbus A220, Airbus A350 and Boeing 787 for Air France. In 2020, Air France ceased operating four-engine aircraft, Airbus A340 and Airbus A380.
- The use of Sustainable Aviation Fuels (SAF), non-fossil fuels produced from industrial or domestic waste that do not compete with the human food chain. Air France has been a pioneer in the use of these sustainable fuels, and this year operated the first ever flight using sustainable aviation fuel produced in France, as well as a Nice-Paris flight powered by 30% sustainable aviation fuel. These fuels can reduce CO_2 emissions by an average of 80% over the entire life cycle and are set to play a key role in decarbonizing air transport. Air France and the Air France-KLM Group are working to make these fuels more accessible in terms of quantity and price, through the creation of a genuine production sector in France and Europe (Air France 2021).

Other governments are passing legislation to ban environmentally harmful energy or mandate green energy. For example, India's "Development of Solar Cities" scheme, which covers up to 60 cities, insists on solar water heating systems in certain categories of buildings, as well as providing financial and technical assistance for increasing renewable energy and energy efficiency. As of the end of 2021, 31 states and the District of Columbia had renewable portfolio standards (RPS) or clean energy standards (CES).

Several governments have taken stringent measures to ban polluting products outright. Plastic, which accelerates climate change by emitting greenhouse gases at every stage of its lifecycle, has been a particular target. Rwanda became the world's first "plastic-free" nation in 2009, 10 years after introducing a ban on all plastic bags and plastic packaging. In May 2021, Canada declared plastic a "toxic" substance, paving the way for its proposed ban on most single-use plastics by the end of 2022 (Atalia et al. 2021).

World Climate Change Initiatives

Global climate is the result of a complex interaction between many different phenomena. Some experts claim that consistent variations in global and regional climates in the past were provoked by relatively modest temperature changes taking place over a few decades or even a few years. Recently, the United Kingdom became the first major economy to legislate bringing all greenhouse gas emissions to Net Zero by 2050. The UK's Net Zero target will need to be reinforced by credible policies across government, inspiring a strong response from business, industry, professional bodies, and society as a whole.

Smart Cities and Inclusive Growth Initiatives

Smart cities are being established all over the developing world and have been used effectively to help build urban areas and have contributed to social and economic growth in India, Indonesia, and other developing countries. In 2020, the Organization for Economic Growth and Development conducted the first Roundtable on Smart Cities and Inclusive Growth (OECD 2020). The Roundtable produced five findings for developing the Smart City:

1) While the digital revolution is offering an unprecedented window of opportunity to improve the lives of millions of urban residents, there is no guarantee that the rapid diffusion of new technologies will automatically benefit citizens across the board. Smart city policies need to be designed, implemented, and monitored as a tool to improve well-being for all people.
2) National Governments should play an enabling role. Building smart cities is not only the business of cities or the private sector, but national governments can also and should play an enabling role to support innovative solution delivery, capacity building, and upscaling.
3) Measuring smart city performance is a complex task but is critically required. Advancing the measurement agenda calls for a comprehensive, multi-sectoral, and flexible framework that is aligned with local and national strategic priorities and embraces efficiency, effectiveness, and sustainability dimensions.
4) Smart cities need smart governance. Business and contractual models need to adapt to rapidly changing urban environments and encompass a more holistic approach, sometimes re-regulate rather than simply de-regulate, and leverage public procurement, including at the pre-procurement stage.
5) Citizens are not only recipients but also actors of smart city policies. Putting people at the centre of smart cities means co-constructing policies with citizens throughout the policy cycle.

Zero Carbon Emissions

Six actions to address Zero Carbon Emissions
1) Provide detailed action plans with clear accountability 2) Be bolder in incentivizing the market and mandating change 3) Boost innovation through increased funding 4) Improve the design and delivery of green initiatives 5) Act as a role model for other parts of the economy 6) Promote a whole-of-society, people-centered approach
Source: Adapted from Atalia (2021).

Global Change and International Cooperation

> I have no doubt in stating that megaprojects dealing with global change issues should have the highest priority.
> — *Umberto Colombo, Former Italian Minister of Universities, Science and Technology*

In recent years, science and technology megaprojects have taken an increasing portion of research budgets, especially governmental research budgets. At a time of scarce resources, they are increasingly judged on their cost and on their direct contribution to responding to near-term societal demand, not so much on long-term strategic relevance (Colombo 2020). In many areas of science, national funding constraints are driving increased interest in international cooperation. The unusually high cost of megaprojects constitutes an insurmountable barrier for smaller countries to undertake them on a national basis. But, increasingly, even the larger countries are finding it difficult to take upon themselves large megaprojects without international cooperation. Distributed facility megaprojects, to which global change belongs, are not associated with a particular geographical location, and therefore require an effective international coordination, in order not to behave as loose networks. Scientists in many institutions and locations participate, and funding often comes from a variety of sources. Researchers from many fields of science are engaged in programs pertaining to global-scale issues, such as World Climate Change.

Global Megaprojects and the Leaders of Tomorrow

> The future of space exploration project management will rely on the innovations and ideas of the people involved in the space program. They will dream what can be possible, and then go about the task of making it happen.
> — *Dorothy Tiffany, Former Chief, Project Management, NASA*

The Excellence and Innovation Office at NASA Goddard Space Flight Center Project Management is evolving in theory and practice. Understanding the future of project management, and how project management processes and tools will be applied, is critical for every project management professional and student. In their scholarly treatise on the future of project management, Project Management Circa 2025, the authors provide important examples of how projects will be utilized to resolve such diverse issues and problems as nanotechnology and energy resources, sustainable manufacturing, conquering new frontiers in space exploration, and monitoring the planet, extreme weather response, and climate control (Cleland and Bidanda 2009).

To understand the types of leaders that will be needed in the world of tomorrow, it is helpful to take a look at the projects of tomorrow and how the changing needs of the world will require a different approach for managing projects (Greiman 2013, p. 408). Some examples of innovative initiatives in the field of project management are highlighted in this section.

United States. The transportation sector accounts for more than 10% of the U.S. gross domestic product, behind only housing, food, and health care. Across the country, taxpayers are pumping billions of dollars into innovative transportation initiatives including the Next Generation Air Transportation System (NextGen) to track aircraft with greater accuracy, integrity, and reliability. NextGen is not one technology, product, or goal. Rather, it is a series of interlinked programs, portfolios, systems, policies, and procedures. It implements advanced technologies and capabilities that dramatically improve the operation of the National Airspace System (NAS).

These investments have improved the safety and efficiency of the nation's system of highways, transit, ports, and airports. Just as important, these projects generated tens of thousands of jobs in transportation and related sectors in a difficult economic environment. Funds have also been designated for transportation infrastructure, including transit capital assistance, high-speed rail, pavement improvements, and bridge repair, as well as the preservation and creation of jobs, to promote economic recovery. Other US-based initiatives include a surge in wind energy production and manufacturing, the development of biofuel technologies, and projects advancing innovative clean coal technology, and the growth of natural shale gas projects across the country.

Developing World. Despite the global economic crisis, megaprojects also continue to be of importance to the developing world. According to the World Bank, in the coming decades the number of international projects will increase exponentially. Current efforts are under way to protect human capital in Latin America and the Caribbean amid COVID-19, to strengthen the social safety net to build resilience and protect human capital in Somalia, to implement digital transformation in Europe and Central Asia to power green, resilient and inclusive development, and to support countries in the Middle East and North Africa toward a green, resilient and inclusive recovery (WB 2021).

International. Megaprojects of the future include the exploration of cyberspace, advancing human understanding of the laws of nature, and the exploration of physics and nuclear fusion research, as demonstrated by the following international projects:

1) **The Large Hadron Collider (LHC)** constructed across Switzerland and France is the world's largest and highest-energy particle accelerator, built by the European Organization for Nuclear Research (CERN) from 1998 to 2008, with the aim of allowing physicists to test the predictions of different theories of particle physics and high-energy physics. It contains six detectors, each designed for a specific kind of exploration. The LHC is expected to address some of the most fundamental questions of physics and advance human understanding of the deepest laws of nature.

Hadron Collider. *Source:* CERN/New Atlas.

2) **The International Thermonuclear Experimental Reactor (ITER)** is presently building the world's largest and most advanced experimental tokamak nuclear fusion reactor in Provence, southern France (ITER 2022). ITER (originally the International Thermonuclear Experimental Reactor) is an international nuclear fusion research and engineering megaproject. This is a true international endeavor run by seven member entities: the European Union, China, India, Japan, Russia, South Korea, and the United States. Overall, 35 countries are participating in the project directly or indirectly. The

project was initiated in 1988 and is expected to start full deuterium-tritium fusion experiments in 2035. (Fusion is the process where two hydrogen atoms combine to form a helium atom, releasing energy). The Manhattan Project to develop the world's first nuclear weapon lasted for 6 years, this one will take 50 years or more and is by far the longest duration international megaproject ever constructed.

3) **The International Space Station (ISS)** has been described as the greatest human endeavor in the history of the world in terms of complexity and meeting the challenges presented. The ISS serves as a microgravity and space environment research laboratory in which crew members conduct experiments in biology, human biology, physics, astronomy, meteorology, and other fields. The station is suited for the testing of spacecraft systems and equipment required for missions to the Moon and Mars. The station has been continuously occupied Since November 2000. As of 2021 it has been visited by astronauts and cosmonauts from 19 different nations.

On 25 May 2021, Space Exploration Technologies Corporation (or SpaceX) became the world's first privately held company to send a cargo load, the Dragon spacecraft, to the International Space Station. In 2020, SpaceX began transporting people to the orbiting laboratory under NASA's Commercial Crew Program.

4) **The National Human Research Genome Institute** is an international effort to sequence the three billion DNA letters in the human genome is considered by many to be one of the most ambitious scientific undertakings of all time, even compared to splitting the atom or going to the moon (NIH 2022). In 2003, an accurate and complete human genome sequence was finished two years ahead of schedule and at a cost less than the original estimated budget. The finished sequence produced by the Human Genome Project covers about 99% of the human genome's gene-containing regions, and it has been sequenced to an accuracy of 99.99%. In addition, to help researchers better understand the meaning of the human genetic instruction book, the project took on a wide range of other goals, from sequencing the genomes of model organisms to developing new technologies to study whole genomes.

Besides delivering on the stated goals, the international network of researchers has produced an amazing array of advances that most scientists had not expected until much later. These "bonus" accomplishments include: an advanced draft of the mouse genome sequence, published in December 2002; an initial draft of the rat genome sequence, produced in November 2002; the identification of more than three million human genetic variations, called single nucleotide polymorphisms (SNPs); and the generation of full-length complementary DNAs (cDNAs) for more than 70% of known human and mouse genes.

Megaprojects and the Growth of the Digital Economy

> People blame technology for humanity's problems. They are much too pessimistic about its power to solve poverty, disease, and pollution in our lifetimes.
>
> — *Ray Kurzweil*

Technological Innovation in Megaprojects: Industry 4.0 Driven Projects

Megaprojects are not confined to large infrastructure or industrial projects but include research and development, cyber infrastructure, defense, food security, biotechnology, and many other fields. Given the increasing number of megaprojects and the movement towards the professionalization of the discipline, project management practice has struggled to develop workable solutions and practices to address complexities that exceed the technical concerns of engineering (Clegg et al. 2017; van Marrewijk et al. 2016). Among the technologies that megaprojects must adopt to advance in a technology driven world are the following:

1. Artificial Intelligence

To overcome these challenges machine learning and intelligent design have become critical to every megaproject. Governments and private industry face an unprecedented challenge in managing AI systems designed for megaprojects that include regulators, markets, and special interests that all play a role in influencing the development of AI in different contexts without a full appreciation of the impact of AI on people's livelihood and other consequences (Greiman and Bernardin 2021). In his 2016 testimony before the Senate Armed Services Committee, Commander of U.S. Cyber Command Admiral Michael Rogers stated that:

> "Relying on human intelligence alone in cyberspace is a losing strategy" (Rogers 2016). "If you can't get some level of AI or machine learning with the volume of activity you're trying to understand when you're defending networks . . . you are always behind the power curve"
>
> *(Lyle 2016).*

This disruptive potential has triggered rapidly growing investments in AI research and development (R&D) as well as speedy uptake in the public and private sectors worldwide. By 2030, AI could contribute as much as $13 trillion to the global economy, a figure that approaches the current annual economic output of China, the world's second-largest economy (Bughin 2018; PwC 2017).

There exists both benefits and challenges of using AI in building megaprojects. AI could play a crucial role in many aspects of human problems including climate change, international conflict, and medical breakthroughs, and could contribute greatly to improving human exploration, scientific knowledge and even quality of life. However, the literature generally has raised a number of areas where AI deployment has created serious concerns about the impact on decisions and outcomes (Raso et al. 2018). Before deployment of AI two questions must be asked:

(1) How will AI help deliver "better" projects?
(2) What are the risks, managerial implications, and challenges in designing and implementing AI in major projects?

Some of our more important issues in recent years such as project security, protecting the health and safety of its workers, project sustainability and value creation, and managing the impact of climate change and a global pandemic involve the application of AI. All of these require a more agile approach to project management and a more sophisticated intelligence that can be generated through Algorithms that are used to generate artificially intelligent systems (Greiman 2020).

Artificial intelligence (AI) is expected to transform economies and impact virtually every aspect of human life over the next couple of decades. This disruptive potential has triggered rapidly growing investments in AI research and development (R&D) as well as speedy uptake in the public and private sectors worldwide (Bartneck et al. 2021). Moreover, as AI applications are expanding to a wide range of sectors, early adopters will be well positioned to reap significant economic and strategic benefits.

In Europe leaders such as former German Chancellor Angela Merkel and French President Emmanuel Macron have stressed the need for Europe to become a leading global player on AI, and the new European Commission has made AI a top priority for the next five years (EC 2020). By declaring AI, a major strategic priority, several member states and EU institutions are taking steps to advance the continent's ambitions for AI leadership. This includes rolling out documents devoted to national and EU-level AI strategy, boosting research and innovation, and exploring new regulatory approaches for managing the development and use of AI (EC 2020).

AI is expected to play a major role in shaping global competitiveness and productivity over the next couple of decades, granting early adopters significant societal, economic, and strategic advantages. As the pace of AI innovation and development picks up – underpinned by advancements in big data and high-performance computing – the United States and China are taking the lead. Europe, meanwhile, despite having certain advantages such as a strong industrial base and leading AI research and talent, is still not sufficiently advanced according to the Carnegie Endowment for International Peace (Brattberg et al. 2020). This state of affairs is especially due to the fragmentation of the EU's digital market, difficulties in attracting human capital and external investment, and the lack of commercial competitiveness.

From the military perspective, Dr. Arati Prabhakar, the former DARPA Director commented, "When we look at what's happening with AI, we see something that is very powerful, but we also see a technology that is still quite fundamentally limited . . . the problem is that when it's wrong, it's wrong in ways that no human would ever be wrong (Pomerlau 2016).

Ethical Concerns in the Deployment of AI   Decisions made by an AI system's human designers can have significant human rights consequences both positive and negative which are informed by the individual life experiences and biases of the designers. The one AI tool with the greatest threat for ethical considerations in megaprojects is the tool designed to help select managers, staff, and team members. It would be a simple task to develop algorithms for compliance with state and federal laws regarding citizenship, age, gender, or specific job requirement skills like educational and work experience. However, there is also a risk of bias being introduced into the algorithms depending on how they are written (Linux 2005). For example, there may be gender-specific job sites or ethnic bias or directives to investigate a candidate's social media sites for political or personality traits deemed to be unacceptable by the algorithm developer. The only way to minimize this risk would be for all data and algorithms to be created through a diverse inter-disciplinary team with regular reviews and testing to ferret out bias.

One of the most important tools for avoiding or mitigating ethical concerns in hiring and promotion is to implement auditing systems that regularly test for bias and errors as well as intentionally designing algorithms to control for human biases. Just as we would test a product before it is released to the market we must also be constantly testing and evaluating the algorithms we are using to ensure that conflicts and bias are eliminated to the fullest extent possible.

2. Machine Learning

The recent advances in AI come under the rubric of "machine learning," (ML) which involves programming computers to learn from example data or past experience (Agrawal et al. 2017, p. 23). Environments with a high degree of complexity are where machine learning is most useful. Predicting the presence of a disease, a risk condition, or the need for repair of heavy equipment are all benefits of machine learning that are applicable to megaprojects. Machine learning and AI was introduced into the business platform in 2020 to create and empower professional business operations. Machine learning has become increasingly embedded in many new technologies and solutions, delivering in-depth insight into business metrics, and improving data-based decision-making.

AI and ML can help megaprojects analyze their data with less manual effort. AI is the development and use of any device that perceives its environment and takes actions that maximize its chance of success of reaching a defined goal (including learning and adapting to its environment). It is not a single technology but a family of technologies. ML is one application of AI. The algorithms that underlay AI rely on inputs of large amounts of data to learn and produce accurate and valuable insights. Based on adoption patterns, studies predict that firms responsible for about 70% of economic output will have adopted at least one type of AI technology by 2030.

Megaprojects rely on data-intensive analytic applications and big data analytics sometimes require that data be processed in different formats and distributed across different locations. These may include cloud computing, bio-inspired computing, or quantum computing. They also require the capacity to store big datasets and to clean them to correct inaccuracies.

Vast complexity is reflected in the sheer volume of data produced by megaprojects. Thousands of reports from designers, contractors, oversight authorities and other stakeholders creates a massive amount of data that the human mind cannot track leading to schedule delays and cost overruns. AI continues to make a contribution in this area as machine learning advances and the human machine relationship is developed.

3. Quantum Computing

Quantum Computing is still a growing innovation, that analysts, businesses, and governments have started working on this century and can be used in future megaprojects. The ability to achieve the first entirely useful, fully functioning quantum computer (additionally called supercomputer) continues to be developed. At present, there are 500 supercomputers in the world with China and the United States accounting for 60% of these computers. As opposed to on-premises machines, quantum PCs will most likely soon be a cloud service with its incredible computational power. IBM now offers Cloud-based quantum computing services.

4. Digital Twins

The concept of building digital twins to empower us to benefit from simulations has taken hold in the industry, and it holds much value in agriculture megaprojects as well. Exact innovation sensors and data collection allow point-by-point and close reality simulations to be developed progressively and carefully, try different things with new food production strategies and increase efficiency and harvest yields. Meanwhile, it is possible to observe and minimize the ecological effect of cultivation by enabling an opportunity to achieve a superior comprehension of the connections between farming and natural habitat which is a positive outcome impacting sustainability of a project.

5. New Nuclear Power

A new opportunity for megaprojects is the construction of new nuclear plants that hold promise to make this power source safer and less expensive. Age IV fission reactors are advancement of conventional methods; small modular reactors; and fusion reactors, an innovation that has seemed to be merely distant without ceasing. For example, Canada's Terrestrial Energy, and Washington-based TerraPower, have entered into R&D partnerships with utilities, aiming for grid supply.

6. Dextrous Robots

Industrial robots are still not capable of absorbing human jobs. A robot can get a segment with incredible accuracy over and over again on a mechanical production system. In any case, while a robot cannot yet be customized to make sense of how to get a grip on any product just by seeing it, as people do, it would now be able to figure out by virtual experimentation how to control the work alone. Robots are being used on megaprojects to conduct routine operations that remove contractors from unsafe situations. Construction companies are also increasingly relying on off-site factories staffed by autonomous robots that piece together components of a building, which are then pieced together by human workers on-site.

7. Blockchain Technology

Blockchain empowers transparency based on value over a variety of business roles. Blockchain is available at the core of business advancement in numerous companies. For example, to make digital transactions, you can use bitcoin, a virtual currency, and blockchain technology in bitcoin trading. Blockchain may be a source of financing megaprojects in the future. Shipping companies like Maersk are already experimenting with blockchain to track cargo and discourage tampering, while farmers (and even mega-chains like Walmart) are using the technology to follow and ensure the quality of livestock transactions. Even diamond-makers are following suit.

8. Supply Chain Management

Machine-learning forecasting engines apply algorithms and hierarchies to predict future need with exceptional accuracy. This is particularly relevant in megaprojects where thousands of suppliers must be coordinated. Food giant Nestlé uses supply chain forecasting to improve forecasting accuracy on a global level, with more than 447 factories operating in 194 countries. This strategy improved Nestlé's sales precision by 9% in Brazil alone (Rogers 2018).

9. Data Analytics

Careful data analysis is crucial for megaprojects to understand performance. This insight is increasingly valuable when coupled with analytical benchmarking, which allows megaprojects to compare themselves with their peers and competitors. Additionally, AI is bridging the gap between operational and predictive reporting. Predictive analytics can foresee everything from employee retention to long-term weather patterns, and machine learning continuously and automatically improves predictions with experience.

10. The Internet of Things

The internet-of-things (IoT) is the inter-networking of physical devices, vehicles, buildings, and other items – embedded with electronics, software, sensors, actuators, and network connectivity that enable these objects to collect and exchange data. When you consider the possibility of connecting any machine, any system, or any site to the Internet to know at any time what is happening, it rapidly becomes clear that the only limit in finding uses for IoT is our imagination. With the large amounts of equipment including software, heavy equipment and utilities needed on megaprojects IoT integration may become an essential component of megaprojects.

11. Natural Language Processing

One of the primary purposes of artificial intelligence in megaprojects is for the determination of risk (Greiman 2020). The number one priority on all projects is on-the-job safety. Natural Language Processing (NLP) is an interdisciplinary field involving humanistic, statistical–mathematical, and computer skills that has been used to understand risk perception. The aim of NLP is to process languages using computers. The human language can be defined as natural because it is ambiguous and changeable. On the contrary, machine language is defined as formal because it is unambiguous and internationally recognized (Di Giuda et al. 2020). The NLP must deal optimally with the ambiguity, imprecision, and lack of data inherent in natural language (Greiman 2020).

AI-based conversational tools have certainly advanced, but when technology relies on artificial language like Java or C++, it is automatically limited to literal translation. Human language is complex and brimming with subtleties, so there is ample opportunity for misunderstanding.

In contrast, natural language processing (NLP) solutions learn to speak organically through practice, just like people do. These tools can even discern a wide range of emotions and recognize the differences among anger, frustration, and fear.

NLP has unlimited potential with language barriers found in global megaprojects. Deloitte recently partnered with Kira Systems to develop NLP models capable of rapidly digesting complex documents and extracting important information for further analysis. This type of solution is likely to have incredible implications for law, finance, and other contract-heavy industries.

12. Drones and the Impact on Human Rights and Privacy

A drone, also known as unmanned aerial vehicle (UAV), is an aircraft without a human pilot on board (Kanellakis and Nikolakopoulos 2017). There has been a rapid development of drones for the past few decades due to the advancement of components such as micro electro-mechanical systems (MEMS) sensors, microprocessors, high energy lithium polymer (LiPo) batteries, as well as more efficient and compact actuators (Fotouhi et al. 2019). Drones are now present in many megaproject infrastructure activities. They are used in many applications such as inspecting pipelines and power lines, surveillance and mapping, military combat, agriculture, delivery of medicines in remote areas, aerial mapping, and many others (Kangunde et al. 2021; Rao et al. 2016).

The emergence of drones challenges traditional notions of safety, security, privacy, ownership, liability, and regulation. With their ability to collect data and transport loads, drones are re-shaping the way we think and feel about our physical environment. However, they also are burdened with the perception of being surveillance equipment, and their commercial use has been criticized by both individuals and activist organizations. In parallel, drones have been legitimized by regulations and licenses from federal agencies, and are used by projects for surveying, inspecting, and imaging, and their technological development are driven by active communities of hobbyists and enthusiasts.

Case Study: Hyperloop: A Fifth Mode of Transportation

The basis for many new megaprojects may be related to Hyperloop. This is a completely new mode of very fast transportation first proposed by Elon Musk and a team of engineers from Tesla Motors and the Space Exploration Technologies Corporation in August 2013. The concept of hyperloop relates to traveling from one place to another in a capsule. The Hyperloop system can propel passenger or cargo pods at speeds of over 1000 km/h. That is three times faster than high-speed rail and more than 10 times faster than traditional rail (Pathare 2020).

In India, for example the possible beginning of a new social technological system is under development. The State Government of Maharashtra is working with the US transportation company Virgin Hyperloop to develop a hyperloop train line between Pune and Mumbai. Hyperloop trains are high speed ground-level transportation systems for passengers and cargo, moved in a hovering pod inside a vacuum tube at speeds as high as 1220 kmh (760 mph). The proposal of Hyperloop Virgin to the Government of Maharashtra included a projection of 150 million passenger trips annually for this route alone. Along with savings of 90 million hours travel time, and potential economic benefits worth of US$55B over the next 30 years. Experts debate vigorously about realistic cost estimates to develop hyperloop systems and their benefits. So, this might be an opportunity to create a new system.

Megaproject Management: Looking Back to Move the Future Forward

There has been a call from the business, management, and project management communities for decades for more research on the history of project management. We need to learn from the past to do better in the future (Söderlund et al. 2017). Though there have been case studies on some of the larger industrial and engineering projects of our time there is a need for more studies to fill the gap of our understanding of best practices for megaprojects. Especially practices that seem to be critical for the success of projects and that are grounded on a contextual understanding of the particular project at hand such as planning techniques, coordination mechanisms, team structures, and interorganizational collaborations (Söderlund and Lenfle 2011).

We can learn from the projects that were built as well as the projects that were never built. One important example is the conventional story of the Atlantic-Pacific sea-level canal, a megaproject that failed to make the transition from idea to reality and thereby enter the pantheon of monumental civil engineering works (Keiner 2020). Unbuilt projects merit attention for many reasons, including the ways in which the planned and improvisational work underlying them influenced decision-making at the time. The resurgence of the sea-level canal proposal at strategic points in the intertwined history of the United States and Panama provides windows into moments of diplomatic, technological, scientific, and environmentalist transformation. Such historical moments in turn remind us of the value of envisioning alternative futures, and of questioning technocratic prescriptions that promise to modernize landscapes and societies without ensuring environmental quality and equal justice for all (Keiner 2020). Today, years on, it is clear that those demanding realistic, comprehensive assessments of the megaproject's environmental effects were as forward-thinking as the other sea-level canal stakeholders. One of the great outcomes of the project was the establishment of the National Environmental Policy Act (NEPA) of 1969 which enabled citizen environmentalists to mobilize the insights of biologists in powerful, unexpected ways (Keiner 2020).

In the coming decades, tunnel and bridge projects will continue to take priority to connect countries, cities, and regions throughout the world. One key decision point with each of these bridge-tunnel projects will be whether to design it strictly for cars and trucks, trains, or whether it should also include provisions for future forms of hyperloop-style tube-transportation.

As an example of the lack of future thinking, the Channel Tunnel between Great Britain and France was not designed to accommodate commercial transport and vehicles between the two countries. This means no cars or trucks, no cargo, and no provision for a potentially much faster tube-train. When the owner Eurotunnel won the contract to build its undersea connection, the firm was obliged to come up with plans for a second Channel Tunnel. . . by the year 2000. Although those plans were published the same year, the tunnel still has not gone ahead. These examples will be critical decision points moving forward.

Leadership for Megaprojects in the Emergent Era

> They don't make plans; they don't solve problems; they don't even organize people. What leaders really do is prepare organizations for change and help them cope as they struggle through it.
>
> — *John P. Kotter (2001)*

In his scholarly work on leadership, Professor Kotter points out that management and leadership are really two distinctive and complementary systems of action, but both are necessary for success in a changing world. He explains that "management is about coping with complexity" through practices and procedures, while "[l]eadership, by contrast, is about coping with change" (Kotter 2001). On the one hand, project managers accomplish their goals by planning and budgeting, setting targets, controlling, monitoring results against the plan, and problem solving. On the other hand, leaders create a vision, develop strategies to achieve the vision, and implement the vision. Project managers use tools and techniques and processes and procedures to accomplish their goals, whereas leaders must impact human emotion, change minds, and motivate and inspire individuals to do things they have never done before or do things in a different way. Since leadership is about change management, the more that change occurs in an organization or project, the more leaders must motivate others in the project to provide leadership as well. Because projects, by their very definition, are about change, good leadership skills, as distinct from managerial skills, are critical to project success.

As described in this chapter, understanding how to manage technological emergence is a critical factor in managing megaprojects due to their high degree of complexity and uncertainty in a constantly changing world. The Baldrige Framework and its Criteria for Performance Excellence sees organizations as always emergent. They live in the "in between." They bridge current organizational systems and leadership with always striving for the next leading edge of leadership (Hertz 2021). How do senior leaders set vision, values, encourage frank two-way communication, and create a focus on action? The following questions should be raised in every megaproject to determine how leadership is fostering innovation:

1) How do senior leaders cultivate innovation and intelligent risk taking?
2) How does your strategic planning process address the potential need for organizational agility, including operational flexibility?
3) How does your strategic planning stimulate and incorporate innovation?
4) How does your strategic planning address key elements of risk, including finding potential blind spots?
5) How do you decide which intelligent risks to pursue?
6) How do you build and manage organizational knowledge, share it, and use it as a platform for innovation?
7) How do you change when your strategy is not working?
8) What is your organization doing to prepare for the emergent era?

The Trends of Megaproject Leadership in the Future

> It is better to lead from behind and to put others in front, especially when you celebrate victory when nice things occur. You take the front line when there is danger. Then people will appreciate your leadership.
>
> — *Nelson Mandela*

Four leading scholars, in the project management field, Nathalie Drouin, Shankar Sankaran, Alfons van Marrewijk, and Ralf Muller (2021) focused upon a question central to megaproject leadership in their book, *Megaproject Leaders: Reflection on Personal Life Stories*. The question they raised is: *What can we learn from the personal view of megaproject leaders in managing their projects* (p. 288)? The interviews were conducted with 16 experienced megaproject leaders in Europe, Asia, Oceania and North America. The analysis in the book revealed four new trends for the advancement of knowledge on the management of megaprojects:

The first trend they found is the increasing attention on human capital and socio-environmental dimensions of megaprojects by their leaders moving the focus away from technological expertise, and making way for a more inclusive, humanistic, and societally oriented approach (p. 289) (Caron and Drouin 2019; Van den Ende and Van Marrewijk 2019).

The second trend revealed from their research is the increasing ability of megaproject leaders to reflect upon their personal management styles. The megaproject leaders interviewed for the research "explained that their learning was not often planned and took them by surprise" (p. 290). This is described as reflection in action and reflection-after-action (Schon 1983) realizing that the opportunity and the people who they were leading shaped their lives during and after the project (p. 290).

The third trend which came out of the life stories is that the experiences of project leaders play a vital and growing role in how they enrich their lives through remembering past experiences (p. 290). This is what Carlsen and Pitsis (2020) call the narrative capital of project managers.

Finally, the authors found a fourth trend, which is the growing awareness of leaders of the issue of project culture and cultural differences in megaprojects (p. 291). Importantly, the research "found that the project culture helped to create a sense of unity during difficult periods in the project, which brought people together from diverse cultural backgrounds to achieve their shared destiny" (p. 291). To support this finding there is a rich literature on the role of culture in project management and how it can help to foster inter-organizational relationships (Clegg et al. 2002; van Marrewijk 2007).

Skills Needed for Megaproject Leaders

As the dynamics and environment of organizations evolve, the challenges of future project managers will follow. With large complex projects, it's becoming a necessity for the project manager to coordinate multidisciplinary knowledge. To effectively accomplish the undertaking, project managers will have to adapt to new technologies and learn which specialized intricate tools will work best for each unique project. Another challenge for project managers will be cutting through the ceaseless sea of information by filtering abundant data and capturing the right information. This will require understanding the big picture and effectively communicating with others.

Having to work across-networks, with people in different countries, and people who come from different cultures will only compound the complexity of the project environment for future project managers. Project managers will also have to acclimate to the changing demographics of the workforce, such as the retirement of baby boomers, immigration flow change, and the number of young professionals entering the workforce. Finally, project managers will also have to adapt to shifting organizational structure. Globalization, limited resources, stakeholders, competition, economics, technology advancements, and many other factors are contributing to the transformation of organizations and the business environment. The one certain thing in the future is that in order to succeed, project managers of the twenty-first century have to be adaptable to constant change, uncertainty, and disruptions (Seymour and Hussein 2014). This requires competency to work through complexity and uncertainty and lead major transformations.

A New Principle Based Approach to Project Management

PMI in its seventh edition *A Guide to the Project Management Book of Knowledge* changes its approach to Project Management and rather than focusing on the 10 knowledge areas previously established moves to a Principle Based Standard through implementation of the 12 Principles shown in Table 12.2.

How Do Megaprojects Produce Value?

How does one produce value in a megaproject as distinct from the traditional focus on cost, schedule, and scope is the endeavor of all projects and will be increasingly important in all megaprojects of the future. Value differs because it is the ultimate success indicator and driver of projects because it focuses on the outcome of the deliverables rather than just adherence to process (PMI 2021). Value is the worth, importance, or usefulness of something which is not always

Table 12.2 Project management principles.

The Project Management Institute® 2021 Project Management Principles
1) Be a Diligent, Respectful, and Caring Steward
2) Create a Collaborative Project Team Environment
3) Effectively Engage with Stakeholders
4) Focus on Value
5) Recognize, Evaluate, and Respond to System Interactions
6) Demonstrate Leadership Behaviors
7) Tailor Based on Context
8) Build Quality into Processes and Deliverables
9) Navigate Complexity
10) Optimize Risk Responses
11) Embrace Adaptability and Resiliency
12) Enable Change to Achieve the Envisioned Future State

Source: Adapted from PMBOK 2021.

measured in monetary terms. Value is subjective because the same concept can have different values for different people and organizations (p. 35). Thus, value is the reason that megaproject success must be viewed holistically in terms of a project's contributions to the project, the performing organization, the environment in which it operates and society at large (Greiman and Sclar 2020). Characteristics of value include:

- Creating a new blue ocean product, service, or result that meets the needs of customers, clients, or users.
- Creating positive social or environmental contributions.
- Sustaining benefits beyond those enabled by previous programs, projects, or business operations.
- Implementing the United Nations 2030 Sustainability Agenda.
- Building resilience into the project's strategic plans.
- Improving efficiency, productivity, effectiveness, or responsiveness.
- Enabling the changes needed to facilitate organizational transition to its desired future state; and
- Exceeding expectations and delivery of added benefits beyond the original scope without impacting cost or schedule.

Megaprojects and System Perspectives

The complex megaprojects of the future will require a systems perspective. This means megaprojects will require a more holistic approach to achieve a project's mission and strive toward a project's vision as shown in Figure 12.1. The Baldrige Performance Program has been adopted by government agencies in the United States and other countries as well as private industry, and non-profits including universities and educational institutions. Each year, the U.S. Department of Commerce recognizes the top performers in the Baldrige Awards for Performance Excellence Programs (NIST 2021).

The Baldrige Performance Excellence Program is built on a set of five measurements that include: (i) Leadership and Governance Results, (ii) Product and Process Results, (iii) Workforce Results, (iv) Customer Results, and (v) Financial, Market, and Strategy Results that contain interrelated core values and concepts. These beliefs and behaviors are embedded in high-performing organizations. Figure 12.1 has adopted in part the Baldrige Performance measurements and added new measurements based on the core values and concepts of megaprojects. The measurements consist of five categories to provide a results-oriented perspective for the megaprojects of the future – strategy, visionary leadership, sustainability and resilience, value delivery and performance and outcome.

Figure 12.1 A System Perspective of Megaproject Management. *Source:* Adapted in part from the Baldrige Performance Excellent Program.

Strategy

The strategy of a megaproject should be clearly defined, mission focused, and should align with the goals of the sponsoring organization and the project owners and users. Strategies should incorporate an ethical and transparent environment that fosters innovation in people, processes, and technological change.

Visionary Leadership

A megaproject should set a vision for the organization, create an owner/user focus, with a partnering approach to its stakeholders both internal and external to the organization, create a culture that enables change, and provide transformational leadership so that the benefits can be passed on to future generations.

Sustainability and Resilience

Sustainability should balance the environmental, social, and economic aspects of megaprojects to meet the current needs of stakeholders without compromising or overburdening future generations. Sustainability is defined broadly to include financial, economic, and social impact. As set forth in the United Nations 2030 Sustainable Development Goals, sustainability should improve the quality of life of the people it impacts, protect the ecosystem, and preserve natural resources for future generations. All of these goals will be advanced through a focus on systems adaptation and resilience.

Value Delivery

Megaprojects should choose and analyze results that help deliver and balance value for the project's stakeholders. Thus, results need to include not just financial results, but also process, product and project results; stakeholder engagement results; and leadership, strategy, and societal performance. Value includes the creation of value well beyond the project itself to include future benefits, knowledge sharing, and a continual collaborative environment even after transition of the project to operations.

Performance and Outcome

From the systems perspective the results should be evaluated and measured for achievement in the project's strategy, leadership, sustainability and resilience, and value delivery. Outcomes should be used to provide accountability, improve governance, and oriented toward success. Measurements should be based on fact and not just hope and inspiration.

Summary

The megaprojects of the future present many challenges including different competencies, a greater understanding of complex adaptive systems and continuous understanding of the ever-changing technological environment. Predicting the future will be difficult while at the same time being prepared for change will require constant vigilance and a new mindset that focuses on opportunities, outcomes, openness, optimization, and blue oceans and the blue economy. The future requires that megaprojects adopt practices that are sustainable and resilient and have an understanding of complex adaptive systems to deliver effective leadership for the grand global challenges that lie ahead. They also require leadership competent to manage the socio environmental dimensions of megaprojects and the growing significance of project culture and cultural differences in megaprojects.

Lessons and Best Practices

Management of the twenty-first century megaprojects will require:

- A focus on success and innovation.
- Skills to manage change in an increasingly complex environment.
- Project environments and culture that is adaptable to change.
- Understanding the organizational culture, systems, processes, and leadership behaviors that support the creation and execution of a Blue Ocean Strategy and making a Blue Ocean Shift.
- Continued knowledge of the evolving technology industrial state and how these new technologies will contribute to greater efficiency and add value to the social and economic growth of the project.
- Creating and implementing visions of the Blue Economy.
- An understanding of the significance of value creation and how it can be measured.
- Transformative leadership that inspires positive change.
- The development of innovative technologies such as renewable energy, carbon capture, waste management, and energy efficiency.
- Building sustainability and resilience to improve the quality of life and protect the global ecosystem.

Discussion Questions

1 The megaprojects of the future will focus on major global issues and the grand challenges that lie ahead that will include the development of new energy resources, the health of the world's population, climate change, and sustainable environmental policy. How do we meet these grand challenges within the context of a megaproject?

2 In 2017, the Project Management Institute estimated that the value of project-oriented economic activity worldwide would grow from $12 trillion in 2017 to $20 trillion in 2027, in the process putting some 88 million people to work in project management-oriented roles. This movement toward project management as a discipline requires new thinking about project management leadership. What will organizations need to think about in structuring their organizations for this new demand?

3 What is meant by value delivery, and should it support or be prioritized over traditional cost, scope, and schedule evaluation?

4 What is meant by transformational leadership and why is it so critical to megaproject management?

5 How can resilience be developed in a megaproject to improve megaproject outcomes?

6 What are some Blue Ocean strategies and Blue Economy strategies that organizations will need to implement in the future?

7 How will the technological industrial state improve the megaprojects of the future?

8 What is the changing role of the megaproject in society?

9 Why is a system perspective so valuable to the megaprojects of the future?

10 What attributes of megaprojects will be the focus of the future? Provide some examples.

11 What are some of the innovative technologies that will be used in megaprojects and how will they be applied?

12 How can repurposing of nuclear sites be used to benefit local communities?

13 What are the leadership skills and the competencies that will be needed for the projects of tomorrow and how will we train these new leaders?

References

Agrawal, A., Gans, J.S., and Goldfarb, A. (2017). What to expect from artificial intelligence. *MIT Sloan Management Review* 58 (3): 22–26.

Air France (2021). *The Air France Group Commits to Setting Science-Based Greenhouse Gas Emission Reduction Targets with the Science-Based Targets Initiative (Sbti)*. Paris: Air France Press.

Association of Project Management (APM) (2019). *APM Body of Knowledge*, 7the. England: Buckinghamshire.

Atalia, G., Mills, M., and McQueen, J. (2021). *Six Ways That Governments Can Drive the Green Transition*. New York: Ernst & Young https://www.ey.com/en_cn/government-public-sector/six-ways-that-governments-can-drive-the-green-transition.

National Institute of Standards and Technology (NIST) (2021). *Proven Leadership and Management Practices for High Performance*. Gaithersburg, MD: Baldrige Performance Excellence Program 2021–2022, U.S. Department of Commerce (DOC).

Bartneck, C., Lütge, C., Wagner, A., and Welsh, S. (ed.) (2021). What Is AI? In: *An Introduction to Ethics in Robotics and AI*, Springer Briefs in Ethics. Cham: Springer https://doi.org/10.1007/978-3-030-51110-4_2.

Clegg, S.R., Pitsis, T.S., Rura-Polley, T., and Marosszeky, M. (2002). Governmentality matters: Designing an alliance culture of interorganizational collaboration for managing projects. *Organization Studies* 23 (3): 317–337. https://doi.org/10.1177/0170840602233001.

Colombo, U. (2020). *A Viewpoint from the Policy Community*, OECD Megascience Forum Workshop on Global Scale Issues. Strengthening the Interaction Between Science and Policy-making. Milan, Italy: Organization for Economic Cooperation and Development (OECD).

Brattberg, E., Csernatoni, R., and Rugova, V. (2020). *Europe and AI: Leading, Lagging Behind, or Carving Its Own Way?* [Working Paper]. Washington, DC: Carnegie Endowment for International Peace.

Bughin, J., Seong, J., Manyika, J. et al. (2018). *Notes from the AI Frontier: Modeling the Impact of AI on the World Economy*. [Discussion Paper]. Washington, DC: McKinsey Global Institute, McKinsey and Company.

Carlsen, A. and Pitsis, T. (2020). We are projects: Narrative capital and meaning making in projects. *Project Management Journal* 51: 357–366.

Caron, M.-C., and Drouin, N. (2019). *A systematic review of the integration of non-financial benefits in major projects: Co-construction of knowledge to monitor complexity*. 19th European Academy of Management (EURAM) Conference, Lisbon, Portugal, 26–28 June.

Clegg, S., Shankar, S., Biesenthal, C., and Pollack, J. (2017). Power and Sensemaking in Megaprojects. In: *The Oxford Handbook of Megaproject Management* (ed. B. Flyvbjerg), 238–258. Oxford, UK: Oxford University Press.

Cleland, D. and Bidanda, B. (2009). *Project Management Circa 2025*. Newtown Square, PA: Project Management Institute.

Daniel, E. and Daniel, P.A. (2019). Megaprojects as complex adaptive systems: The Hinkley Point C case. *International Journal of Project Management* 37 (8): 1017–1033.

Di Giuda, G.M., Locatelli, M., Schievano, M. et al. (2020). Natural Language Processing for Information and Project Management. In: *Digital Transformation of the Design, Construction and Management Processes of the Built Environment*.

Research for Development (ed. B. Daniotti, M. Gianinetto and S.D. Torre). Cham: Springer.

Drouin, N., Sankaran, S., van Marrewijk, A., and Müller, R. (2021). *Megaproject Leaders: Reflections on Personal Life Stories*, New Horizons in Organization Studies Series. Cheltenham, UK: Edward Elgar.

Economic Commission for Africa (ECA) (2016). *Africa's Blue Economy: A Policy Handbook*. Ethiopia: United Nations.

European Commission (EC) (2020). *On Artificial Intelligence: A European Approach to Excellence and Trust*. https://ec.europa.eu/info/sites/info/files/commission-white-paper-artificial-intelligence-feb2020_en.pdf.

Flyvbjerg, B. (2014). What you should know about megaprojects and why: An overview. *Project Management Journal* 45 (2): 6–19. https://doi.org/10.1002/pmj.21409.

Folke, C. (2006). Resilience: The emergence of a perspective for social–ecological systems analyses. *Global Environmental Change* 16 (3): 253–267.

Folke, C., Carpenter, S.R., Walker, B. et al. (2010). Resilience thinking: Integrating resilience, adaptability and transformability. *Ecology and Society 15* (4): 20. http://www.jstor.org/stable/26268226.

Fotouhi, A., Qiang, H., Ding, M. et al. (2019). Survey on UAV cellular communications: Practical aspects, standardization advancements, regulation, and security challenges. *IEEE Communications Surveys & Tutorials* 21 (4): 3417–3442.

Frey, T. (2016). *Megaprojects set to explode to 24% of global GDP within a decade*. Davinci Institute http://www.futuristspeaker.com/job-opportunities/megaprojects-set-to-explode-to-24-of-global-gdp-within-a-decade.

G20 Countries (2021). *Energy Policy Tracker Website*. https://www.energypolicytracker.org/region/g20/.

Gillin, K. (2019). *A Sustainability-based approach to Nuclear Back-End Management*. 4th Canadian Conference on Nuclear Waste Management and Environmental Restoration, Ottawa, Canada, September 8–11.

Greiman, V.A. (2013). *Megaproject Management: Lessons on Risk and Project Management from the Big Dig*. Hoboken, NJ: John Wiley & Sons, Inc.

Greiman, V.A. (2020). *Artificial Intelligence in Megaprojects: The Next Frontier*. Published in the Proceedings of the 19th European Conference on Cyber Warfare and Security, University of Chester, UK, June 25–26, 2020.

Greiman, V.A. and Bernardin, E.M. (2021). *Cyber Resilience: A Global Challenge*. Reading, UK: Academic Conferences and Publishing International (ACPI). ISBN-10 1914587022.

Greiman, V. and Sclar, E. (2020). Mega-infrastructure as a dynamic ecosystem: Lessons from America's interstate system and Boston's Big Dig. *Journal of Mega Infrastructure and Sustainable Development* https://doi.org/10.1080/24724718.2020.1742624.

Gunderson, L.H. (2000). Ecological resilience: In theory and application. *Annual Review of Ecology and Systematics* 31: 425–439.

Haenlein, M. and Kaplan, A. (2019). A brief history of artificial intelligence: On the past, present, and future of artificial intelligence. *California Management Review* 64 (4): 5–14.

Hertz, H. (2021). *Looking Back to Learn for the Future: The Emergent Organization, Strategy, and Innovation*. Washington, DC: The Baldridge Cheermudgeion, The National Institute for Standards and Technology (NIST), U.S. Department of Commerce.

Hirschman, A.O. (1995). *Development Projects Observed*, 2nde. Washington, DC: Brookings Institution.

Holland, J.H. (2014). *Complexity: A Very Short Introduction*. Oxford, UK: Oxford University Press.

Holland, J.H. (1999). *Emergence: From Chaos to Order*. New York, NY: Basic Books.

Holling, C.S. (1973). Resilience and stability of ecological systems. *Annual Review of Ecology and Systematics* 4: 1–23.

Holling, C.S. (1996). Engineering Resilience Versus Ecological Resilience. In: *Engineering Within Ecological Constraints* (ed. P. Schulze), 31–44. Washington, DC: National Academy Press.

INSEAD. Blue Ocean Strategy Institute (2022). Fontainebleu, France. https://www.insead.edu/centres/blu Our oceans are under attack by climate change, overfishing e-ocean-strategy. ISBN 9780128037263.

International Trade Administration (ITA) (2021). *Egypt – Commercial Country Guide*. Washington, DC: U.S. Department of Commerce https://www.trade.gov/country-commercial-guides/egypt-electricity-and-renewable-energy.

ITER Organization (2022). *What Is Iter?* https://www.iter.org/proj/inafewlines.

Johnson, L. (2010). *Science & Technology Innovation as a Complex Adaptive System: Applying the Natural Processes of Complexity to Policymaking*. [Paper] Annual Meeting of the American Political Science Association (APSA), September 2–5, 2010. https://ssrn.com/abstract=1657193.

Kanellakis, C. and Nikolakopoulos, G. (2017). Survey on computer vision for UAVs: Current developments and trends. *Journal of Intelligent and Robotic Systems* 87 (1): 141–168.

Kangunde, V., Jamisola, R.S., and Theophilus, E.K. (2021). A review on drones controlled in real-time. *International Journal of Dynamics and Control* 9 (4): 1832–1846. https://doi.org/10.1007/s40435-020-00737-5.

Keiner, C. (2020). *Deep Cut Science, Power, and the Unbuilt Interoceanic Canal*. Athens, Georgia: The University of Georgia Press.

Kim, W.C. and Mauborgne, R. (2015). *Blue Ocean Strategy, Expanded Edition: How to Create Uncontested Market Space and Make the Competition Irrelevant*. Brighton, MA: Harvard Business Review Press.

Kim, W.C. and Mauborgne, R. (2017). *Blue Ocean Shift: Beyond Competing – Proven Steps to Inspire Confidence and Seize New Growth*. New York, Boston: Hachette Books. ISBN: 0316314048.

Kotter, J.P. (2001). What leaders really do. *Harvard Business Review* 79: 85–98. Harvard Business School Publishing, Brighton, MA.

Linux (2005). *Algorithms a Very Brief Introduction*. San Francisco, CA: The Linux Foundation, July 29.

Lovejoy, B. (2021). Battersea Power Station Tube opens ready for Apple's London campus. Apple, 9to5mac. https://9to5mac.com/2021/09/20/apples-london-campus-tube-station/.

Lyle, A. (2016). *National Security Experts Examine Intelligence Challenges at Summit*. U.S. Department of Defense, September 9. https://www.defense.gov/News/News-Stories/Article/Article/938941/national-security-experts-examine-intelligence-challenges-at-summit/.

Maher, H., Chraïti, A., Laabi, A. et al. (2021). *Igniting Innovation-Based Growth in Africa*. Boston, MA: Boston Consulting Group https://www.bcg.com/publications/2021/innovation-in-africa.

National Institute of Health (NIH) (2022). *International Consortium Completes Human Genome Project*. National Human Genome Research Institute, Department of Health and Human Services and Office of Science and U.S. Department of Energy https://www.genome.gov/11006929/2003-release-international-consortium-completes-hgp.

Organization for Economic Cooperation and Development (OECD) (2020). *OECD Roundtable on Smart Cities and Inclusive Growth*, Building on the outcomes of the 1st OECD Roundtable on Smart Cities and Inclusive Growth. Korea: OECD and The Ministry of Land, Infrastructure and Transport.

Outa, G.O., Osano, P.M., Muchiri, M. et al. (2021). *Science, Research, and Innovation for Harnessing the Blue Economy*. Nairobi: Government of Kenya & Stockholm Environment Institute - Africa Centre.

Pathare, K.S. (2020). Hyperloop: Fifth mode of transpotation. *International Journal of Engineering Applied Sciences and Technology* 4 (11): 274–278.

Pomerlau, M. (2016). *DARPA Director Clear Eyed and Cautious on AI*. Government Computer News, May 10.

Price Waterhouse Coopers (PwC) (2017). *Sizing the Prize What's the Real Value of AI for Your Business and How Can You Capitalize?* London: Price Waterhouse Coopers.

Project Management Institute (PMI) (2021). *A Guide to the Project Management Body of Knowledge*, 7the. Newtown Square, PA: PMI.

Raso, F.A., Hilligoss H., Krishnamurthy V., Bavitz, C., and Kim, L. (2018) Artificial intelligence & human rights: Opportunities & risks. Berkman Klein Center Research Publication No. 2018-6. http://dx.doi.org/10.2139/ssrn.3259344.

Rao, B., Gopi, A.G., and Maione, R. (2016). The societal impact of commercial drones. *Technology in Society* 45: 83–90.

Rodriques, S.S., Almeida, P.J., and Almeida, N.M.C. (2020). *Mapping, Managing, and Crafting Sustainable Business Strategies for the Circular Economy*. Hershey, PA: IGI Global.

Rogers, A. (2018, April 13). *Innovation Case Studies: How Companies Use Technology to Solidify a Competitive Advantage*. Jersey City, NJ: Forbes.

Rogers, M. (2016). *Testimony of Michael Rogers, Senate Armed Service Committee*. Hearing to Receive Testimony on Encryption and Cyber Matters, United State Senate Committee on Armed Services, September 13, 2016. https://www.armed-services.senate.gov/hearings/16-09-13-encryption-and-cyber-matters.

Scheffer, M. (2009). *Critical Transitions in Nature and Society*. Princeton, NJ: Princeton University Press.

Schindler, S., Fadaee, S., and Brockington, D. (2019). Contemporary megaprojects. *Environment and Society* 10 (1): 1–8. https://doi.org/10.3167/ares.2019.100101.

Schon, D.A. (1983). *The Reflective Practitioner: How Professionals Think in Action*. New York: Basic Books.

Seymour, T. and Hussein, S. (2014). The history of project management. *International Journal of Management & Information Systems* 18 (4): 233–240. https://doi.org/10.19030/ijmis.v18i4.8820.

Söderlund, J. and Lenfle, S. (ed.) (2011). Special issue: Project history. *International Journal of Project Management* 29 (5): 491–493.

Söderlund, J., Sankaran, S., and Biesenthal, C. (2017). The past and present of megaprojects. *Project Management Journal* 48 (6): 5–16.

Sullivan, T. (2011). Embracing complexity. *Harvard Business Review* 89 (9): 89–92.

United Nations (UN) (2015). *The 2030 Agenda for Sustainable Development*. New York, NY: The United Nations Department of Economic and Social Affairs (ESA) https://sdgs.un.org/2030agenda.

United Nations (UN) (2018). *Official Aide-mémoire for the SBEC 2018 Conference*. Sustainable Blue Economy Conference (SBEC) 2018. Nairobi, Government of Kenya, Ministry of Foreign Affairs. United Nations Economic Commission for Africa, United Nations Economic and Social Council, United Nations.

Van den Ende, L. and Van Marrewijk, A.H. (2019). Teargas, taboo and transformation: A neo-institutional study of public resistance and the struggle for legitimacy an Amsterdam subway project. *International Journal of Project Management* 37 (2): 331–346. https://doi.org/10.1016/j.ijproman.2018.07.003.

van Marrewijk, A.H. (2007). Managing project culture: The case of Environ Megaproject. *International Journal of Project Management* 37: 331–346.

van Marrewijk, A., Ybema, S., Smits, K. et al. (2016). Clash of the titans: Temporal organizing and collaborative dynamics in the Panama Canal megaproject. *Organizational Studies* 37 (12): 1745–1769.

Vardoulakis, S. and Kinney, P. (2019). Frontiers grand challenges in sustainable cities and health. *Frontiers in Sustainable Cities* 1: 7. https://doi.org/10.3389/frsc.2019.00007.

Visbeck, M. (2018). Ocean science research is key for a sustainable future. *Nature Communications* 9: 690. https://doi.org/10.1038/s41467-018-03158-3.

Walker, B.H., Holling, C.S., Carpenter, S.R., and Kinzig, A. (2004). Resilience, adaptability, and transformability in social–ecological systems. *Ecology and Society* 9 (2): 5. http://www.ecologyandsociety.org/vol9/iss2/art5.

Wang, L., Müller, R., Zhu, F., and Yang, X. (2021). Collective mindfulness: The key to organizational resilience in megaprojects. *Project Management Journal* 52 (6): 592–606.

White House (WH) (2022). *National Strategic Plan for Quantum Information Science and Technology Workforce Development*. A product of the National Science and Technology Council Subcommittee on Quantum Information Science (SCQIS), Office of Science & Technology Policy and U.S. National Science Foundation.

World Bank (WB) (2017). *The Potential of the Blue Economy*. Washington, DC: International Bank for Reconstruction and Development.

World Bank (WB) (2021). *World Development Report: Data for Better Lives*. Washington, DC: World Bank.

Glossary

This glossary of terms is intended as a reference aid and should not be considered an exhaustive or complete list of all the terms set forth in this book or in the project management literature on the topics discussed herein.

A

Accountability: Broadly defined, accountability is the acknowledgment and assumption of responsibility for actions, projects, decisions, and policies including the administration, governance, and implementation within the scope of the role in a project or employment position.

Adaptation Governance: Sensitivity to climate change is a challenge to humanity that requires an integrated, anticipatory, and strategic approach to governance.

Agile Project Management: Agile methods promote a process that encourages development iterations, teamwork, stakeholder involvement, objective metrics, and effective controls. While the term was originally associated with software development projects, Agile approaches can apply to any project and has been used successfully in megaprojects.

Alternative Dispute Resolution: Provides procedures for settling disputes by means other than litigation and includes such mechanisms as negotiation, mediation, arbitration, and mini trials.

Ambiguity: Subject to more than one interpretation.

Architecture Governance: The principles, standards, guidelines, contractual obligations, and regulatory framework within which goals are met at an enterprise-wide level.

Global Megaprojects: Lessons, Case Studies, and Expert Advice on International Megaproject Management, First Edition. Virginia A. Greiman.
© 2023 John Wiley & Sons, Inc. Published 2023 by John Wiley & Sons, Inc.

Association of Project Management (APM): The APM is a U.K.-based organization committed to developing and promoting project and program management.

Assumption: Something taken as true without proof. In planning, assumptions regarding staffing, complexity, learning curves, and many other factors are made to create plan scenarios. These provide the basis for estimating. Assumptions are not facts. Alternative assumptions should be made to get a sense of what might happen in a given project.

Australian Institute of Project Management (AIPM): The AIPM is the primary body for project management in Australia. Formed in 1976 as the Project Managers' Forum, the AIPM has been instrumental in progressing the profession of project management in Australia in the decades since.

B

Baldrige Performance Excellence Program: This program provides some key elements for analyzing an organization's governance system. These elements include organizational governance, legal and ethical behavior, and societal responsibilities and community support.

Bankable Projects: Lenders are willing to finance the project because it is financially, environmentally, economically, technically, and socially feasible.

Basel III: Basel III is an internationally agreed set of measures developed by the Basel Committee on Banking Supervision in response to the financial crisis of 2007–2009. The measures aim to strengthen the regulation, supervision and risk management of banks.

Baseline: A point of reference. The plan used as the comparison point for project control reporting. There are three baselines in a project: schedule baseline, cost baseline, and product (scope) baseline. The combination of these is referred to as the *performance measurement baseline.*

Best Value Procurement: The owner considers both qualifications and cost when selecting the design-build team.

Bilateral Investment Treaty: These Treaties are entered into by countries seeking protection for investments made by investors from their own state from wrongful acts of host countries including expropriation, nationalization, and discrimination.

Blended Concessional Financing: Explicitly uses scarce concessional financing to enable mobilization of private investment into areas it would not have otherwise gone.

Blue Economy Strategy: The blue economy is a globally emerging concept for oceans governance that seeks to tap the economic potential of the oceans in environmentally sustainable ways.

Blue Ocean Strategy: A focus on differentiating a project or product from its competitors through innovation.

Bonds: A financial security, bearing a fixed interest rate, issued by private businesses or governments as a means of raising money and long-term funds (i.e. borrowing). When an investor buys bonds, he is lending money to the issuer. Bonds are repaid by the issuer at maturity.

Building Information Model (BIM): A digital representation of physical and functional characteristics of a facility. It serves as a shared knowledge resource for information about a facility, constituting a reliable basis for decisions during its life cycle, from inception onward.

Build-Operate-Transfer (BOT): A popular form of project delivery whereby the project is transferred back to the owner or party granting the concession after it has been built and operated for a period of time stated in the concession.

Business Case: The information that describes the justification for a project. The project is justified if the expected benefits outweigh estimated costs and risks. The business case is often complex and may require financial analysis, technical analysis, organization impact analysis, and a feasibility study.

Business Continuity Institute (BCI): Along with the Disaster Recovery Institute International (DRII), BCI formulates the common body of knowledge that provides a structured and systematic approach to business continuity.

Business Continuity Management: Business continuity management seeks to identify potential risks or threats to an organization and allows it to plan and develop ways to react and recover from major risk events. Business continuity management is tied closely to crisis management that systematically deals with a disaster or a risk event as it arises.

C

Catastrophic Loss: One or more related losses whose consequences are extremely harsh in their severity, such as total loss of assets or loss of life.

Central Artery/Tunnel Project: The Central Artery/Tunnel Project in Boston, often referred to as the "CA/T Project" or the "Big Dig," was the country's largest publicly funded construction project, costing $14.8 billion. It was known for its technological advancement of slurry wall construction, ground freezing, and the world's widest cable-stayed bridge.

Change: Difference in an expected value or event. The most significant changes in project management are related to scope definition, availability of resources, schedule, and budget.

Change Management: The process of identifying, documenting, approving, and implementing changes within a project. It is a structured and systemic approach to achieve a sustainable change in human behavior within an organization.

Change Request: A documented request for a change in scope or other aspects of the plan.

Charter: A high-level document usually issued by the project initiator or sponsor that describes the purpose of a project, the manner in which it will be structured, and how it will be implemented, and that provides the authorization for the project.

Client: The person or organization that is the principle beneficiary of the project, sometimes called the "project owner." Generally, the client has significant authority regarding scope definition and whether the project should be initiated and/or continued.

Cohesion Fund: An instrument of the EU's regional policy which supports investments in the field of environment and trans-European networks in the area of transport infrastructure.

Collective Action Perspective: Megaprojects are vast actor-networks formed to develop a new large-scale designed artefact: the infrastructure system. High-order decision-making within these networks is driven by the need to build interorganizational consensus at the core of the network.

Collective Mindfulness: Collective mindfulness refers to organizational processes or practices that help organizations detect, categorize, and respond to unexpected events and errors. Collective mindfulness is about both the quality of attention and conserving scarce attention.

Complex Adaptive Systems: They are characterized by feedback loops, tipping points and emergent properties, and their future is impacted by their past, and surprise is inevitable.

Complexity: Complexity arises from the interdependency of thousands of moving parts associated with funding, managing, and governing social and organizational relations and is particularly significant in megaprojects. Importantly, as noted by scholars in the field of project management, it is, not the cost but the complexity that marks out a megaproject.

Communications Management: The process of identifying, creating, reviewing, and distributing communications to stakeholders within a project, as well as receiving feedback and information.

Conceptual Phase: A period of time in which the description of how a new product will work and meet its performance requirements is formulated.

Concession Agreement: The agreement with a government body that entitles a private entity to undertake an otherwise public service.

Configuration Management: Adopted by the U.S. Air Force (USAF) in the 1950s as a technical management discipline, it is a process for establishing and maintaining consistency of a project's performance, both functional and physical attributes of its design and operational requirements throughout the project's life cycle.

Conflict: Competing interest and cultural divides.

Consensus: Unanimous agreement among the decision makers that everyone can at least live with the decision (or solution). To live with the decision, one has to be convinced that the decision will adequately achieve objectives. As long as someone believes that the decision will not achieve the objectives, there is no consensus.

Consortium: A consortium is an association of two or more individuals, companies, organizations, or governments (or any combination of these entities) with the objective of participating in a common activity or pooling their resources for achieving a common goal.

Constraints: The factors that must be considered during the life of the project that cannot be changed. These may include deadlines, regulatory requirements, and dependencies on other projects to deliver.

Construction Manager at Risk: The construction manager (CM) begins work on the project during the design phase to provide constructability, pricing, and sequencing analysis of the design. The CM becomes the design-build contractor when a guaranteed maximum price is agreed upon by the project sponsor and the CM.

Contingency Reserve: A designated amount of time and/or budget to account for parts of the project that cannot be fully predicted such as the risk of schedule and cost slippage due to unknown knowns. For example, it is relatively certain that there will be some rework, but the amount of rework and where it will occur in the project (or phase) are not known.

Control: The process of monitoring, measuring, and reporting on progress and taking corrective action to ensure project objectives are met.

Corporate Governance: The system by which an organization is overseen and controlled by its shareholders.

Corporate Social Responsibility (CSR): CSR is defined by the World Bank as the commitment of business to contribute to sustainable economic development, working with the employees, their families, the local community, and society at large to improve quality of life in ways that are both good for business and good for development.

Cost-Benefit Analysis: Used to show that the expected benefits of a project are sufficient to warrant the cost of carrying it out. Monetary units are usually used for the comparison. A CBA requires an investigation of a project's net impact on economic, and social welfare. Benefits and costs are typically evaluated for a period that includes the construction period and an operations period ranging from 20 to 50 years after the initial project investments are completed.

Country Partnership Frameworks: Partnering with other International Government Organizations' (IGOs) to fund projects in the poorest of countries.

Credit Rating Agency: Evaluates the likelihood of success of a public financing undertaking. Project debt ratings address default probability or the level of certainty with which lenders can expect to receive timely and full payment of principal and interest according to the terms of the financing documents.

Critical Infrastructure: As defined by the U.S. National Infrastructure Protection Plan (NIPP), critical infrastructure includes systems and assets, whether physical or virtual, so vital that the incapacity or destruction of such may have a debilitating impact on the security, economy, public health or safety, environment, or any combination of these matters across any federal, state, regional, territorial, or local jurisdiction.

Critical Path: The critical path is the sequence of activities that must be completed on time for the entire project to be completed on schedule. It is the longest-duration path through the project and the shortest amount of time in which a project can be completed.

Crossrail Project: The Crossrail Project, in London, is presently the largest rail network expansion project in Europe.

Culture: Socially learned behaviors and assumptions of social interaction and problem solving.

D

Debt Service: Required payments on borrowings including state bonds and notes.

Debt Service Cover Ratio (DSCR): The DSCR tells you how much equity will be needed to timely pay off the short-term debt commitments in the project financing agreement. Typically, equity sponsors may include the project owner (typically a public entity), the project developer or main contractor, and often institutional investors that do not play an active role in the project and are known as passive investors.

Design (Conceptual): The period during which public hearings are held, financing is authorized, environmental approvals are obtained, and right-of-way plans are developed to describe how a new project will be structured and how it will meet its performance requirements.

Design (Final): Requires multiple separate design contracts ranging from less than $1 million to over $50 million. Activities can include right-of-way (ROW) acquisitions, traffic control plans, utility drawings, permits and licensing, final cost estimates, and contractor bid solicitation.

Design-Bid-Build: The traditional project delivery method used whereby design and construction are sequential steps in the project development process.

Design-Build: It is a method to deliver a project in which the design and construction services are contracted by a single entity known as the design-builder or design-build contractor. In contrast to design-bid-build, design-build relies on a single point of responsibility contract and is used to minimize risks for the project owner and reduces the delivery schedule by overlapping the design phase and construction phase of a project.

Design Development: Design development is the evolution of design during the life of the project. There are many potential reasons for design changes, including environmental, risk, quality, cost, and schedule issues. Design professionals must balance process and structural considerations with regulatory, maintainability, and human factors.

Developing Country: Defined by the World Bank in terms of gross national income per capita.

Differing Site Conditions: These are conditions that can occur on a project that were not anticipated in the project's design and drawings. Differing site conditions can also shift the risk on a project from the contractor to the owner if they are materially different than represented in the contract drawings.

Disaster Recovery Institute International (DRII): Along with the Business Continuity Institute (BCI), the DRII formulates the common body of knowledge that provides a structured and systematic approach to business continuity.

Dispute Review Board (DRB): A panel of experienced, respected, and impartial reviewers that takes in all the facts of a dispute and makes recommendations on the basis of those facts and the board's expertise.

E

Earned Value: An approach whereby you monitor the project plan, actual work, and work-completed value to see if a project is on track. Earned value shows how much of the budget and time should have been spent for work done, and what are the variances from the budget, the estimate to complete, and the budget at completion.

Electronic Identification: The program that provided for notification to contract management upon entry or exit on the work site.

Emergency Preparedness: The project's plan for responding to emergencies, including force majeure events, disasters, catastrophes, shutdowns, and fatalities and serious injuries.

Eminent Domain: Sometimes called compulsory purchase, expropriation or simply a taking. It refers to the power of the government to take private property and convert it into public use.

Empirical Research: A way of gaining knowledge by means of direct and indirect observation or experience. Empirical evidence (the record of one's direct observations or experiences) can be analyzed quantitatively or qualitatively. Through quantifying the evidence or making sense of it in qualitative form, a researcher can answer empirical questions, which should be clearly defined and answerable with the evidence collected (usually called *data*). Research design varies by field and by the question being investigated.

Enterprise Governance: The entire accountability framework of the organization.

Environmental Impact Assessment: A project review to assess the impact a project will have on the local environment including the community, its businesses, and the natural environment by evaluating the project through the EIA, a determination can be made as to whether the project is safe to move forward or if alternative options should be explored.

Equator Principles: A financial industry benchmark for determining, assessing, and managing social and environmental risk in project financing. The Equator Principles incorporate the International Finance Corporation (IFC) and World Bank environmental performance standards and guidelines.

Errors and Omissions (E&O) Insurance: Professional liability or mal-practice insurance, which covers the professional negligence of design professionals.

Escalation of Commitment (EOC): Escalation of commitment (EOC) is a common behavior among investors who receive negative feedback (NF) in public-private partnership (PPP) projects, and this behavior typically leads to sizable losses

Ethics: The basic concepts and fundamental principles of human conduct. It includes study of universal values such as the essential equality of all men and women, human or natural rights, obedience to the law of the land, concern for health and safety, and, increasingly, concern for the natural environment.

European Bank for Reconstruction and Development (EBRD): Located in London, the EBRD serves as the primary lender to the developing world, particularly Eastern Europe and Russia, by virtue of their membership.

European Regional Development Fund: Finances programs in shared responsibility between the European Commission and national and regional authorities in Member States.

Eurotunnel: The company that was formed on 13 August 1986, to finance, build, and operate a tunnel between Britain and France. Groupe Euro-tunnel S.A. presently manages and operates the car shuttle services and earns revenue on other trains passing through the tunnel. It is listed on both the London Stock Exchange and Euronext Paris.

Events and Causal Factors Analysis (ECFA): An integral and important part of the management oversight and risk tree (MORT-based) accident investigation process.

Ex Ante Evaluations: In the case of project finance, ex ante means assessment of a project during the early conception and planning phases to test its viability and feasibility. The predicted outcome serves as a basis for comparing the prediction to the actual results (ex-post).

Ex Post Evaluations: Represents the actual results attained by the project, which is the return on investment that the project yielded. Ex-post evaluations are used throughout the European Commission to assess whether a specific intervention was justified and whether it worked (or is working) as expected in achieving its objectives and why. Ex-post evaluations also look for unintended effects (i.e. those which were not anticipated at the time of the initial project feasibility study.

Export Credit Agencies: Government departments or financial institutions that benefit from government guarantees or direct funding, which provide financing as a means of supporting exports from their countries.

Expropriation: The taking over by a state of a company or investment project, with compensation usually being paid. Creeping expropriation occurs when a government gradually takes over an asset by taxation, regulation, access, or change in the law.

F

Fast-Track Construction: Involves the commencement of construction before all of the design is completed.

Feasibility Study: A document that confirms the likelihood that a range of alternative solutions will meet the requirements of the customer.

Federal Highway Administration (FHWA): The entity within the U.S. Department of Transportation that oversees state-level projects that receive federal-aid highways funds.

Finance Plan: A project's report to the public on the project's design and construction status, cost center status, audits, insurance and safety and health program, and the project financing and budget.

Financial Structure: The manner in which the project is funded, whether through public or private financing, equity, debt, bonds, or revenue streams.

Force Majeure: An event that is not foreseeable and is beyond the control of the parties, such as a hurricane, earthquake, act of war, or change in the law that will excuse the parties from fulfilling their contractual obligations. Projects generally obtain insurance to protect against these events.

Foreign Direct Investment: A business decision to acquire a substantial stake in a foreign business or project to invest in or buy it outright in order to expand its operations to a new region.

Funder: The person or group, often called *sponsor*, which provides the financial resources, in cash or in kind, for a project.

G

General Obligation Bonds: Debt instruments issued by state and local governments to fund highway and infrastructure projects.

Global Megaproject: Global megaprojects consist of the creation of a special organizational structure for the purpose of financing a unique multi-billion-dollar investment involving numerous stakeholders and complex interdependencies and interorganizational relations in diverse regions of the world that will provide sustainable and long-term benefits to local communities and the larger eco system of which the megaproject is a part.

Globalization: The spread of free-market capitalism to virtually every country in the world.

Golden Age of Globalization: The period 1870–1940 when international trade, international migration flows and international mobility of financial capital reached historical peaks.

Governance: A set of relationships between a company's management, its board, its shareholders, and other stakeholders. Project governance is defined as the framework of authority and accountability that defines and controls the outputs, outcomes and benefits from projects, programs, and portfolios.

Governance Structure: An oversight and control function that can change to adapt to the emerging context of the project. Megaprojects are unique in that traditional hierarchical structures are replaced by a unique blending of vertical and horizontal engagements that require coordination.

Grand Challenges: Ambitious but achievable goals that harness science, technology, and innovation to solve important national or global problems.

Great Megaproject Era: Defined by Altshuler and Luberoff (2003) as the period between the late 1990s when the federal Interstate Highway Project got underway and the late 1960s when construction peaked.

Green Bond: A type of fixed-income instrument that is specifically earmarked to raise money for climate and environmental projects. These bonds are typically asset-linked and backed by the issuing entity's balance sheet, so they usually carry the same credit rating as their issuers' other debt obligations.

Greenfield Sites: An area of agricultural or forest land, or some other undeveloped site, earmarked for commercial development or industrial projects.

H

Hazards: The real or potential conditions (natural, accidental, or human) that cause injury to people or loss or damage to property, or which interfere with the operations of a project or the government owner.

I

Imprévision theory: The right to reject a contract that has become uneconomic.

Improvisation: Improvisation has been defined as the practice of reacting and of making and creating. Improvisation is linked with aspects of time, and, particularly, pressure to achieve to a demanding or compressed timetable, which is a typical attribute of most megaprojects. Improvisation is a developing theory of project management and is not recognized universally by all the professional bodies.

Infrastructure Report Card: A Report card issued by the American Society of Civil Engineers (ASCE) revealing the condition of infrastructure in the United States. In 2021, America's Infrastructure GPA is a C-.

Initiating (Project): The process of describing and deciding to begin a project (or phase) and authorizing the project manager to expend resources, effort, and money for those that are initiated.

Innovation: The successful exploitation of new ideas.

Innovation Megaproject: Innovation has been defined in the context of megaprojects as a new product, process, or service that has a step change and creates value including financial value, environmental value, societal value, and job creation.

Innovative Finance: Innovative methods of financing construction, maintenance, or operation of transportation facilities. The term *innovative finance* covers a broad variety of nontraditional financing, including the use of private funds or the use of public funds in a new way, for example, grant anticipation notes or special tax districts.

Institutional Learning: Institutional learning is proposed as a process through which adaptations can be made to accommodate shortcomings in the prevailing institutional environment.

Integrated Change Control: A centralized organization within a project to manage all claims and changes that arise during the course of the project. This centralized system provides many benefits, including more informed decision making, consistency in decision making, important data for the prevention and mitigation of risk, and finance and budget controls.

Integrated Project Delivery (IPD): Integrated project delivery (IPD), as defined by the American Institute of Architects (AIA), is a project delivery approach that integrates people, systems, business structures, and practices into a process that collaboratively harnesses the talent and insights of all participants to optimize project results, increase value to the owner, reduce waste, and maximize efficiency through all phases of design, fabrication, and construction (AIA).

Integrated Project Organization (IPO): An organization where both the owner's employees and the management consultant's employees work under one organization structure.

Integration: The coming together of primary participants (which could include owner, designer, constructor, design consultants, and trade contractors, or key systems suppliers) at the beginning of a project, for the purpose of designing and constructing the project together as a team. Also includes the integration of processes and programs such as integrated change control, quality control, and risk management.

International Finance Corporation (IFC): Established in 1956 as the private sector arm of the World Bank Group to advance economic development by investing in strictly for-profit and commercial projects which reduce poverty and promote development.

International Project Finance Association (IPFA): An international, independent, not-for-profit association established in 1998. The IPFA aims to raise awareness and understanding about project public–private partnerships (PPPs), and their crucial role in infrastructure and economic development.

International Project Management Association (IPMA): The IPMA is a nonprofit, Swiss-registered organization for the promotion of project management internationally. The IPMA is a federation of more than 50 national and internationally oriented project management associations with over 120 000 members worldwide as of 2012.

International Space Station: An international partnership representing 15 countries that developed a modular space station in low Earth orbit to study the Earth's environment and the universe.

Islamic Finance: Finance that is structured to be compliant with the principles of Islamic law (known as Shariah law in Arabic). Key principles require financing without interest and uncertainty and the sharing of profits and losses.

L

Leaders: Those who create a vision, develop strategies to achieve the vision, and implement the vision. Leaders must impact human emotion, change minds, and motivate and inspire individuals to do things they have never done before, or do things in a different way.

Lenders: The entities providing debt contributions to the project company.

Life Cycle Costs: The costs of a project over its entire life – from project inception to the end of a transportation facility's design life.

Loss Control: An organized and usually continuous effort to help decrease the possibility of unforeseen losses and the impact of those that do occur. Loss control can be applied to all kinds of losses on a construction project such as those caused by contractor negligence, design error, fires, electrical surges, hurricanes, or just about anything that results in unexpected harm, injuries, or damage.

M

Management Consultant: The party responsible for overseeing the project construction. Sometimes called a delivery team.

Management Reserve: A designated amount of time and/or budget to account for parts of the project that cannot be predicted. These are sometimes called "unknown unknowns." Use of the management reserve generally requires a change to the total budget.

Megaproject: The Federal Highway Administration (FHWA) characterizes megaprojects as any projects costing $1 billion or more, and they are commonly distinguished by size, duration, uncertainty, and significant political and external influences.

Metrics: Metrics are quantitative measures, such as the number of on-time projects. They are used in improvement programs to determine whether improvement has taken place or whether goals and objectives have been met.

Subordinated Debt/Mezzanine Financing: A combination of financing instruments, with characteristics of both debt and equity, providing further debt contributions through higher-risk, higher-return instruments, and sometimes treated as equity.

Mobilization of Private Finance: The Multilateral Development Bank's ability to crowd-in capital from private creditors as well as other nongovernmental organizations such as the export credit agencies.

Modular Innovation: A significant change within a component but does not require change in the interacting set of components.

MORT Risk Factor Analysis: MORT (management oversight and risk tree) was developed for the Department of Energy in the 1970s by Bill Johnson. The chart consists of 1500 items arranged into a large/complex fault tree, which is used primarily for accident investigation.

Multilateral Development Bank: Agency or institution created by international agreement among multiple countries funded by its members. The major global multilateral, the World Bank Group consists of the IBRD and the IDA, the primary lenders to the developing world. The IFC encourages the investment of private capital.

N

Net Present Value (NPV): Net present value (NPV) is an estimate that helps organizations determine the financial benefits of long-term projects. NPV compares the value of a pound today to the value of that same pound in the future, taking inflation and returns into account.

Nonrecourse Loans: refers to the lenders' inability to access the capital or assets of the sponsor to repay the debt incurred by the project company. In cases where project financings are limited recourse as opposed to truly non-recourse, the sponsor's capital or assets may be at risk only for specific purposes and in specific (limited) amounts set forth in the project financing documentation.

O

Occupational Safety and Health Administration (OSHA): OSHA is a federal government agency that continuously monitors and studies the reasons for project failures. Each year, OSHA identifies the most frequently cited violations of its own standards.

Optimism Bias: Refers to biased estimations that managers form based on delusional optimism instead of on a rational cost-benefits analysis.

Optimization: Managed change and continuous improvement in project operations including governance, finance, technology, economic, and social dimensions.

Organizational Resilience: Defined as the "capacity of a social system to proactively adapt to and recover from disturbances that are perceived within the system to fall outside the range of normal and expected disturbances."

Organizational Structure: An enterprise environmental factor that can affect the availability of resources and how projects are conducted. Organizational structures can range from functional to projectized, with a variety of matrix structures between them.

Organization for Economic Cooperation and Development (OECD): A multilateral organization based in Paris and focused on the harmonization of the international trade laws and the advancement of international trade and development.

Oversight Coordination Committee (OCC): Established by the Massachusetts state legislature to coordinate the oversight of the Big Dig among the Commonwealth's major oversight agencies, including the Office of the Inspector General, the State Auditor's Office, and the Office of the Attorney General.

Owner: Usually the owner of a project's assets in a public project that is commonly the government agency responsible for funding the project.

Owner-Controlled Insurance Program (OCIP): A comprehensive, project-specific insurance program obtained by the owner and intended to cover all key project participants commonly used on megaprojects. An OCIP is managed by the project owner and has the benefits of economies of scale and centralized control.

P

Paris Club Investments: The Paris Club is an informal group of creditor nations whose objective is to find workable solutions to payment problems faced by debtor nations. The Paris Club has 22 permanent members, including most of the western European and Scandinavian nations, the United States, the United Kingdom, and Japan.

Participatory Governance: A structure in which stakeholders or external groups or committees are involved in the project's decision making or oversight.

Partnering: Establishing a long-term win–win relationship based on mutual trust and teamwork and sharing of both risks and rewards. Partnering arrangements can be between labor and management, government owners and management consultants, subordinates and executives, suppliers and customers, designers and contractors, and contractors and contractors. The objective is to focus on what each party does best, by sharing financial and other resources, and establishing specific roles for each participant.

Performance Bond: A bond payable if a project is not completed as specified.

Phase: A set of project activities and tasks that usually result in the completion of one or more project deliverables.

Political Risk: Risks associated with cross-border investing usually comprising currency inconvertibility, expropriation, war and terrorism, nongovernment activists, and legal approvals. In the United States, it is often referred to as *regulatory* or *financial risk*.

Polycentric Governance: An intuitive approach to structure large arenas of consensus-oriented collective action including ecological, social, and environmental endeavors.

Portfolio: A combination of projects and programs both related and unrelated and other matters managed under the organization's strategic plan.

Power: The ability to influence the actions of others. Power may come from formal delegation of authority, reference power, subject matter expertise, the ability to influence, rewards and penalties, as well as other sources.

Power Purchase Agreement: A form of guarantee for the project sponsors and lenders that the purchaser will agree to purchase the output of the project at an agreed upon price and timeline.

Procurement Management: A component of a project or program management plan that describes how a project team will acquire goods and services from outside the performing organization.

Program: A group of related projects managed as a whole to obtain benefits not available from managing them individually.

Program Governance: The structure by which related projects and other work are integrated, coordinated, and managed among all stakeholders in alignment with the strategic goals of the parent organization.

Program Management Office (PMO): Provides support and oversight for the projects under its jurisdiction. The role of the PMO depends on the responsibility it assumes.

Project: In the project management literature, *project* is generally defined as being temporary in nature; undertaken to create a unique project, service, or result; and completed when the goals are achieved or when the project is no longer viable.

Project Finance: The financing of long-term infrastructure, industrial projects, and public services based upon a nonrecourse or limited-recourse financial structure whereby project debt and equity used to finance the project are paid back from the cash flow generated by the project.

Project Governance: The system by which projects are managed to ensure benefits are received and requirements are met in alignment with the organization's and/or the program's goals.

Project Management: The discipline of planning, organizing, and managing resources to bring about the successful completion of specific project goals and objectives.

Project Management Body of Knowledge (PMBOK): The Project Management Institute's Body of Knowledge comprises the sum of knowledge within the profession of project management that is generally recognized as good practice.

Project Management Institute (PMI): PMI is the world's leading not-for-profit membership association for the project management profession, with more than 600000 members and credential holders in more than 185 countries. PMI issues standards and guideline publications developed through a voluntary consensus standards development process.

Public-Private Partnership (PPP): The public and the private sector work together to design, construct, finance, operate, and maintain infrastructure projects. PPPs are usually, but not always, funded in part by the private sector and part by the public sector. While the public sector usually retains ownership in the facility or system, the private party will be given additional decision rights in determining how the project or task will be completed.

Q

Qualitative Risk Assessment: As described in the *Project Management Body of Knowledge* (*PMBOK*), the process of qualitative risk analysis is to assess the likelihood that a specific risk will occur and the impact on cost, quality, or performance, including both negative effects for threats and positive effects for opportunities if it does occur.

Quality: The extent to which the final deliverable conforms to the customer requirements.

Quality Assurance: A structured review of the project, usually by an external resource, to determine the overall project performance (e.g. against schedule and budget) and conformance (i.e., to the management processes specified for the project).

Quality Control: The internal monitoring and control of project deliverables, to ensure that they meet the quality targets set for the project.

Quality Planning: The process of identifying and scheduling quality assurance and quality control activities to improve the level of quality within a project.

R

Rating: An evaluation of creditworthiness provided by a rating agency such as Standard & Poor's Corporation, Fitch Group, or Moody' Investor Service.

Renewable Energy: Energy that is collected from renewable resources that are naturally replenished on a human timescale. It includes sources such as sunlight, wind, hydropower, biomass, and geothermal heat.

Resident Engineer: The individual assigned as the authorized representative for the owner's construction contracts on a project and interagency agreements.

Resilience: The ability to return to a prior state after a perturbation. It has been used in the context of a cyber attack to survive, adapt, recover and return to a new normal after a cyber attack.

Resilience Thinking: Resilience thinking addresses the dynamics and development of complex social–ecological systems (SES).

Return on Investment (ROI): Every project is analyzed from a return-on-investment (ROI) perspective with profitability often a key consideration. ROI tries to directly measure the amount of return on a particular investment, relative to the investment's cost.

Risk: Any event that is likely to adversely affect a project's ability to achieve the defined objectives or an event that creates an opportunity.

Risk Allocation: The process of allocating responsibility for a risk to the party best able to manage the risk. Sometimes the best allocation is to share a risk.

Risk Identification: Determining what risks or hazards exist or are anticipated, their characteristics, remoteness in time, duration period, and possible outcomes.

Risk Intelligence: Is "the organizational ability to think holistically about risk and uncertainty, speak a common risk language and effectively use forward-looking risk concepts and tools in making better decisions, alleviating threats, capitalizing on opportunities and creating lasting value."

Risk Management: The process of identifying, quantifying, mitigating, responding to, and controlling risks throughout a project's life.

Risk Mitigation: The actions taken to avoid, transfer, or mitigate risks within a project.

Risk Monitoring and Control: Defined in PMI's *Project Management Body of Knowledge* (*PMBOK*) as the process of identifying, analyzing, and planning for newly arising risks, keeping track of the identified risks and those on the watch list, reanalyzing existing risks, monitoring trigger conditions for contingency plans, monitoring residual risks, and reviewing the execution of risk responses while evaluating their effectiveness.

Risk Planning: The identification and scheduling of actions needed to reduce the level of risk within a project.

Risk Response: Action that can be taken to address the occurrence of a risk event. Contingency plans are collections of risk responses. Typical risk response methodologies include avoidance, loss prevention, loss reduction, separation, duplication, and risk transfer.

Risk Sharing: A risk management method in which the cost of the consequences of a risk is distributed among several participants in an enterprise, such as in syndication.

Root-Cause Analysis: The Department of Energy defines root-cause analysis as any analysis that identifies underlying deficiencies in a safety management system that, if corrected, would prevent the same and similar accidents from occurring, and to identify the lessons to be learned to promote the achievement of better consequences.

S

Safety and Health Awards for Recognized Excellence (SHARE): An awards program used on the Big Dig to incentivize workers to exercise safe behaviors resulting in reduced lost time incidents and reduced Occupational Safety and Health Administration (OSHA) recordables.

Shared Values: Those principles or beliefs that the project participants agree are the most important and will be given priority over all other principles that may arise as the project evolves.

Special Purpose Vehicle (SPV): An independent legal entity created to accomplish a specific project. Often a limited-liability company or a limited-liability partnership, the SPV generally is dissolved once the project is completed and its financial goals achieved.

Specialized Funding: Used for Climate Change and other sustainable development goals by governments and non-profit organizations.

Stakeholder: Anyone, internal or external to an organization, who has an interest in a project or will be affected by its deliverables. The International Finance Corporation (IFC) of the World Bank Group defines partners and stakeholders to include "a wide range of groups that have a stake in their projects, are affected by their work, or help strengthen impact on sustainable private sector development."

Standard of Care: A designer's normal standard of care is reasonable care required of members of one's profession.

Strategic Misrepresentation: Strategic misrepresentation is a behavioral bias that consists in the tendency to deliberately and systematically distort or misstate information for strategic purposes. It is sometimes also called political bias, strategic bias, power bias, or the Machiavelli factor.

Step-In Rights: The Right of a Lender under certain conditions usually a breach of the contract to step in and replace the existing contractor with a replacement contractor.

Strategy: A direction in a project that contributes to the success and survival of the project in its environment and aligns with the goals of the project's parent organization. Strategy is all about gaining (or being prepared to gain) a position of advantage over adversaries or best exploiting emerging possibilities. As there is always an element of uncertainty about the future, strategy is more about a set of options (*strategic choices*) than a fixed plan.

Supply Chain: A network between the megaproject and its contractors and the contractors and sub-contractors to produce and distribute a specific project, product or service. The entities in the supply chain include producers, vendors, warehouses, transportation companies, testing labs, and distribution centers.

Sustainable Development: The meeting of present needs without compromising the ability of future generations to meet their own needs.

Sustainable Repurposing: The management of nuclear decommissioning in a sustainable way by repurposing the site to a new sustainable use such as parklands, community infrastructure, or small nuclear reactors.

System Engineering: An interdisciplinary approach and means to enable the realization of successful systems.

Systems Integration: Refers to the work undertaken across organizational boundaries in interorganizational projects to integrate the systems that these projects deliver.

System of Systems: Multiple systems whose elements are managerially and/or operationally independent systems but are integrated collections of constituent systems that usually produce results unachievable by the individual systems alone (Source: System Engineering Handbook. International Council on Systems Engineering).

Systems Thinking: The process by which one attempts to look at the whole rather than the individual parts to gain a better understanding of how the parts interact and are interdependent within the larger system.

T

Take or Pay Contract: A contractual term whereby the buyer is unconditionally obligated to take any product or service that is being offered (and pay the corresponding purchase price), or to pay a specified amount if he refuses to take the product or service.

Transition Economies: One that is changing from central planning to free markets. Since the collapse of communism in the late 1980s, countries of the former Soviet Union, and its satellite states, including Poland, Hungary, and Bulgaria, sought to embrace market capitalism and abandon central planning. Although many eastern and central European countries completed the transition by 2000, there are more than 20 countries still in transition.

Transparency International: An organization that supports a global movement working in over 100 countries to end the injustice of corruption.

U

Uncertainty: Information inadequacy when too many variables interact – unknowable, unmeasurable, and uncontrollable.

United Nations 2020 Agenda for Sustainable Development Goals: Among the most ambitious of global initiatives consisting of 17 sustainable goals including no poverty, zero hunger, clean water and sanitation, climate action, life below water and life on land implemented in investment projects through the multilaterals and other organizations globally.

V

Value Capture: A type of public financing where increases in the private land values generated by public transportation investments are "captured" to repay the cost of the public investment. Using value capture mechanisms to finance new or existing transportation infrastructure connects the benefit of the infrastructure investment with the cost to provide it.

Value Creation: Value creation in megaprojects has been approached from outcome-based and system lifecycle-based perspectives. From the outcome-based perspective, a megaproject creates value after the project's completion for the organizations participating in it, when it achieves the desired outcomes set initially in the strategic front-end phase.

Value Engineering (VE): VE is a systematic method for improving the value of goods or products and services. It was developed at General Electric Corporation during World War II and is widely used in industry and government, particularly in areas such as defense, transportation, construction, and health care. VE is an effective technique for reducing costs, increasing productivity, and improving quality.

W

World Bank (WB): A multilateral agency based in Washington, DC, whose primary mission is to lend to developing countries to assist in reducing the poverty level. The bank consists of the following divisions: International Bank for Reconstruction and Development (IBRD), International Development Association (IDA), the International Finance Corporation (IFC), the Multilateral Guarantee Association (MIGA), and the International Center for the Settlement of Investment Disputes (ICSID).

Acronyms

ABC:	Artery Business Committee
ADR:	Alternative Dispute Resolution
AfDB:	African Development Bank
AG:	Attorney General
AI:	Artificial Intelligence
AIA:	American Institute of Architects
AIIB:	Asian Infrastructure Investment Bank
AIPM:	Australian Institute of Project Management
ANSI:	American National Standards Institute
APM:	The U.K. Association for Project Management
ASCE:	American Society of Civil Engineers
ASQ:	American Society for Quality
AUKU:	Australia-UK-US Security Pact
AUM:	Assets Under Management
BIT:	Bilateral Investment Treaty
BMZ:	German Federal Ministry of Economic Cooperation and Development
BOT:	Build-Own-Transfer
BOOT:	Build-Own-Operate-Transfer
BVP:	Best Value Procurement
CA/T:	Central Artery/Tunnel Project
CHSR:	California High Speed Rail
CMR:	Construction Manager at Risk
CPS:	Construction Phase Services
CSR:	Corporate Social Responsibility
C/SU:	Cost/Schedule Update
DB:	Design-Build

Global Megaprojects: Lessons, Case Studies, and Expert Advice on International Megaproject Management, First Edition. Virginia A. Greiman.
© 2023 John Wiley & Sons, Inc. Published 2023 by John Wiley & Sons, Inc.

DBB: Design-Bid-Build
DOE: Department of Energy
DRBF: Dispute Review Board Foundation
DRII: Disaster Recovery Institute International
EIB: European Investment Bank
EBRD: European Bank for Reconstruction and Development
ECA: Export Credit Agency
ECFA: Events and Causal Factors Analysis
EHSGs: The World Bank's Environmental, Health and Safety Guidelines
EIA: U.S. Energy Information Administration
EIB: European Investment Bank
EIS: Environmental Impact Statement
EMDE: Emerging Markets and Developing Economies
ENE: Early Neutral Evaluation
EOC: Escalation of Commitment
EPA: Environmental Protection Agency
EPC: Engineering Procurement and Construction Contract
ESA: European Space Agency
ESF: Environmental and Social Framework
EV: Earned Value
FAO: United Nations Food and Agriculture Organizations
FE: Field Engineer
FEIS: Final Environmental Impact Statement
FDI: Foreign Direct Investment
FIDIC: The International Federation of Consulting Engineers
GANs: Grant Anticipation Notes
GAO: [U.S.] Government Accountability Office
GERD: Grand Ethiopian Renaissance Dam
GIH: Global Infrastructure Hub, A G20 Initiative
GOBs: General Obligation Bonds
HRO: High Reliability Organization
IBRD: International Bank for Reconstruction and Development, World Bank
ICE: Institute of Civil Engineers
ICSID: International Center for the Settlement of Investment Disputes
IDA: International Development Agency, World Bank
IDB: Inter-American Development Bank
IFAD: International Fund for Agriculture Development
IFC: International Finance Corporation (of the World Bank Group)
IHS: Interstate Highway System
IFS: International Financial Institutions
IMF: International Monetary Fund

IoT:	Internet of Things
IPF:	Investment Policy Financing
IPFA:	International Project Finance Association
IPMA:	International Project Management Association
ISO:	International Standards Organization
ISS:	International Space Station
ISTEA:	Intermodal Surface Transportation Efficiency Act of 1991
ITER:	The International Thermonuclear Experimental Reactor
JBIC:	Japan Bank for International Cooperation
JV:	Joint Venture
LDCs:	Lesser Developed Countries
LHC:	Large Hadron Collider
MDB:	Multilateral Development Bank
ML:	Machine Learning
MSR:	Megaproject Social Responsibility
MTP:	Mega Transport Projects
NAO:	National Audit Office UK
NASA:	National Aeronautics and Space Administration
NEC:	New Engineering Contract
NIH:	National Institute of Health
NIST:	National Institute of Standards and Technology
NLP:	Natural Language Processing
OCC:	Oversight Coordination Commission
OCIP:	Owner-Controlled Insurance Program
OECD:	Organization for Economic Cooperation and Development
OIG:	Office of the Inspector General
OSA:	Office of the State Auditor
OSHA:	Occupational Safety and Health Administration
PCA:	Potential Change Allowance
PF:	Potential Forecast
PMBOK:	Project Management Body of Knowledge
PMI:	Project Management Institute
PMM:	Project Management Monthly
PMO:	Project Management Office
PPI:	Private Participation in Infrastructure
PPP:	Public–Private Partnership or P3
QA:	Quality Assurance
QC:	Quality Control
QCP:	Quality Control Plan
RE:	Resident Engineer
RFP:	Request for Proposal

RM:	Risk Manager
RCA:	Root Cause Analysis
ROI:	Return on Investment
SDC:	Section Design Consultant
SDGs:	Sustainable Development Goals
SHARE:	Safety and Health Awards for Recognized Excellence
SNWTP:	South North Water Transfer Project
TQM:	Total Quality Management
TRB:	Transportation Research Board
UNCTAD:	United Nations Commission on Trade and Development
UNECE:	United Nations Economic Commission for Europe
USDOT:	United States Department of Transportation
VE:	Value Engineering
VECP:	Value Engineering Change Proposal
WB:	World Bank
WBG:	World Bank Group
WBS:	Work Breakdown Structure
WCD:	World Commission on Dams
WEF:	World Economic Forum
WESP:	World Economic Situation and Prospects Report (WESP)
WWTP:	Wastewater and Treatment Projects

Index

a

Accountability   172, 183, 187, 209
Actors in the project organization
    178–179
Adaptation governance   180, 465
African Development Bank   115–116
Agile project management   173–174,
    479, 491
Airbus 380   362
Allocation of risk   407–408
Alternative dispute resolution
    285–289
America's Interstate Highway System
    6, 14–15, 30
Aqueducts of Rome   8
Ariane 5 Rocket Disaster   332
Artificial intelligence   459, 469,
    478–479, 483
Asia, innovation in   449
Asian Development Bank   116
Asian Infrastructure Investment
    Bank   117
Asphalt road   30
Association of Project Management
    (APM)   465
Azerbaijan Rural Investment
    Project   195

b

Baldrige Performance Excellence
    Program   487, 490–491
Bangladesh   192–193
Battersea Power Station, sustainable
    repurposing   466
Bechtel   13, 172–173, 292–293, 397
Belt and Road Initiative   30–31
Benefits Assessment   351
Benin   107
Benin Agricultural Productivity and
    Diversification Project   107,
    196–197
Big Bertha   398
Big Dig. See Central Artery
    Tunnel Project
Bilateral investment treaties
    415–417
Blockchain technology   482
Blue ocean strategy   444–445
  blue economy thinking   461–462
  blue ocean thinking   460–461
Boeing 787 Dreamliner   445–446
Borrower Credit Conditions   61–62
BRICS Bank   116–117
British Petroleum (BP) Deepwater
    Horizon Water Spill   339

Global Megaprojects: Lessons, Case Studies, and Expert Advice on International Megaproject Management, First Edition. Virginia A. Greiman.
© 2023 John Wiley & Sons, Inc. Published 2023 by John Wiley & Sons, Inc.

c

California High-Speed Rail System   25
Catastrophic loss   336–340
Central Artery Tunnel Project (Big
        Dig)   18–19, 73
    claims and changes   291
    cost and schedule   255, 268–275, 280
    innovation   431–432
    integration   221–224
    risk management   397
Change and Conflict Management   159
Civil engineers   425–427
Claims and changes   283–284
Climate initiatives   473–474
Collaboration   412
Collaborative contracting   414
Collective mindfulness   370–372
Complex Adaptive Systems   464–465
Complex Global Projects   135–156
Complex megaprojects
    case studies of   146–151
    characteristics of   141–143,
        144–146
    compared to uncertainty, ambiguity,
        conflict, risk   138–141
    definition of   136–138
    sources and impact of   145
Concession Agreement   260
Contract checklist   417–419
Contracts
    challenging clauses   391–403
    negotiation   391
Corporate governance   169
Corruption in megaproject governance
        190–191
    impact   369
Cost and schedule tracking
    claims and changes   283–284
    lessons learned   292–294
    project controls   278–279
    quality   281–282

    regulations and oversight   279–281
    safety and health   282
Country Partner Frameworks   105–106
Creative Economy and
        Sustainability   450
Credit Rating Agencies   82–84
Crossrail Project London   24
    cost and schedule   269–279
    innovation   433–434
    integration   223–224
    value capture   76
Culture and governance   189–190

d

Data analytics   483
    Internet of things   483
    natural language processing
        483–484
Delhi Airport Express   233–234
Delivery Methods   262–262
Developing countries   99
Developing countries investment   49–51
Dexterous Robots   482
Differing site conditions   398–399
Digital economy   478–484
Digital road   31
Digital twins   481
Disaster lessons learned   340
Dispute Resolution and Choice of Forum
        403–407
Dynamic structural complexity
        152–153

e

Ecosystem Megaprojects   28–30
Eiffel Tower   10–11
Emerging issues in contracts   414–415
EPC contracts   184
Equator Principles   84–85
Equity Investment   62–64
Era of emergence   468–469

Era of Great Tunnel, Energy, and Pipeline Projects   17–20
Escalation of Commitment   366–367
European Bank of Reconstruction and Development   116
European Investment Bank   53–54
European Motorway   14
Eurotunnel   17, 60, 381, 486
Evaluation Criteria for Project Finance   70–71
Export Credit Agencies (ECAs)   65–67

f

Failed projects
  characteristics of   356–360
  reasons for   360–369
Force majeure   409–411
Fukushima Daiichi nuclear disaster   339
Future of megaprojects
  challenges   463–469
  introduction   459–460

g

Globalization, definition of   3
Golden Age of Globalization Projects   9–12
Gotthard Base Tunnel   6, 23, 25
Governance failure   191–195
Governance literature   170–171
Governance of megaprojects   167–169
  frameworks   174–177
  governance literature   169–171
  roles and responsibilities   177–179
  structure   177–178
Governance performance   187–188
Grand challenges   28, 469–473
Great Depression Projects   12–14
Great Megaproject Era   14–17

Great Wall of China   7–8
Green energy initiatives   472–473

h

Hadron Collider   6, 25, 461, 476
Heathrow Terminal   5, 219–221
High Reliability Organizations   157–158, 335–336
Holistic view of success   355–356
Hong Kong PPP Development   236–237
Hong Kong Zhuhai-Macao Bridge   430–431
Hoover Dam   12–13
Human Rights and Privacy   485–486
Hurricane Katrina   338
Hyderabad Metro Rail   362

i

Implied duties in contracts   394–396
Indonesia Lucie Mudslide   339
Infrastructure demand   96–97
Infrastructure Financing
  European Union   52–53
  United States   51–52
Integrated project organizations   207–211, 221–224
International Bank for Reconstruction and Development   121–123
International contracting   387–391
International Development Association   121–123
International development case studies   118–121
International Finance Corporation   108, 184
International Space Station   3, 216–218, 216–219, 302, 359, 477
International Thermonuclear Experimental Reactor (ITER)   476–477
Iridium case study   354

j

James Bay Project   16–17
James Webb Space Telescope   193–194
Joint Strike Fighter F-35   157
Joint Ventures   413–414

l

Labor   301, 410
Laerdal Tunnel   17–18
Law and Contracts   381–384
Leadership in Megaprojects   474–477
Legal framework for governance   183–187
Legal Framework Project Agreements   59–61
Lender key questions   59–61
Life Cycle Evaluation   85–86
Litigation
   arbitration   284–286
   dispute review boards   288–289
   mediation   286–288
   partnering   288
London Olympics Integration   221

m

Machine learning   480–481
Mandatory law   409
Megaproject cost and budgets   276–277
   contingency reserves   278
   management reserves   278
Megaproject evaluation   77–84
   ex ante evaluation   78–82
   ex-post evaluation   79, 83–84
Megaproject implementation   255–256
   development   256–257
   implementation   263–266
   initiation   257–258
   scope development   268–270
   transition to operations   289–290

Megaproject innovation and resilience   425–431
   Asia   448
   best practices   443–450
   enablers and challenges   439–443
   impediments   439
   index   446–448
   Kazakhstan   450
   programs   431–435
   strategy   435–439
   Thailand   448–450
Megaproject management systems
   perspective   490–492
Megaproject risk structure   306–307
Megaprojects
   characterization of   3–5
   definition of   5–6
   Global Timeline   6
   Why Study   26–28
Megaproject schedules   270–276
Megaproject social responsibility   179–180
Megaproject troubles and triumphs   349–350
Millennium Projects   23–26
Mobilization of Capital   107
Multilateral Development Banks   67–68
Multilaterals   95
   financing   67, 115–119
   role of   101–115

n

National Aeronautics Space
   Administration (NASA)   333–334, 365, 470, 474–475, 477
National Audit Organization UK   277
National Human Genome Institute   477
Netherlands (The)   8

New nuclear power   483
Normalization of deviance   333–335
Nuclear Power   21–23
Nuclear power decommissioning   340, 365, 465

o
Oresund Bridge   6, 25, 29, 214
Organizational and Social Complexity   156–157
Organization for Economic Cooperation and Development (OECD)   47, 308, 354–355

p
Panama Canal   6, 9, 301
Power purchase agreement   235–236
Private investment   97–99
Privatization of Water Supply   237
Procurement of megaprojects   385–387
Program governance   168
Project Bonds   64–65
Project Company structure and legal agreements   55–59
Project finance   45–51
Project Management Institute (PMI)
  governance   173–174
  principle based approach   489–490
  quality   281
  value creators   71
Projects of the future   460
Public private partnerships   225–231
  advantages and challenges   231
  economics   234–236
  global implementation   231–233
  legislation and regulation   243–246
  procurement   239–241
  social considerations   241–243
  transit   239

Public-Private Partnership Structures   69–70
Pyramid of Giza   7

q
Quality at entry   369–370
Quantum computing   481

r
Relational governance   181
Renewable energy projects   471
Resident engineers, manual   267–268
Resilience   281, 370, 425, 466–468
Return on Investment   68–69
Risk and megaprojects   301–342
  business continuity   327–330
  characterization of   330–332
  classification of risks   325–327
  definition of   305
  emerging risks   302–305
  frameworks   313–321
  opportunities   321–325
  resilience   308–309
  risk intelligence   310–312
  structure   306–307
Root cause evaluation   327–330

s
Safety and health   268, 287–288
Science Parks   448
Seikan Tunnel   18
Shoreham Nuclear Power Plant   365
Singapore case study   209–2011
Small modular reactors (SMRs)   482
Smart Cities and Inclusive Growth   473–477
Sources of Funding for Project Finance   54–59, 108–109

South-North Water Transfer Projects
126–127
Space Hornet Project   214–215
Space shuttle challenger   333
Special Purpose Vehicle   183–184
Stakeholders
  community consultation of
    158–159
  engagement   371
  public outreach   185–187
Step-in-rights   408
St. Lawrence Seaway   16
Strategies for learning from failures
    357–360
Successful megaproject failures   350
Successful projects   353–356
  characteristics of   370–372
Suez Canal   10
Supply chain management   482
Sustainability   415, 463
  environmental   461–462
  innovation   440
  repurposing   465–466
Sustainability pillars   352
Sustainable development frameworks
    124–127, 448
Sustainable governance models
    179–183
Sydney Opera House   15–16
Systems engineers   209
Systems Integration and complexity
    151–154
Systems integration on the
    battlefield   213
Systems of systems   211–213

t
Temporary institutional structures
    172–173
Tennessee Valley Authority Projects
    13–14
Three Gorges Dam   258–259
Trans Alaskan Pipeline   19–20
Transition economies   100–101
Trans-Siberian Railway   11–12
Triborough Bridge   13

u
Uncertainty   143–144
Unforeseen subsurface conditions
    396–399
United Nations 2030 Agenda   95–96
Unknown unknowns   154–155

v
Value capture   73–77
Value creation   71–73
Value delivery   492
Value engineering   443
Value of megaprojects   123–124

w
Water privatization   237–238
Work Breakdown Structure (WBS)
    268–270
World Bank Environmental and Social
    Framework (ESF)   392–394
World Bank Funding Structure
    107–109
World Bank Group   47, 50, 55, 57, 67,
    77, 79, 82, 701